ANNUAL REVIEW

OF CHINESE ARCHITECTURAL DESIGN WORKS · VOLUME 16

中国建筑设计作品年鉴·第十六卷

《中国建筑设计作品年鉴》编委会　北京主语空间文化发展有限公司　编

江苏凤凰美术出版社

图书在版编目(CIP)数据

中国建筑设计作品年鉴. 第十六卷 /《中国建筑设计作品年鉴》编委会, 北京主语空间文化发展有限公司编. -- 南京：江苏凤凰美术出版社, 2023.6
 ISBN 978-7-5741-1063-2

Ⅰ.①中… Ⅱ.①中… ②北… Ⅲ.①建筑设计—作品集—中国—现代 Ⅳ.①TU206

中国国家版本馆CIP数据核字(2023)第113406号

出版统筹	王林军
责任编辑	孙剑博
责任设计编辑	韩 冰
装帧设计	杨建辉
责任校对	王左佐
责任监印	唐 虎

书　　名	中国建筑设计作品年鉴·第十六卷
编　　者	《中国建筑设计作品年鉴》编委会　北京主语空间文化发展有限公司
出版发行	江苏凤凰美术出版社（南京市湖南路1号 邮编: 210009）
总 经 销	天津凤凰空间文化传媒有限公司
总经销网址	http://www.ifengspace.cn
印　　刷	北京世纪恒宇印刷有限公司
开　　本	787 mm×1092 mm　1/8
印　　张	65
版　　次	2023年6月第1版　2023年6月第1次印刷
标准书号	ISBN 978-7-5741-1063-2
定　　价	598.00元（精）

营销部电话　025-68155675　营销部地址　南京市湖南路1号
江苏凤凰美术出版社图书凡印装错误可向承印厂调换

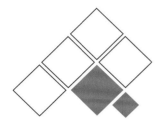

中国建筑设计作品年鉴·第十六卷

谨以此年鉴献给为中国建筑设计行业发展而辛勤付出的设计师们!

《中国建筑设计作品年鉴·第十六卷》(以下简称《年鉴》)收录了国内外五十余家设计机构的近五百件优秀建筑设计作品。这些不同类型、不同设计理念的作品基本反映了当前建筑设计行业多元化的发展状况和实践成果,从中可以直观地领略到中国建筑设计融汇古今、贯通中西的风采。

作为忠实记录中国建筑设计行业发展状况的出版物,《年鉴》对于各地建设主管部门和城市规划、建筑设计、景观设计、工程建设、房地产开发的从业人员以及相关科研教育机构人员掌握建筑设计行业发展趋势、了解全新的设计理念具有很好的参考、借鉴作用。

谨向为《年鉴》提供帮助和支持的单位和个人表示衷心的感谢!

《中国建筑设计作品年鉴》编辑部
2022年11月

主　　办:中国建筑文化中心
编　　者:《中国建筑设计作品年鉴》编委会
　　　　　北京主语空间文化发展有限公司
网络服务:阿特莱斯(北京)网络科技有限公司
编委会主任:曾少华
编委会副主任:李　剑
主　　编:朱　辉
副　主　编:肖　措　张中平
执行主编:肖　峰
编辑部主任:张　辉
编　　辑:张竹村　王　蓉　章丽君　宫　超　张　澜　王　伟　刘笑艳　冷娜英
美术编辑:文　夏

编辑委员会(按姓氏首字母排序)
陈继松　陈晓丽　程志毅　房庆方　冯志成　胡柏龄　李光荣　李建新　李振东　李子青
林坚飞　刘　鲲　刘永富　刘志昌　柳　青　田　明　屠锦敏　王西明　王正刚　夏恩恕
向世林　谢家谨　谢志平　熊建平　杨洪波　杨焕彩　杨建华　查　敏　张发懋　张建民
郑应炯　周　游　朱正举

编辑部地址:北京市海淀区三里河路13号
　　　　　　中国建筑文化中心807室
电　　话:010-88151966
监督电话:13910120811
邮　　箱:zgjsmail@126.com

编委题词 Epigraphy by Committee

设计年鉴，推陈出新，
是我们的良师益友。
唐国安

愿《年鉴》迈上更高的台阶
愿中国建筑师铸就新的辉煌
马辛

吴良镛先生言，一切艺术
综合于建筑，而礼乐诗书画
剧之表演，必与建筑背景
调成一气，美好生活必每一
元的强盛时代莫不有伟
大的建筑计划。容纳和表
现这一些丰富之生命、
先生此言实为千古名论。
以新世纪我们更应努力这
一和谐的美的章早日
来临。
吴良镛

敬业 诚信 创新
与"年鉴"及同仁的共勉
JY 俞锦文
二〇一三年八月

以铜为鉴，可正衣冠；
以古为鉴，可知兴替；
以人为鉴，可明得失；
以众筑为鉴，可知未来。

中国建筑设计作品
年鉴 是在中国
做设计最好的学习
工具与展示！
饶及人
2013.9.12

建筑是一瓶陈年老酒
唯有时间+空间+人素材为
可调配其绝妙

《中国建筑设计作品年鉴》不仅
记录了建筑师们一年来辛勤耕耘的
成果，更记载了祖国建设一年来迈
出的巨大步伐！
刘光

祝《中国建筑设计作品年鉴》
越办越好！
2013

祝《中国建筑设计作品年鉴》
越办越好！
2013.8.21

做世界级的中国设计
迟亨忠
2014.11.12.

设计更好的新生活！
天大设计学院

特邀编委 Contributing Editors

(按姓氏首字母排序)

姓名	所属机构
安日红	航天规划设计集团有限公司
蔡沪军	彼印建筑工作室
蔡爽	启迪设计集团股份有限公司
曹辉	中国建筑东北设计研究院有限公司
陈剑飞	哈尔滨工业大学建筑设计研究院
陈霖峰	佛山市顺德建筑设计院股份有限公司
陈雄	广东省建筑设计研究院有限公司
程鹏	贵州省建筑设计研究院有限责任公司
程万海	天津大学建筑设计规划研究总院有限公司
褚冬竹	重庆市设计院有限公司
丁永君	天津大学建筑设计规划研究总院有限公司
董丹申	浙江大学建筑设计研究院有限公司
董明	贵州省建筑设计研究院有限责任公司
多庆巴珠	拉萨市设计集团有限公司
冯高磊	山西省建筑设计研究院有限公司
冯智新	广州冯智新建筑设计有限公司
高明磊	北方工程设计研究院有限公司
高庆辉	东南大学建筑设计研究院有限公司
高正华	香港道和国际设计有限公司
韩艳红	深圳市联合创艺建筑设计有限公司
何嘉欣	中国中轻国际工程有限公司
黄会明	华汇集团
黄俊华	广东省建筑设计研究院有限公司
姜兴兴	法国AREP设计集团
金旭炜	中铁二院工程集团有限责任公司
李韬	方舟国际设计有限公司
李晓梅	晋思建筑设计事务所
梁昆浩	佛山市顺德建筑设计院股份有限公司
林世彤	西迪国际	CDG国际设计机构
刘伟平	香港道和国际设计有限公司
刘存发	天津华厦建筑设计有限公司
刘磊	北京行者匠意建筑设计咨询有限责任公司
刘卫兵	四川省大卫建筑设计有限公司
刘玉龙	清华大学建筑设计研究院有限公司
龙卫国	中国建筑西南设计研究院有限公司
卢子敏	广州珠江外资建筑设计院有限公司
马树新	包钢集团设计研究院(有限公司)
庞波	华蓝设计(集团)有限公司
蒲净	太原市建筑设计研究院
钱方	中国建筑西南设计研究院有限公司
乔丛	北京东方华脉工程设计有限公司
饶红	昆明新正东阳建筑工程设计有限公司
盛开	Archilier Architecture建筑设计顾问咨询公司
孙建超	中国建筑科学研究院建筑设计院
谭东	上海砼森建筑规划设计有限公司
汤朔宁	同济大学建筑设计研究院(集团)有限公司
王慧琳	拉萨市设计集团有限公司
王平	GWP建筑事务所
王元新	新疆印象建设规划设计研究院(有限公司)
吴强	安徽建筑大学·建筑与规划学院
	安徽建筑大学·城乡规划设计研究院
夏平	苏州城发建筑设计院有限公司
肖诚	深圳华汇设计有限公司
谢璇	上海鼎实建筑设计集团有限公司
熊承志	昆明新正东阳建筑工程设计有限公司
许可	中电光谷建筑设计院有限公司
薛巍	中煤科工重庆设计研究院(集团)有限公司
杨书林	杭州市城建设计研究院有限公司
杨竖	广州珠江外资建筑设计院有限公司
杨旭	深圳市建筑设计研究总院有限公司
尹碧涛	中电光谷建筑设计院有限公司
于飞	航天规划设计集团有限公司
袁大昌	天津大学建筑设计规划研究总院有限公司
查金荣	启迪设计集团股份有限公司
张国威	GWP建筑事务所
张凯	江苏中大建筑工程设计有限公司
	英国UKLA太平洋远景国际设计机构
张宁	ppas德国佩帕施城市发展咨询
张琦	中电光谷建筑设计院有限公司
张羽	北京中鸿建筑工程设计有限公司昆明分公司
赵松林	广州市设计院集团有限公司
钟洛克	高驰国际设计有限公司
朱洪宇	贵州省建筑科研设计院有限公司

目录 Contents

■ 主要建筑设计作品   1

◆ 境外
- 晋思建筑设计事务所   2
- 西迪国际 | CDG国际设计机构   12
- GWP建筑事务所   22
- ppas德国佩帕施城市发展咨询   28
- Archilier Architecture建筑设计顾问咨询公司   30
- 江苏中大建筑工程设计有限公司   36
- 英国UKLA太平洋远景国际设计机构
- 法国AREP设计集团   44
- 美国华贝设计有限公司（中国）   48

◆ 香港
- 香港道和国际设计有限公司   50

◆ 安徽
- 安徽建筑大学·建筑与规划学院   56
- 安徽建筑大学·城乡规划设计研究院

◆ 北京
- 清华大学建筑设计研究院有限公司   62
- 中国建筑科学研究院有限公司建筑设计院   70
- 航天规划设计集团有限公司   84
- 中国中轻国际工程有限公司   94
- 北京行者匠意建筑设计咨询有限责任公司   102
- 北京东方华脉工程设计有限公司   106

◆ 重庆
- 重庆市设计院有限公司   116
- 中煤科工重庆设计研究院（集团）有限公司   126
- 高驰国际设计有限公司   134

◆ 广东
- 广东省建筑设计研究院有限公司   140
- 广州市设计院集团有限公司   150
- 广州珠江外资建筑设计院有限公司   158
- 广州冯智新建筑设计有限公司   164
- 广州市天作建筑规划设计有限公司   170
- 深圳市建筑设计研究总院有限公司   176
- 深圳华汇设计有限公司   184
- 深圳市联合创艺建筑设计有限公司   200
- 深圳市地平匠造规划设计有限责任公司   206
- 深圳大学景观设计研究所
- 佛山市顺德建筑设计院股份有限公司   208

◆ 广西
- 华蓝设计（集团）有限公司   218

◆ 贵州
- 贵州省建筑设计研究院有限责任公司   228
- 贵州省建筑科研设计院有限公司   238

◆ 湖北
- 中电光谷建筑设计院有限公司   252

◆ 河北
- 北方工程设计研究院有限公司   264

◆ 黑龙江
- 哈尔滨工业大学建筑设计研究院   268
- 方舟国际设计有限公司   276

◆ 江苏
- 东南大学建筑设计研究院有限公司   288
- 启迪设计集团股份有限公司   300
- 苏州城发建筑设计院有限公司   314

◆ 辽宁
- 中国建筑东北设计研究院有限公司   322

◆ 内蒙古
- 包钢集团设计研究院（有限公司）   328

◆ 山东
- 山东省建筑设计研究院有限公司   334

◆ 山西
- 山西省建筑设计研究院有限公司   342
- 太原市建筑设计研究院   354

◆ 上海
- 同济大学建筑设计研究院（集团）有限公司   360
- 彼印建筑工作室   368
- 上海砼森建筑规划设计有限公司   376
- 上海鼎实建筑设计集团有限公司   384

◆ 四川
- 中国建筑西南设计研究院有限公司   398
- 四川省大卫建筑设计有限公司   406
- 四川华西建筑设计院有限公司   414
- 中铁二院工程集团有限责任公司   418

◆ 天津
- 天津大学建筑设计规划研究总院有限公司   426
- 天津拓城建筑设计有限公司   440
- 天津华厦建筑设计有限公司   444

◆ 新疆
- 新疆印象建设规划设计研究院（有限公司）   450

◆ 西藏
- 拉萨市设计集团有限公司   460

◆ 云南
- 昆明新正东阳建筑工程设计有限公司   470
- 北京中鸿建筑工程设计有限公司昆明分公司   476

◆ 浙江
- 浙江大学建筑设计研究院有限公司   482
- 杭州市城建设计研究院有限公司   494
- 华汇集团   506

■ 设计作品索引   512

主要建筑设计作品
Main Architectural Design Works

境外

Gensler 晋思建筑设计事务所

晋思建筑设计事务所(以下简称Gensler)成立于1965年,是一家全球性的建筑、设计与规划公司,在全球拥有50间办公室,分布在亚洲、欧洲、大洋洲、中东以及美洲地区,为各行业3 500多家活跃客户提供专业的设计服务。Gensler的设计专家们致力于打造精彩的、可持续的、有影响力的居住、办公、休闲场所。

Gensler is a global architecture, design, and planning firm with 50 locations across Asia, Europe, Australia, the Middle East, and the Americas. Founded in 1965, the firm serves more than 3 500 active clients in virtually every industry. Gensler designers strive to make the places people live, work, and play more inspiring, more resilient, and more impactful.

地址:上海市新闸路669号博华广场3楼
电话:021-61351900
传真:021-61351999
邮箱:gc_info@gensler.com
网址:www.gensler.com

Add:3rd Floor, Bohua Building, No.669 Xinzha Road, Shanghai
Tel:021-61351900
Fax:021-61351999
Email:gc_info@gensler.com
Web:www.gensler.com

银科金融中心
Yintech Finance Center

项目地点:上海
建筑面积:117 756 m²
用地面积:13 515 m²
设计时间:2016年8月—2018年11月
竣工时间:2021年4月

Location: Shanghai
Building Area: 117 756 m²
Site Area: 13 515 m²
Design Time: August 2016 to November 2018
Completion Time: April 2021

设计团队运用极简的设计手法勾勒出具有视觉冲击力的几何立方造型,以"连接"为设计主旨,充分融入自然空间,为五千多位金融科技从业者打造"共享、灵活、创新、健康"的总部办公空间。

位于主体建筑中心的12层挑高中庭空间,充分引入自然采光,并与建筑外部景观形成通透的视线走廊。中庭的连廊设计,以及内部的连通楼梯,建立了水平以及垂直层面上的联系,真正实现了"互联互通",鼓励部门内部以及跨部门之间的共享、协作,从而有效地提升团队创造力。

银科金融中心通过严谨、高效及经典的设计,体现公司愿景价值的同时,营造一体化协作的办公空间,并且和谐地融入了城市肌理。

The design for Yintech Finance Center consists of an office "Cube" perched strategically above a low-rise podium. The "Cube" will contain the major business units of this financial technology company for more than 5 000 employees with a naturally daylit atrium at the center flanked by linear office components.

The atrium, with its deep respect for natural light, provides a space that encourages collaboration by connecting the flanking office wings through a series of bridges spanning this twelve-story void. Together, the atrium along with interconnecting stairs at strategic locations within the office space, create a visually intuitive navigation path providing each employee connectivity throughout the "Cube" with direct access to natural light and views to the city beyond.

The Yintech Finance Center is an example of responsible architecture rooted in timelessness that communicates its values through its rigor, rationality, and precision. The geometric form of the project communicates trust and stability for this financial technology giant while creating an integrated collaborative workplace and outdoor landscaped spaces fitting seamlessly into their urban context.

东方美谷JW万豪酒店
JW Marriott Hotel, Shanghai Fengxian

项目地点：上海	Location: Shanghai
建筑面积：64 551.9 m²	Building Area: 64 551.9 m²
用地面积：39 889.6 m²	Site Area: 39 889.6 m²
设计时间：2017年12月—2019年12月	Design Time: December 2017 to December 2019
竣工时间：2021年1月	Completion Time: January 2021

酒店的整体造型由顶部的两条蜿蜒的曲线及底部的3个几何体形成有机的组合，既规避了狭长的地块带来的设计局限性，又打造了柔和、动感的建筑造型，在和谐地融入周边自然环境的同时，建筑所具有的视觉冲击力也使之成为整个区域的标志建筑。

建筑设计摒弃传统酒店的双廊客房的设计模式，打破单一的狭长的建筑体量，以两条曲线造型取而代之，采用单廊客房与双廊客房相结合的平面布局，确保所有客房均享有湖水景观，并形成内部围合庭院空间，充分利用弹性设计策略，将自然采光以及通风引入首层公共区域。

为打造最优的用户体验，Gensler在项目的规划中认真处理各功能之间的关系，实现人车分流、动静分离的规划策略。瞬时客流相对较大的会议功能被设置在项目的入口处，而酒店大堂则被布置在隐私性较强的内部区域，为酒店住客提供安静的入住体验。交通线路布局实现完全的人车分离，车行线路布局在项目的北面，南部空间则为精心规划的步行系统。

Gensler的设计团队充分利用了湖水景观以及周边的森林公园，整体规划中将外部水景引入项目中，形成内外呼应的一体化水系景观，为用户打造不同层次的亲水开放空间，旨在建造延续当地文脉、忠于用户体验并引领未来设计趋势的酒店建筑。

The smooth, curving form of the east and west wings are structured atop three geometric bases, which support the structure and provide space for the ground-floor amenities that include convention center, dining, lobby and amenities. The dynamic form of the building also breaks the design limitation of the long and narrow site, and enables it the focal point of an emerging art, entertainment, and tourism-oriented neighborhood.

Flowing over the supporting structure is the residential section of the hotel, shaped to ensure views of the water from every room. Instead of a traditional hallway separating the suites, the design of the hotel creates a courtyard through its split flow design, inviting light and air into the project, while providing further connection with nature through use of green space.

The circulation plan for the project confines automotive traffic to the northern shore of the peninsula, while creating space for the pedestrian environment in the south, exposing the guests of the hotel to vast, uninterrupted views of the lake. The convention center, which has busy peak traffic, is planned at the entrance of the site and the hotel lobby is planned in the end to provide a more private guest experience.

The design introduces water elements on site that connect with adjacent water resources and landscaped parks - creating a unique water transit system to augment the hotel and conference experience. The movement of water is also reflected in several hotel terraces with clustered, outdoor gardens. The intent is to create several smaller enclaves where hotel guests can enjoy their natural setting and vistas in this unique site.

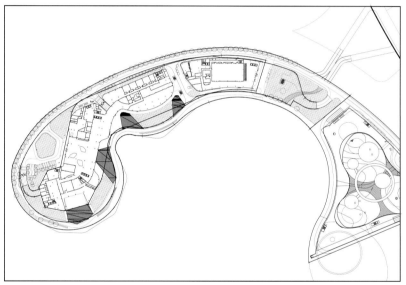

临港新片区创新魔坊
Lingang Sci-Tech City

项目地点：上海	Location: Shanghai
建筑面积：78 000 m²	Building Area: 78 000 m²
用地面积：43 831.5 m²	Site Area: 43 831.5 m²
设计时间：2017年9月—2019年9月	Design Time: September 2017 to September 2019
竣工时间：2020年11月	Completion Time: November 2020

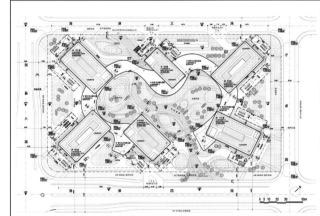

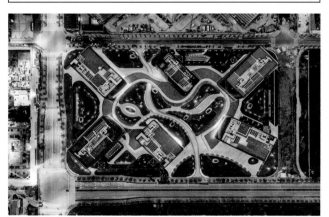

Gensler将中国古代"曲水流觞"的传统引入临港新片区创新魔坊的设计理念，旨在激发灵感创意，推进园区内的协同创新。为了营造园区内鼓励协同创新的氛围，Gensler的设计团队引入了连廊的设计元素，以一条蜿蜒曲折的连廊将园区内6栋楼连接起来，象征着园区内创意的萌发与汇聚，增强园区内协同功能。连廊内部设有大厅、会议及休闲空间，集视觉性与功能性于一体。

出于环境可持续性考虑，连廊顶棚以耐腐蚀性、持久性强的赤色陶土管铺就，与地面的红砖呼应，营造出温暖放松的户外氛围。在连廊的环绕之下，园区内形成一个别具新意的开放式空间，在6栋大楼和园区中心添设绿色景观，让身处其中的人获得花园式体验。

建筑的玻璃外墙时尚而动感，幕墙上点缀的百叶排列成波浪形状，作为点睛之笔，使园区和谐融入其地理环境——东海之滨。或铜色、或银色的百叶在幕墙上形成一道道柔和的波浪线，营造出诗意而动感的视觉效果。

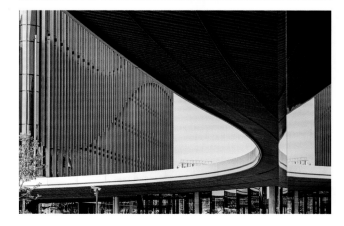

To build an international collaborative zone that welcomes synergy and sharing, the Gensler team proposed the idea of a loop. Thousands of years ago, the streams were common gathering spots for Chinese poets to drink and share ideas. Echoing such a graceful tradition, the loop connects the ground floors of six buildings in the area, creating a space that promotes sharing and innovation. In the loop are pavilions programmed for lobby, meeting, and amenity functions.

Designed with sustainable approach, the space of the loop is characterized by its tactile material with a ceiling of red terracotta rots and a floor made of red bricks. The warming, nurturing color and texture of the tactile material elevates outdoor experience. The loop also creates a large green garden inside the community, which extends the meeting space with natural surroundings.

Looking from distance, one will find the glassy curtain walls of the buildings both modern and dynamic, featuring poetic wave pattern fins. This delicate design is to heighten the community's connection to its geographic location—a reclaimed land facing East China Sea. When night falls, some of the fins would be illuminated and shine by the oceanside.

前海世茂中心
Qianhai Shimao Tower

项目地点：深圳	Location: Shenzhen
建筑面积：194 000 m²	Building Area: 194 000 m²
用地面积：12 746.66 m²	Site Area: 12 746.66 m²
设计时间：2014年7月—2016年12月	Design Time: July 2014 to December 2016
竣工时间：2019年12月	Completion Time: December 2019

近300 m高的前海世茂大厦，是深圳前海地区的地标，也是世界上最高的结构扭转建筑之一，它采用自下而上的收分式设计，呈现为外幕墙和结构外框柱同时扭转的形态。作为前海地区高密度的城市建筑群中最高的建筑，前海世茂大厦位于6座高层建筑之中。它的设计旨在实现"和而不同"的设计哲学，既和谐地融入整个建筑群，又凭借独特的形态脱颖而出。

塔楼的两个造型体块从地面向天空旋转上升，在上升的过程中以优雅的形态紧密相拥，通过旋转相拥的造型，代表了对于共同发展的美好向往。塔楼的朝向设计力求最大化海湾和山丘的景观视野，同时使各建筑之间的视野影响最小化。考虑到场地的高密度，塔楼在上升的同时逐渐旋转，通过释放邻近空间，与周边建筑形成更加友好的相邻关系。

塔楼严谨的几何造型基于一分为二的正方形平面，从不规则的基地开始，建筑造型向上逐渐收分的同时旋转了45度。建筑体量在塔冠处分开，展示轴线关系，隐喻地表达了两座城市的地理联结。最终形成的螺旋形几何造型既符合建筑美学的独特性，又兼具实现可持续目标的实用功能。这种旋转造型有效降低了风荷载，并减少了总体重量和所需的结构量，这在台风频繁登陆的前海地区是十分必要的。旋转的建筑造型和嵌入的凹槽大大降低了侧向风荷载，从而缩小了结构尺寸并有效降低了隐含碳排放。前海世茂大厦已获得LEED金级预认证，以及中国绿色建筑三星认证。

At nearly 300 m tall, Shimao Qianhai Center stands as one of Qianhai's landmark towers and one of the world's tallest structurally twisting buildings. The building tapers and simultaneously rotates, creating a striking imagery with twisting curtain wall and structural columns. The tower sits as the tallest building within a high-density urban block surrounded by six towers. It is designed to both harmonize the cluster of buildings into a wholistic composition and stand out as an iconic structure.

Two volumes rise gracefully from the ground towards the sky, each representing its own identity yet inextricably intertwined in a virtual dance. The building's soaring, spiraling form represents aspirations toward a harmonious future. It stands as a symbol of innovation and optimism. The positioning of the tower maximizes the views to the bay and the surrounding hills while minimizing views between towers. Given the density of the site, the rising tower's gradual twist responds well to its context by relaxing spatial adjacencies and freeing up views.

The tower's rigorous geometry is based on a perfect square split into two distinct volumes. It tapers and rotates 45 degrees as it rises from its extruded rectangular base to the apex. At the building's crown, the volumes cleave apart to reveal an axial connection, a metaphor for the connection between the two geographies and a link between mountain and sea. The resulting spiraled geometry is both aesthetically distinctive and technically functional for environmental resiliency goals. The rotation mitigates typhoon-level wind loads prevalent in Qianhai and reduces the overall weight and amount of structure required. The twisting building form and recessed corners reduce the lateral wind-loads significantly, downscaling the structural members and embodied carbon. The project has achieved LEED Gold precertification and China Three Star GBEL precertification.

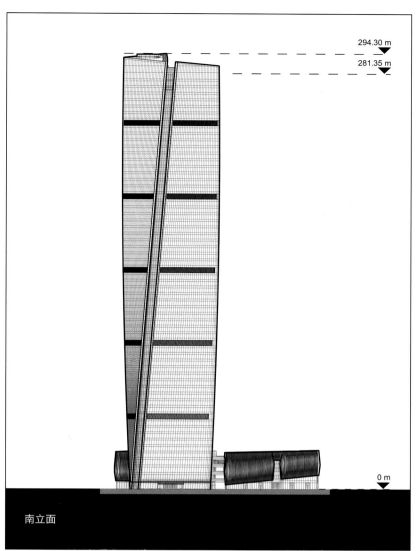

294.30 m
281.35 m
0 m

南立面

星扬西岸中心
Lumina Shanghai

项目地点：上海	Location: Shanghai
建筑面积：254 498 m²	Building Area: 254 498 m²
用地面积：18 622 m²	Site Area: 18 622 m²
设计时间：2015年8月—2018年5月	Design Time: August 2015 to May 2018
竣工时间：2021年11月	Completion Time: November 2021

星扬西岸中心坐落于上海西岸开发的中心位置，与前滩板块隔江相望，与陆家嘴金融区遥相呼应。其280 m的地上总高度为所在区域内最高地标建筑，引领整体规划的天际轮廓线。塔楼的造型灵感来自于上海市市花白玉兰。竖向的线条感配合四角柔和的曲线，以一种流畅而挺拔的姿态从地面向上生长，象征着萌芽的玉兰花含苞待放，丰富且精致的竖向线条设计也使建筑形体在视觉上显得更加纤细。塔楼顶部，塔冠大尺寸的竖向装饰线条使得塔楼犹如绽放的繁花，点缀在滨江半空。星扬西岸中心为上海西岸的文化创意、科技创新、创新金融三大产业提供了大体量、高标准的办公及商业配套空间，助力将上海西岸打造成世界领先的高品质城市核心中央活动区。

The 280 m high Lumina Shanghai by Henderson Land is located in the bustling Xuhui Riverside District, joining the urban skyline as a prominent part of the West Bund Development master plan. It overlooks Shanghai's Qiantan district and Huangpu River, and echoes the tall buildings of Lujiazui across the Bund. Inspired from white magnolia buds, representing beauty and strength, the tower's form is reminiscent of budding and blossoming branches. As a landmark tower, rising from the ground, with a delicate curved form, its crown is as if a new beginning, marking a new horizon along the riverbank, and defining a new height point for the riverside. Lumina Shanghai features a prime office space with retail amenities, creating an environment of commerce, culture, and creativity for professionals, visitors, and locals alike.

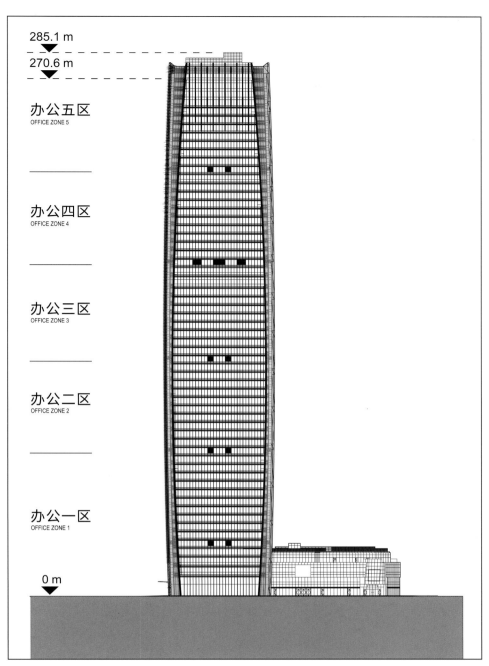

285.1 m	
270.6 m	
办公五区 OFFICE ZONE 5	
办公四区 OFFICE ZONE 4	
办公三区 OFFICE ZONE 3	
办公二区 OFFICE ZONE 2	
办公一区 OFFICE ZONE 1	
0 m	

CONCORD 西迪国际®
DESIGN GROUP
西迪国际|CDG国际设计机构

CDG国际设计机构(以下简称CDG)于2001年在加拿大BC省首府维多利亚注册成立,其宗旨为针对中国市场整合设计组合和提供综合性设计服务。机构成立至今,历经21年耕耘,CDG已成为在中国业界受人尊敬的知名设计品牌,连续多年荣获机构及个人多项行业内奖项。

CDG关注城市的整体性与项目的开放性共存共生,秉承"建筑助力城市风貌,在平凡中助力城市肌理的完善"的城市设计者责任,为社会真正意义的改良提供支持。多年来,CDG在区域规划、城市核心区改造、城市更新、文化保护区以及综合居住区、商业、娱乐、办公、酒店等领域打造了大量优秀的作品,作品遍布中国各地及海外多个区域。

CDG国际设计机构在加拿大及澳大利亚参与过当地多个项目的投资开发及规划设计。对中西方不同的生活、文化及市场的了解,使CDG为中国客户提供服务时更有针对性和前瞻性。以国际化视野打造高品质的产品和服务,一直是CDG每一个设计作品成功的关键,也是与每一位客户长期合作的基础。多年来,CDG国际设计机构与万科、华夏、华润、中建、金茂、金隅、远洋、亚泰、龙湖、天地控股、中金、康桥等知名企业形成了长期的战略合作伙伴关系。

2012年,CDG国际设计机构推出其中文品牌"西迪国际",也正是对公司文化的本土化注解。希望更有亲和力的中文名称,伴随着西迪国际每一个成功作品的面世,能够更加深入人心。目前机构在北京、雄安、深圳、上海、沈阳、重庆、杭州、香港等地设有公司,规划、建筑、景观等各领域专业设计人员共200余人。

CDG International Design Ltd. was established in Victoria, BC, Canada in 2001 with the aim of providing integrated design services for the Chinese market. 21 years after its foundation, CDG today has become a well-known and respected design brand in the Chinese industry, and has been the recipient of various awards and honors in the preceding years.

CDG aims to focus on the symbiosis of city integrity and project openness, taking care to uphold the urban designer responsibility of "buildings that improve the city, helping the city establish perfection in the ordinary". Over the years, we have created a large number of works in the areas of regional planning, urban transformation, urban renewal, cultural preservation and comprehensive residential areas, commerce, entertainment, office, hotels, etc., all which are spread throughout China and various overseas regions.

CDG International is involved in the investment, development and planning of many projects located in Canada and Australia. An understanding of the different lifestyles, cultures and markets in China and the West enables CDG to become more targeted and forward-looking in its service to Chinese clients. Creating high-quality products and services from an international perspective has always been the core focus of every design work for CDG and the basis for a long-term cooperation with every customer. Over the years, CDG International has formed a long-term partnership with clients including but not limited to Vanke, CFLD, CR Land, CSCEC, JINMAO, BBMG, Sino-Ocean, Yatai, Longfor, Universe, CICC, and Kang Qiao.

In 2012, CDG International launched its Chinese brand - "Sidi international". The new Chinese name hopes to have more affinity, establishing a deeper root in the hearts of the people. At present, the organization has branches located in Beijing, Xiong'an, Shenzhen, Shenyang, Chongqing, Shanghai, Hangzhou, and Hong Kong. The company currently consists of more than 200 professional designers working in planning, architecture, landscape and other fields.

沈阳万科·东第
Vanke New Oriental Mansion Shenyang

项目地点：辽宁 沈阳　　Location: Shenyang, Liaoning
建筑面积：154 038.13 m²　Building Area: 154 038.13 m²
用地面积：60 134.13 m²　Site Area: 60 134.13 m²

CDG® 国际设计机构
城市区域规划 建筑设计 环境景观设计

CDG及**西迪国际**均为CDG建筑设计(北京)有限公司在国家工商行政管理总局商标局注册的中国境内专利商标，范围包含建筑设计及景观设计，任何公司及个人未经允许在相关领域使用上述名称，将被视为侵权行为并追究相应法律责任。

服务号：CDGCANADA
订阅号：CDG-NEWS

北京
地址：北京市朝阳区望京街10号望京SOHO T3B座9层
电话：010-57076855/6955　邮箱：cdg@cdgcanada.com
BEIJING
Add: Level 9, Building B T3, Wangjing SOHO, Chaoyang District, Beijing
Tel: 010-57076855/6955　Email: cdg@cdgcanada.com

深圳
地址：深圳市南山区蛇口兴华路6号南海意库1栋209
电话：0755-21616094/6294
SHENZHEN
Add: 209 Building 1, Nanhai Yiku, 6 Xinghua Road, Shekou, Nanshan District, Shenzhen
Tel: 0755-21616094/6294

沈阳
地址：沈阳市和平区南堤西路901号中海国际中心B座701
电话：024-81054583
SHENYANG
Add: 701, Building B, Zhonghai International Center, 901 Nanti West Road, Heping District, Shenyang
Tel: 024-81054583

武汉华润半岛九里
China Resources Between the Lake Wuhan

项目地点：湖北 武汉　　Location: Wuhan, Hubei
建筑面积：167 800 m²　Building Area: 167 800 m²
用地面积：144 098 m²　Site Area: 144 098 m²

长春嘉惠九里
Jiahui Jiuli Changchun

项目地点：吉林 长春	Location: Changchun, Jilin
建筑面积：109 800 m²	Building Area: 109 800 m²
用地面积：24 200 m²	Site Area: 24 200 m²

郑州康桥东麓园
Kangqiao Donglu Garden Zhengzhou

项目地点：河南 郑州　　Location: Zhengzhou, Henan
建筑面积：138 668.03 m²　　Building Area: 138 668.03 m²
用地面积：55 582.98 m²　　Site Area: 55 582.98 m²

深圳勤诚达·云邸
KEENSTAR Cloud Mansion Shenzhen

项目地点：深圳
用地面积：16 987.4 m²

Location: Shenzhen
Site Area: 16 987.4 m²

沈阳万科新都心
Vanke Shenyang New Urban Center

项目地点：辽宁 沈阳
建筑面积：960 000 m²
用地面积：600 000 m²

Location: Shenyang, Liaoning
Building Area: 960 000 m²
Site Area: 600 000 m²

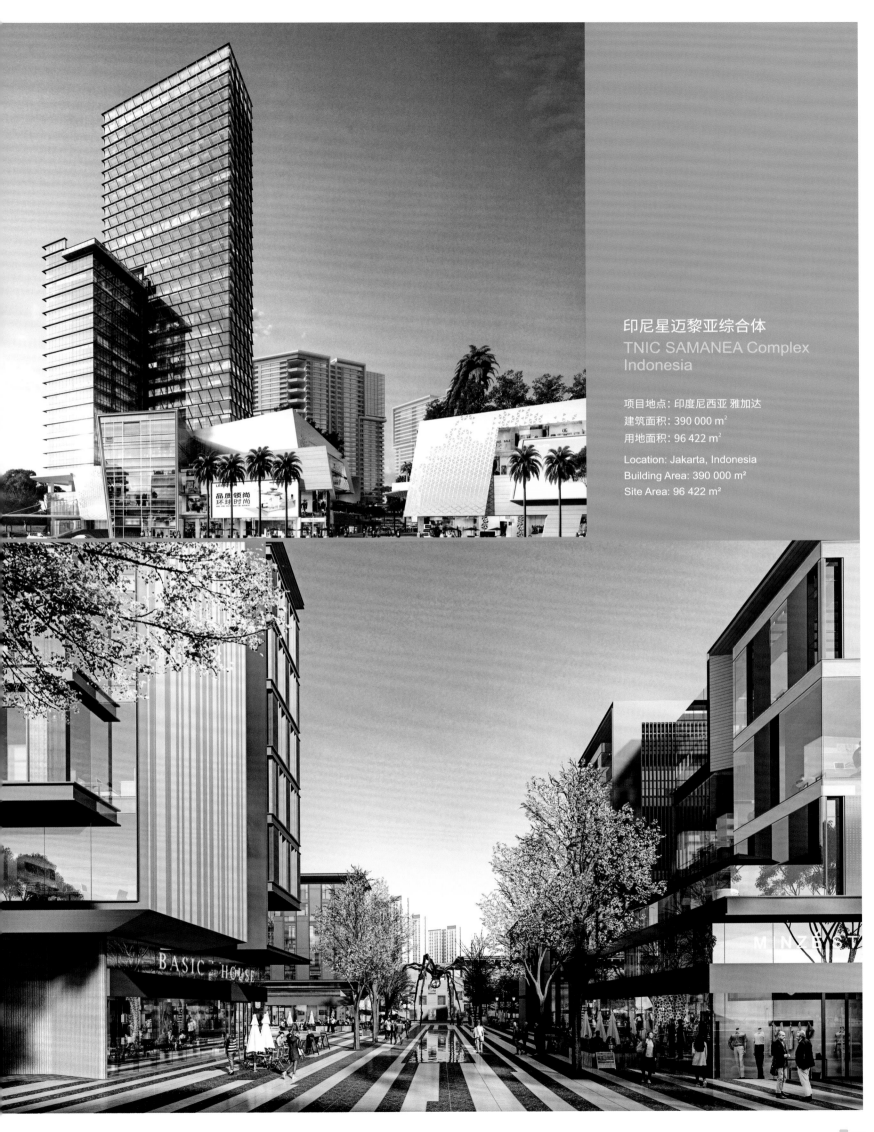

印尼星迈黎亚综合体
TNIC SAMANEA Complex
Indonesia

项目地点：印度尼西亚 雅加达
建筑面积：390 000 m²
用地面积：96 422 m²

Location: Jakarta, Indonesia
Building Area: 390 000 m²
Site Area: 96 422 m²

越南胡志明市莲花东
Lotus East Ho Chi Minh City Vietnam

项目地点:越南 胡志明
建筑面积:140 000 m³
用地面积:270 000 m²

Location: Ho Chi Minh City, Vietnam
Building Area: 140 000 m²
Site Area: 270 000 m²

沈阳汇置·峯
HuiLand Sky Villa Shenyang

项目地点:辽宁 沈阳
建筑面积:1 046 787 m²
用地面积:171 334 m²

Location: Shenyang, Liaoning
Building Area: 1 046 787 m²
Site Area: 171 334 m²

天津金隅·云筑
BBMG Aerial Palace Tianjin

项目地点：天津
建筑面积：291 449.17 m²
用地面积：132 346.3 m²

Location: Tianjin
Building Area: 291 449.17 m²
Site Area: 132 346.3 m²

济南市中万科城
Vanke City Ji'nan

项目地点：山东 济南
建筑面积：366 476.13 m²
用地面积：1 117.74 m²

Location: Ji'nan Shandong
Building Area: 366 476.13 m²
Site Area: 1 117.74 m²

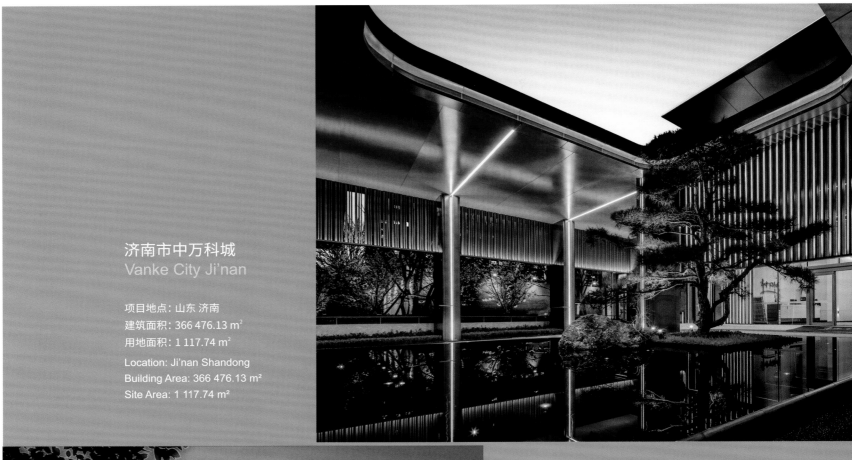

北京万科观承别墅
Vanke Cathy Villa Beijing

项目地点：北京
建筑面积：210 000 m²
用地面积：110 000 m²

Location: Beijing
Building Area: 210 000 m²
Site Area: 110 000 m²

GWP 建筑事务所
GWP Architecture Co., ltd

GWP建筑事务所（以下简称GWP）始创于美国波士顿，在中国广州、中国香港设立中国区域总部以及海外项目分部，参与研究和运作的项目遍及全球，以国际化视野、思维和专业技术开展有关城市规划和建筑设计一体化等专业领域的项目实践。GWP秉持着"全球智慧联盟"的理念，汇聚多方人才，以"创意与实用的平衡"为设计宗旨，凭借多维度思考赋能提升项目综合价值。业务范围主要涵盖超高层总部、商业综合体、公建地标、生态文旅、文化艺术中心以及品质居住等多种类型建筑及空间设计，设计作品荣获百余项国内外知名设计奖项。

GWP团队擅长从聆听分析客户和项目的需求出发，以研究引领设计、技术赋能实践、管理把控品质的实践方法论，与客户携手并进，共同创造具有影响力的地标项目。其核心设计及管理团队由毕业于哈佛大学、加利福尼亚大学、新加坡国立大学、伦敦大学、华南理工大学、东南大学等国内外知名学府，在城市规划、建筑设计、景观设计、室内设计等领域有着卓越的专业和学术教育背景，以及在专业领域实践多年有着丰富经验的项目顾问、设计总监、资深设计师、工程师组成。有扎实过硬的技术总工团队全过程指导，把控每一个项目"从创新设计到高效品质落地"的流程。

Founded in Boston, USA, GWP (GWP Architecture Co., ltd) has set up China regional headquarters and overseas project branches in Guangzhou and Hong Kong, China. The projects involved in research and practice have spread all over the world. With international vision and expertise, GWP has carried out project practices in such specialized fields as urban planning and architectural design integration. GWP adheres to the concept of "Global Wisdom Partnership", brings together various talents, takes "balance between creativity and practicality" as the design purpose, and enhances the comprehensive value of projects from multi-dimensional thinking. The business scope mainly covers super-tall headquarters, commercial complex, public landmarks, ecology tourism, culture and art center, and quality residential, and other types of architecture and space design. The works of design won more than hundred domestic and foreign well-known design awards.

GWP team starts from listening and analyzing the needs of clients and projects, leading design with research, enabling practice with technology, controlling quality with management, and working hand in hand with clients to create influential landmark projects. Its core design and management team composes of Harvard University, University of California, National University of Singapore, University of London, South China University of Technology, Southeast University and other well-known universities from both China and overseas, which have excellent professional and academic education background in urban planning, architectural design, landscape design, interior design and other fields, as well as project consultants, design directors, senior designers and engineers with rich experience in the professional field. We have a solid technical team to guide and control the whole process of each project from "innovative design to quality landing with high efficiency".

地址：广州市天河区珠江新城花城大道68号环球都会广场32层
电话：020-23339010
邮箱：gwpbusiness@gwp-architects.com
网址：www.gwp-architects.com

Add: 32F, 68 Huacheng Avenue, International Metropolit
Tel: 020-23339010
Email: gwpbusiness@gwp-architects.com
Web: www.gwp-architects.com

广州丰盛101超高层综合体
Guangzhou Fengsheng 101 Highrise Mixed-Use

项目地点：广东 广州
设计时间：2019年
项目阶段：在建中，将于2023年竣工
建筑面积：81 000 m²
用地面积：7 000 m²
建筑高度：200 m
服务范围：规划设计、建筑设计、景观设计

Location: Guangzhou, Guangdong
Design Time: 2019
Project Stage: Under construction, will be completed in 2023
Building Area: 81 000 m²
Site Area: 7 000 m²
Building Height: 200 m
Design Scopes: Master Planning, Architectural Design, Landscape Design

项目设计以扬帆起航的画面为概念灵感，意在用凝固的建筑去描述这种动态的趋势，以更亲近的尺度拉近市民与社区、自然的距离，使之成为融入城市的场所。GWP设计团队希望在丰盛101超高层综合体的设计中本着以人为本的出发点，注重采光、通风、与自然生态景观相融合，引入多层次的交互开放空间，在建筑形式与结构美学相得益彰的同时，赋予项目丰富的个性和独特的魅力。

The project is inspired by the image of sailing, hoping to depict the dynamic gesture with a static architectural form that can reach to the people, community and nature at a much more intimate distance. GWP always advocates that people at the center of the concept, the design team hopes to merge natural light, natural ventilation, and natural landscape in this high-rise landmark by introducing multiplelayered interactive open spaces, which promotes both formal and structural elegance and meanwhile endows the project with rich characteristics and unique charms.

广州三七互娱全球总部大厦
Guangzhou 37 Interactive Entertainment Headquarters

项目地点：广东 广州
设计时间：2020年
项目阶段：在建中，将于2024年竣工
建筑面积：95 315 m²
用地面积：6 427 m²
建筑高度：172.5 m
服务范围：规划设计、建筑设计

Location: Guangzhou, Guangdong
Design Time: 2020
Project Stage: Under construction, will be completed in 2024
Building Area: 95 315 m²
Site Area: 6 427 m²
Building Height: 172.5 m
Design Scopes: Master Planning, Architectural Design

GWP团队深度解读琶洲西区城市设计导则与三七互娱企业文化对于未来美好生活与工作环境发展愿景的需求，以"玩心创造世界"为设计思考起源，通过具有标识性与功能性相融合的设计，传递一种有趣和谐、开放创新、凝聚共享的企业文化。该项目的建成，不仅会提升三七互娱企业员工工作的幸福感和归属感，其开放性的首层大堂设计还会成为促进建筑与城市周边对话的标志性场所，为总设计师制度在琶洲片区的实践提供新的解题思路。

The GWP team deeply interprets the urban design guidelines of Pazhou West District and the corporate culture of Sanqi Interactive Entertainment for the development vision of a better life and working environment in the future, and takes "Playing the heart to create the world" as the origin of design thinking. The design conveys a corporate culture of fun, harmony, openness and innovation, cohesion and sharing. The completion of the project will not only enhance the happiness and sense of belonging of the employees of Sanqi Enterprise, but also the open lobby design on the first floor will become a landmark place to promote the dialogue between the building and the surrounding city, providing the chief designer system in Pazhou area The practice provides new problem-solving ideas.

城市绿洲
Urban Oasis

大参林医药集团荔湾运营中心
Dashenlin Pharmaceutical Group Liwan Operation Center

项目地点：广东 广州
设计时间：2019年
项目阶段：已竣工
建筑面积：112 324 m²
用地面积：15 487 m²
建筑高度：100 m
服务范围：规划设计、建筑设计

Location: Guangzhou, Guangdong
Design Time: 2019
Project Stage: Completed
Building Area: 112 324 m²
Site Area: 15 487 m²
Building Height: 100 m
Design Scopes: Master Planning, Architectural Design

由GWP主持设计的大参林医药集团荔湾运营中心项目，位于荔湾区海龙国际科技创新产业园区起步区内。作为广州荔湾十大重点项目之一、广州首批新型产业用地（M0）示范用地，承载着打造广州工程模范标杆的重要意义。该项目以"城市绿洲"为设计概念，旨在满足混合产业功能的同时，充分发挥建筑空间的灵活性，以其独特魅力与坚定姿态，链接多元产城，创造一个传统与未来的对话空间。

The Dashenlin pharmaceutical group Liwan operation center designed by GWP, it is located in the Hailong International Science and Technology Innovation Industrial Park in Liwan District. As one of the ten key projects in Guangzhou Liwan and the new industrial land (M0) demonstration land in Guangzhou, it carry the significance of building in Guangzhou projects. The design concept of "Urban Oasis", the project aims to satisfy the functions of mixed industries while was used to the flexibility of the building space. With its unique glamour and firm attitude, it links the multi-industry city and creates a dialogue space between tradition and future.

无限流动
Infinite Flow

东莞 CBD 示范性停车楼
Dongguan CBD Parking Building

项目地点：广东 东莞
设计时间：2019年
项目阶段：已竣工
建筑面积：20 000 m²
用地面积：4 844 m²
服务范围：建筑设计、景观设计

Location: Dongguan, Guangdong
Design Time: 2019
Project Stage: Completed
Building Area: 20 000 m²
Site Area: 4 844 m²
Design Scopes: Architectural Design, Landscape Design

　　GWP团队从项目的实际功能需求出发，以"无限流动"为设计概念把停车、人车动线和商业布局巧妙匹配，功能和优美的建筑形态完美融合，将创新型的设计理念与城市新地标的整体形象相结合，力图在解决该区域停车拥堵问题的同时为都市通勤者构建一种便捷舒适的交通体验，创建一个令人驻足停留的城市新场所。

　　Starting from the project's actual requirements in function. And uses the concept of "Infinite Flow" to perfectly solve the relationship between parking efficiency, circulation and supporting programs, creating a beautiful architectural form. While merging the innovative design concepts into the comprehensive image of a new urban landmark, GWP strives to resolve the local traffic issue and improve the experience on the road for urban commuters, as well as to create a new urban 'place' for people to gather and stay.

日本北海道温泉度假酒店
Hokkaido Spa Resort

项目地点：日本 北海道	Location: Hokkaido, Japan
设计时间：2021年	Design Time: 2021
项目阶段：深化设计	Project Stage: Deepening Design
建筑面积：600 m²	Building Area: 600 m²
用地面积：13 000 m²	Site Area: 13 000 m²
服务范围：建筑设计、室内设计	Design Scopes: Architectural Design, Interior Design

"以极简主义演绎建筑恒久之美，让纯粹和质朴成为生活的主旋律。"由GWP建筑室内专业一体化设计的日本北海道温泉度假酒店坐落在以滑雪观光业著称的日本城镇——新雪谷町。设计团队将雪山经过光影洗礼呈现优雅美感的动态过程作为设计灵感的来源，结合简洁流畅的线条赋予室内空间流转的动感，给每一位邂逅者留下犹如居于云山间的视觉印象。该项目的设计既表达了人们对未来美好生活的信念，同时也是GWP团队在对创新设计、品质落地的理念坚持下的又一个国际化高品质设计输出的代表作。她将在世界的关注中展现东方文化底蕴中的勃勃生机。

The Hokkaido Spa Resort transmits the eternal beauty of architecture via minimalism, making purity and rusticity the main theme of life. This project is located in Niseko Town, Hokkaido, Japan. It is a ski resort. The GWP team created Hokkaido SPA resort hotel from architecture and interior. The team took the beautiful dynamics presented by the light and shadow of the snow-capped mountains as the design inspiration, and used simple and smooth lines to give the interior space, Every viewer with the feeling of living in the mountains surrounded by clouds. It is not only an expression of faith in a brighter future, but also another exemplary piece of work with international influence by GWP Architects, a team committed to designing with creativity and delivering with quality. The project will reveal the vitality embedded within the cultural background of East Asia.

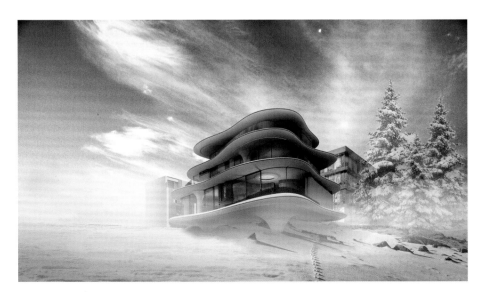

杭州苕溪公园商业街
Hangzhou Tiaoxi Park Commercial Street

项目地点：浙江 杭州
设计时间：2019年
项目阶段：已竣工
建筑面积：41 523 m²
用地面积：15 928 m²
服务范围：规划设计、建筑设计、景观设计

Location: Hangzhou, Zhejiang
Design Time: 2019
Project Stage: Completed
Building Area: 41 523 m²
Site Area: 15 928 m²
Design Scopes: Master Planning, Architectural Design, Landscape Design

　　GWP团队从商业思维的角度出发，融合临安文化与商业环境，团队力图通过区别于"开市即营，散市即归"的传统商业模式，将全新的生活方式融入商业休闲业态。建筑形态以烟雨江南山水画为灵感进行提炼升华，沿着市民日常的生活脉络，总结归纳出酒店与商业、文化、娱乐等多功能业态互补共荣的商业环境。在移步中，感知一店一景、一店一特色，将苕溪公园文化休闲商业街打造成杭州市的商业新地标。

　　From the perspective of business thinking, GWP team integrates the Lin'an culture and commercial environment. By try to integrate a new life into the commercial leisure model through a traditional business model that is different from the traditional commercial pattern of "Opening the market and returning to the market". The architectural form is refined with the inspiration of the Jiangnan painting in the daily life of the citizens, it summarizes the multifunctional business such as commerce, culture and entertainment. In the process of moving, you can perceive the view around the store, special feature in different store, and create Hangzhou Tiaoxi park commercial street as a new commercial landmark in City.

ppa|s ppas德国佩帕施城市发展咨询
Pesch & Partner, Architekten & Stadtplaner

境外

"以富有远见的设计促进城市的美好转变",是ppas一贯秉承的专业理念。

ppas德国佩帕施城市发展咨询(简称ppas)在1982年于德国北威州成立,是德意志制造联盟成员企业,现已成为德语区最富影响力的规划咨询机构之一。在德国规划泰斗弗兰茨•佩西教授领导下,公司于1995年和2012年在巴符州首府斯图加特和上海市静安区分别设立了南德和大中华区总部。得益于来自发展咨询、区域与城市规划、建筑与景观设计等跨领域多层次的专业团队,ppas致力于在城市发展不同层面,提供我们这个时代最富创造力的设计解决方案。

ppas的设计涵盖区域发展规划、城市设计、实施规划、高质量中小建筑设计、公共空间设计5个领域,在欧洲与中国参与的重要项目有:联邦直辖市不来梅内城规划、巴符州首府斯图加特市城市发展规划、瑞士巴塞尔IBA 2020德国区城市设计、科隆市环城大道沿线城市设计、雄安新区启动区城市设计、长三角一体化示范区江南水乡客厅城市设计、上海市嘉定新城远香湖中央活动区城市设计、上海廊下郊野公园核心区设计等。

"Promoting the beautiful transformation of a city with visionary design" is the philosophy that ppas has always adhered to.

ppas, founded in 1982 in North Wales, Germany, is a member enterprise of the Deutscher Werkbund and has emerged as one of the most influential planning consulting firms in German speaking areas. Under the leadership of Professor Franz Pesch, a respected German architect and urban planner, ppas has set up South Germany Headquarters and Greater China Headquarters in Stuttgart, the capital of Baden-Württemberg, and Jing'an District of Shanghai in 1995 and 2012, respectively. Boasting diverse professionals from a wide rage of disciplines involving development consulting, regional and urban planning, architecture and landscape design, ppas is committed to providing creative design solutions in our era at different dimensions of urban development.

The designs of ppas cover five fields, namely regional development planning, urban design, implementation planning, high-quality design for small and medium-sized buildings, and public space design. The major European and Chinese projects that ppas has played a part in are: inner city planning of Bremen, a German municipality; urban development planning of Stuttgart, the capital of Baden-Würt temberg; urban design of German exhibition area of IBA Basel 2020; urban design of regions along the Cologne Ring; urban design of the pilot region of Xiong'an New Area; urban design for "Watertown Living Room" Project of the Demonstration Area for Yangtze River Delta Integrated Development; urban design of the Yuanxiang Lake Central Activity Zone in Jiading New City, Shanghai; design of the core area in Langxia Country Park, Shanghai, etc.

上海	Shanghai	多特蒙德	Dortmund	斯图加特	Stuttgart
地址:上海市中山东二路88号D幢501	Add: 501, Building D, No. 88, Second East Zhongshan Road, Shanghai	地址:多特蒙德市霍尔德堡街11号	Add: 11 Horder BurgstraBe, Dortmund	地址:斯图加特市莫里克街1号	Add: 1 MorikestraBe, Stuttgart
邮编:200002	P.C: 200002	邮编:44263	P.C: 44263	邮编:70178	P.C: 70178
电话:021-52127848	Tel: 021-52127848	电话:+49(0)231.477929-0	Tel: +49(0)231.477929-0	电话:+49(0)771.2200763-10	Tel: +49(0)771.2200763-10
网址:http://ppas.cn/	Web: http://ppas.cn/	网址:http://ppas.cn/	Web: http://ppas.cn/	网址:http://ppas.cn/	Web: http://ppas.cn/

嘉定新城远香湖地区城市设计
Urban Design of Yuanxiang Lake Area in Jiading Xincheng

项目地点:上海
建筑面积:5 250 000 m²
用地面积:4.56 km²

Location: Shanghai
Building Area: 5 250 000 m²
Site Area: 4.56 km²

嘉定市示范图片
Model photos of Jiading City

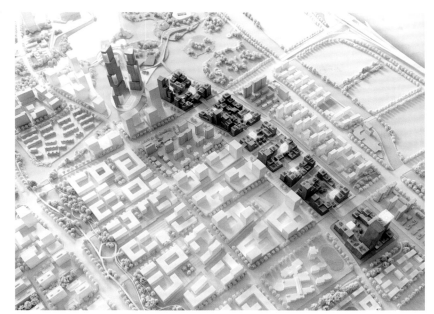
"光明谷"——创新与城市生活的融合
"Bright Valley"-The integration of innovation and urban life

"文化环"为市民提供独特的滨水体验
"Culture Loop" providing unique waterfront experience for citizens

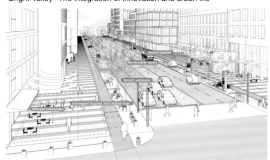
市民智能车站
Smart-station for citizens

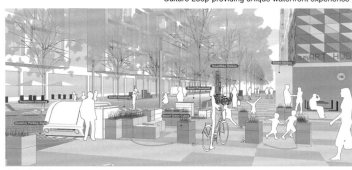
人车智能街道
Smart streets for people and vehicles

在上海最新的规划中,嘉定新城将作为独立新城带给大都市区更具活力的未来。在ppas的规划中,新城核心区通过将文化环、垂直花园塔与光明谷三个关键要素富有魅力地凝聚一体,并通过"一体枢纽、创造赋能、活力空间、智慧共生、低碳韧性"5项相互关联的规划战略,为这个"嘉昆太"区域的创造中心塑造前瞻型的发展模式。

In terms of the latest city planning of Shanghai, the construction of an independent new city, Jiading New City, will enliven the metropolitan area. In PPAS's planning, the core area of the new city is composed of three attractive key elements: Culture Loop, Vertical Garten Tower, and Bright Valley. A forward-looking development model for the creation center of Jiading-Kunshan-Taicang Cooperative Zone will be shaped through five interdependent planning strategies, i.e. hub-based integration, creation to empower, dynamic space, smart co-existence, and climate resilience.

"光明谷"创造活力空间
"Bright Valley" creates vitality spaces

城市中心的智能体验
Smart experience in city center

一座绿色可持续发展的城市
A green and sustainable city

Archilier Architecture 建筑设计顾问咨询公司
archilier architecture

Archilier Architecture 建筑设计顾问咨询公司（以下简称AA）总部位于美国纽约，并在上海设有工作室。作为一个充满活力的国际建筑设计创作公司，在全球拥有广大的客户群。AA一直致力于为客户提供优质的城市酒店、度假酒店、商业办公综合体和大型度假区与小区规划等各种类型项目的建筑设计及咨询服务。AA关注每一个项目的独特性，立足于尊重当地的特色文化、民俗与建筑传统，诠释客户的独到理念，帮助实现他们的独特愿景。

公司的国际设计师团队曾在美国和亚洲的多国成功设计过许多重大项目，特别是豪华酒店、度假酒店、办公大楼及大型商业综合体的设计。他们有着从总体规划、建筑设计、室内设计、BIM设计到绿色设计的丰富经验。不论项目在哪里，设计都着力打造成功的建筑形象和群体氛围，促进城市的内部连接和环境美化。

在设计方法上AA注重与客户、内部团队和各顾问方之间的沟通协作。Archilier是一个由"Archi"与法语后缀"lier"合成自创的词，在法语中，"lier"的意思是"连接"。设计需要将多股协作力量凝聚成一体，最终打造出完美的设计作品。AA团队不是只专注于形式的能手，而是极富经验的工艺师。从手绘草图、数位视觉化研究到建模，设计过程中的每一步都旨在通过空间、形式、灯光、材料等媒介创造令人难忘的客户体验。

设计理念

每个项目都独一无二：AA从不先入为主提供一个固定的设计风格。每一个设计都根据当地独特的地理、人文、经济与技术参数量身定制。

精心营造空间场所：大至公共广场，小到酒店客房，创造令人难忘的空间与体验一直是我们不懈的追求。

坚持原创设计：我们力图在设计中再现场所的时代感与永恒感。

承载社会责任：在自然环境面临严峻挑战的时代，AA恪守本职、保护环境、尽责尽力。

竭诚服务客户：为实现客户愿景全力以赴，提供超值高效的设计服务。

Headquartered in New York, Archilier Architecture is a dynamic architectural design consultancy group serving a wide range of clients, from developers to design firms, focused on hospitality and mixed-use commercial projects. Wherever our projects are located, we strive to design projects that respect the distinct cultural, social and architectural traditions of their communities while also reflecting our client's unique vision. Our intention is to create masterful architectural environments that inspire urban connectivity and elegant placemaking within our cities.

Our international team of experts has a broad range of experience designing and leading projects in the United States and Asia, most notable hotels, resorts, retail centers, office complexes and large scale mixed-use developments. The core members of Archilier Architecture together have extensive experience in the various design fields of architecture, master planning, interior design, BIM practice and green design.

Our approach to design is one of Open Collaboration – collaboration with clients, colleagues and consultants. In French, the "lier" in Archilier means "to link". By connecting our client's goals with intelligent, creative, and beautiful design we are able to weave the multiple strands of collaborative input into a unified whole, producing great work.

Our Values

Every project is unique: Archilier does not have a signature style or any preconceptions. Every design emerges from the unique physical, cultural, economic, and technological parameters that define it.

Place making is paramount : Creating memorable places & experiences is our raison d'etre, whether at the scale of a public plaza or a hotel guestroom.

Authenticity is crucial: We always look to provide authentic storytelling; we understand a design needs to connect with the area it is located. Whether designing skyscrapers in dense metropolis or a small boutique retreat by the lakeside, we always focus on creating unique and coherent user experience, both tangible and emotional. We use the culture, climate, and the physical characteristics of a given site to inspire us, we strive to capture both the timeless spirit of a place as well as the spirit of the times.

Responsible design is required: We believe in being good global citizens. Even if when we're not working within a ratings system like LEED, our work is informed by a sensitivity to issues of climate, resilience, and sustainability.

Exemplary service is mandatory: We take our responsibility to our clients seriously. Deadlines will be met. Goals will be satisfied. Expectations will be exceeded.

地址：上海市杨浦区国权北路1688弄78号湾谷科技园A4座6楼
邮编：200438
电话：021-61803512
邮箱：info@archilier.com

Add: Building A4, 6th Floor, No.78,1688 North Guoquan Road, Yangpu, Shanghai
P.C: 200438
Tel: 021-61803512
Email: info@archilier.com

青岛海天中心
Qingdao Haitian Center

项目地点：山东 青岛	Location: Qingdao, Shandong
建筑面积：490 000 m²	Builidng Area: 490 000 m²
启动时间：2012年	Begin Time: 2012
竣工时间：2021年	Completion Time: 2021
主创建筑师：盛开	Chief Architect: Kai Sheng
摄影：章勇、陈辰、傅兴、刘颖	Photographer: Yong Zhang, Chen Chen, Xing Fu, Ying Liu

海天中心位于山东青岛CBD沿海地区，共包含3幢超高层塔楼（自编号T1、T2、T3），最高的T2塔楼高369 m，截至目前，是青岛市也是山东省的最高建筑。

海天中心的前身，诞生于1988年的海天大酒店，是山东省最早的涉外五星级酒店之一，作为改革开放的见证者，在青岛市民心中占有举足轻重的地位。设计试图将老海天的美好记忆糅合进新海天的生命中，在保留原有历史底蕴的同时，用新的语言创造出了新的形式。

Sitting at the beautiful seashore in the new CBD of Qingdao, a major seaport and commercial center in China, Haitian Center features three multi-functional towers with the tallest tower in both Qingdao and Shandong province.

The site was home to the original Haitian Hotel, the first 5-star hotel in Qingdao City. Having once served as the city's hospitality hot spot, the site holds a special place in the city's history and its citizens' memories. We sought to carry the city's fond memory of the original Haitian hotel on the same site into our design while creating new forms with languages that represent the vigor and aspiration of an emerging new "first-line" coastal city in China.

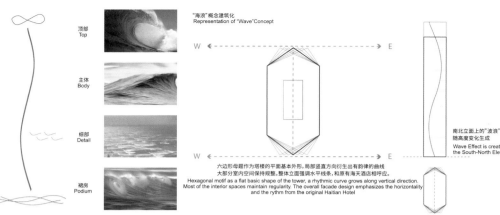

承袭场地历史文脉和滨海环境,建筑设计提取了"海之韵"的理念来表现大海天然的有机曲线和动态能量,以六边形平面作为形体出发点,将六边形南北两个端点随高度升高逐层位移,使建筑体量沿竖向旋转生长,从而在主立面上创造出一条有韵律的竖向曲线,巧妙地把海洋元素转化为设计语言,在形态上形成了与大海相呼应的优雅柔和又充满张力的形体。起伏的海浪通过裙楼蜿蜒的墙体得以表达,柔和的涟漪则通过幕墙的排列得到体现,错落有致的玻璃幕墙,在阳光的照耀下,如海风吹起的千层浪,远远观之,与塔身的曲线、塔冠的穹顶一同形成了"大浪、小浪、千层浪"的垂直景观。波光荡漾,碧波万顷,是玻璃在阳光的折射下形成的诗意效果。

Drawing from the site's rich history and proximity to the seashore, we formulated the "Rhythm of the Sea" as the project's design narrative to reflect the movement and energy of ocean waves and to represent Qingdao's unique coastal character. We used the original building's hexagonal geometry as our tower plates' starting point to pay homage to the site's history. Then we shift the hexagon's tips at the north and south ends progressively as the towers rise, creating a vertical wave along the building's main elevations and thus metaphorically bringing the sea into our site. Arching crests and surging waves are expressed at the podium as continuous undulating walls, while the sea's gentler movements are articulated in the tower's skin with triangular-shaped panel systems creating a rippling effect. These panels shimmer as sunlight bounces off the building's facade.

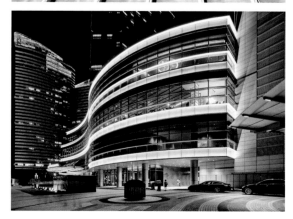

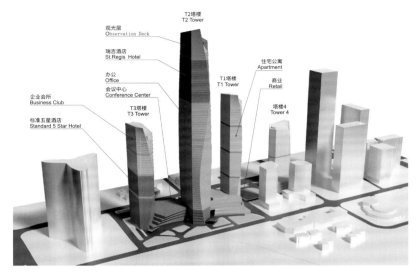

项目含两个五星酒店、国际会议中心、5A甲级写字楼、豪华公寓和商业中心等主要业态，最高的T2塔楼还设有艺术中心、城市观光厅和VIP俱乐部。

位于T2塔楼58层的瑞吉酒店空中大堂拥有全景观海视野，空中大堂下层的高空无边泳池将天空与海景融为一体。最顶层的云端钻石俱乐部拥有城市至高视点，位于81层的城市观光平台周围的360度落地环绕玻璃将城中美景尽收眼底，观光平台西侧设有3个全玻璃结构的三角形盒子，踏在透明的玻璃地板上，可从336 m的高空俯瞰城市。

The overall project contains two five-star hotels, a conference center, Class-A office space, a luxury apartment tower, and a 5-story retail center. T2 contains the St Regis hotel and high-end offices.

The hotel is on top with its sky lobby at floor 58 and sweeping views. The hotel's infinity pool is located on the floor below - reflecting the sky, the water in the pool visually merges with the sea below. T2's special function space, the "Diamond Club" sits at the top of the building under an electrochromic glass dome. It is the crown jewel of the tower providing the highest viewpoint in the whole city. An observation deck and museum "Art Above the Clouds" are right below the dome. With a 360-degree view over sea and city, this unique guest experience is enhanced by the three glass prisms projected outside the building which invite daring people to stand on a glass floor 336 m above-ground.

 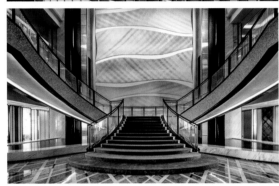 

设计不仅希望将项目打造成城市地标，还希望将它作为公共客厅有机地融入城市肌理，在多个层面与城市发生联系：地下空间连接轨道交通，迎来送往大量乘客；地面街角广场和裙楼通廊保证空间的连续性和通透性，为居民留下充分的步行空间；平行于海岸线的漫步大道贯穿3栋塔楼，与城市道路和景观融为一体；裙楼屋顶花园则打通室内外空间，为宾客提供亲海观景点。

This project was established not only as an icon but also as a public "living room" woven into the dynamic urban fabric. It is connected to the city at multiple levels with a basement link to subway lines and two-level urban plazas at both seaside and city side. A promenade, connecting the three towers across the full length of the property, parallels the oceanfront and include a promenade park with a series of landscaped terraces linking the site's gardens to the district's pedestrian network. With multiple vantage points of sea views, the development's podium roof terraces, just like the ground floor plazas, extend the buildings' interiors to the outside.

中海国际中心与环宇荟办公商业综合体
China Overseas International Center

项目地点：北京	Location: Beijing
建筑面积：76 210 m²	Builidng Area: 76 210 m²
用地面积：15 242 m²	Site Area: 15 242 m²
启动时间：2013年	Begin Time: 2013
竣工时间：2017年	Completion Time: 2017
建筑高度：100 m	Builidng Height: 100 m

项目包含两幢15层高的办公塔楼和5层高的商业裙楼，平面规划通过贯穿裙楼的通道将地块东南侧的元大都城垣遗址公园和西北侧的奥体公园有机地连接起来，面积约3 000 m²的广场包括了绿化、商业入口、落客停车以及通往地下商业和地铁接驳站的入口。

The concept for the design was inspired by the idea of glass crystals rising from the earth in the form of the office towers that merge with glass crystals falling from the sky, collecting to form the retail podium. The program is comprised of two 15 story office towers totaling 48 000 sm which sit above a five story podium totaling 28 510 sm of retail, restaurant, and entertainment uses. The design of Site 4 is defined by two major boulevards and occupies a prominent corner in the urban design of the commercial district. Site 4 faces Yuan Dynasty Capital Heritage Garden to the southeast and the new Olympic Center Park to the northwest of the property. Archilier Architecture's design creates a dynamic connection through the podium that unites the parks. Public plazas totaling 3 000 sm while providing the required green spaces serve as retail entrances, vehicular drop-offs and access to the network of below grade retail and subway concourses. The two office towers form a diagonal axis and serve as a gateway to the interior of the new business district.

天目湖 WEI 精品酒店
WEI Retreat Tianmu Lake

项目地点：江苏 溧阳
建筑面积：8 998 m²
用地面积：7 385 m²
启动时间：2013年
竣工时间：2016年
房间数：35

Location: Liyang, Jiangsu
Builidng Area: 8 998 m²
Site Area: 7 385 m²
Begin Time: 2013
Completion Time: 2016
Keys: 35

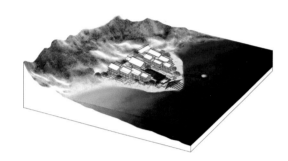

本项目位于自然景色环绕的天目湖中半岛上，朝南面湖，朝北面山，拥抱极致景色，极为隐秘脱俗。建筑夹在山与水之间，以自然为主体，概念以山形层次，堆叠延续，将中国古建筑之屋、檐、瓦、墙与现代简洁利落的建筑语言相结合，营造出宅院深度与文化质感。材料的选择体现了江南建筑的灵魂，灰砖、清瓦、白墙融合贯通了草木山水的气息。设计过程中对砖墙与瓦片的颜色、接缝、深浅进行了多次比对，对屋檐弧度与山峦起伏的和谐进行了多次推敲，最终使其风格纯粹，定位清晰，独具一格。

This project is located on a freshwater peninsula surrounded by beautiful natural scenery. It has mountains to the north and a lake to the south, providing ultimate privacy and natural views. The architectural design integrates traditional Chinese elements with modern style inspired by the nearby layering mountain ranges. The project features curved roofs with large overhanging eaves, stone tiles, white walls, wood details, and garden courtyards. The exclusive private house reflects traditional culture with clever and simple design. With its pure style and scenic location, the Tianmu Lake project will be a very unique boutique hotel in the region.

境外

扫描查看更多信息

ZDDI 中大国际
江苏中大建筑工程设计有限公司
JIANGSU ZHONGDA ARCHITECTURAL ENGINEERING DESIGN Co.,LTD.
英国UK LA太平洋远景国际设计机构
U.K PACIFIC LONG-RANGE INTERNATIONAL DESIGN INSTITUTION

中大国际由江苏中大建筑工程设计有限公司和英国UKLA太平洋远景国际设计机构强强联合组建。总部位于美丽的山水洲城、六朝古都南京，在徐州设立分公司，并于香港、上海及郑州设立分支机构，先后通过ISO9001质量管理体系认证、ISO14001环境管理体系认证、ISO45001职业健康安全管理体系认证，并被评定为AAA级信用企业。

江苏中大建筑工程设计有限公司由东南大学建筑设计研究院与台北启大设计顾问公司于1993年创办，持有国家建筑行业（建筑工程）甲级设计证书：甲级A232060221，是全国首批合资甲级设计公司之一。中大建筑成立近30年来，在江苏省和周边省份完成了大量优秀业绩。

英国UKLA太平洋远景国际设计机构是一家国际专业设计公司，公司致力于城市与建筑的功能规划和空间设计研究，在国内在建项目达数千万平方米，为江苏规模最大的境外设计机构之一，先后被评为中国最具影响力设计机构、中国最具品牌价值设计机构、建设部《中国建筑设计作品年鉴》特邀理事单位。

ZDDI Is Jointly Established By Jiangsu Zhongda Architectural Engineering Design Co., Ltd and U.K Pacific Long-Range International Design Institution. The headquarter is located in the beautiful landscape city, Nanjing, the ancient capital of Six Dynasties, with branch offices in Xuzhou, and branches in Hong Kong, Shanghai and Zhengzhou. It has passed ISO9001 quality management system certification, ISO14001 environmental management system certification, ISO45001 occupational health and safety management system certification, and was rated as AAA credit enterprise.

Jiangsu Zhongda Architectural Engineering Design Co., Ltd. was founded in 1993 by Southeast University Architectural Design Institute and Taipei Qida Design Consulting Co., LTD., holding the NATIONAL Architectural industry (Architectural engineering) Class A design certificate: Class A AW132020712, is one of the first joint venture Class A design companies in China. Since its establishment nearly 30 years ago, CUHK has accomplished a lot of outstanding achievements in Jiangsu Province and neighboring provinces.

U.K Pacific Long-Range International Design Institution is an international professional design company, the company is committed to urban and architectural functional planning and spatial design research, in China under construction projects of tens of millions of square meters. It is one of the largest overseas design institutions in Jiangsu. China's most influential design institutions, China's most brand value design institutions, the Ministry of Construction "Annual Review of Chinese Architectural Design Works" invited director units.

地址：南京市建邺区河西大街198号同进大厦五单元11层
电话：025-84739678
邮箱：UKLA2000@126.com
网站：www.chinesearch.com

Add: 11th Floor, Unit 5, Tongjin Building, No. 198 Hexi Street, Jianye District, Nanjing
Tel: 025-84739678
Email: UKLA2000@126.com
Web: www.chinesearch.com

郑州格拉姆国际中心
Zhengzhou Gramm International Center

项目地点：河南 郑州　　　Location: Zhengzhou, Henan
建筑面积：88 660 m²　　　Building Area: 88 660 m²
用地面积：0.94 hm²　　　 Site Area: 0.94 hm²

格拉姆国际中心由意大利格拉姆财团和罗马市政府委托设计，是郑东唯一一家由外商独资开发的商业地产项目。河南是中华文明的发源地，意大利是欧洲文明的摇篮，格拉姆国际中心的设计，希望体现出中国和意大利文化，项目采用超高层双子楼设计，并将哥特式风格与象征财富的钻石元素和中国文化中象征吉祥的莲花造型有机融合在一起，成为郑东新区一颗耀眼的明珠。

Zhengzhou Gramm International Center is commissioned by the Italian Rome, Italy Gramm consortium and Roman municipal government. And it is the only project in commercial property developed independently by a wholly foreign owned company Currently. Italy is the cradle of European civilization while Henan is the birthplace of China's Yellow River civilization. The design is full of culture blend and collision between Italia and China, and a mixture of architectural language; it helps to establish a modern, efficient landmark building with epochal characteristics, which reflects both Italian and Chinese cultures. The higher, the narrower, that's the style of the building. It means "progress at every step". The top is like a sparkling diamond. This project will finally become a dazzling pearl in Zhengzhou East CBD.

连云港新港城几何中心总体空间规划及核心区城市设计
The Overall Spatial Planning of the Geometric Center of Lianyungang Port New City and the Urban Design of Its Core Area

项目地点：江苏 连云港
项目规模：总体空间概念规划193 km²，核心区城市设计范围28.6 km²
获奖情况：国际招标以第一名中标

Location: Lianyungang, Jiangsu
Site Area: The overall space concept planning is 193 km², and the urban design area of the core area is 28.6 km²
Awards: The first place in the international bidding won the bid

方案提出"资源整合、土地注能、生态先行、因地制宜、有机更新、经济可行"的设计理念，通过"多维纵横的HUB"实现"连山、连水、连文、连轨、连城、连产"6大核心要素的串接，实现打造"区域协同发展的高站位战略核心区、产城绿色发展的优品质示范先行区、多元创新发展的强活力要素聚集区"的发展目标。

方案发挥"新港城几何中心"在交通赋能、产业服务、版块联动方面的优势，建构未来新港城城市中心体系和整体功能板块布局结构，彰显"六脉皆通海，青山半入城"的城市风貌特色，勾画未来新港城"山海港城互望、蓝绿网脉融城"的城市空间结构。

Plan put forward "resource integration, land can note, ecology first, adjust measures to local conditions, organic update, economically feasible" design concept, through "HUB" of the multidimensional and realize "lianshan, even water, even together, even the rail, and even produce" six core elements of concatenated, implementation to build the coordinated development of regional high positioning strategy core; Excellent quality demonstration pilot area for green development of the city; The development goal of the strong dynamic element gathering area of diversified innovation and development.

Play a "new harbor city geometric center" in the traffic assignment can linkage advantage, industry services, section, construct the new harbor city urban center system and the overall function section layout structure, reveal "six vein is sea, castle peak half into the city," city landscape features, shaping the new harbor city "shanhai harbour city looked at each other, blue green net vein into the" urban spatial structure.

中华药港核心区建筑设计
Architectural Design of the Core Area of China Pharmaceutical Port

项目地点：江苏 连云港
建筑面积：215 000 m²
用地面积：8.13 hm²

Location: Lianyungang, Jiangsu
Building Area: 215 000 m²
Site Area: 8.13 hm²

中华药港核心区总体定位为"山海连云、药港智谷"。规划从连云港山海城市特征入手，以层层退台的建筑为山、引水拟海，打造山水中的城市和城市中的山水；同时引入垂直森林和天街系统，将会议中心、运动中心、中央厨房、生活广场及专家公寓等功能串联为整体，为整个中华药港区域提供综合性服务。

The overall positioning of the core area of China Pharmaceutical Port is "mountain sea Lianyun and Pharmaceutical Port Zhigu". Starting from the characteristics of Lianyungang mountain and Sea city, the plan takes the buildings that retreat from the platform layer by layer as the mountains and draw water from the sea to create the city in the landscape and the landscape in the city; at the same time, it introduces the vertical forest and Tianjie system to connect the functions of the conference center, sports center, central kitchen, living square and expert apartment as a whole, so as to provide comprehensive services for the entire China Pharmaceutical Port area.

东台高新区邻里中心
Dongtai High-tech Zone Neighborhood Center

项目地点：江苏 东台
项目规模：150 000 m²
获奖情况：第一名中标

Location: Dongtai, Jiangsu
Site Area: 150 000 m²
Awards: First Place Winner

规划以"海潮云起、高山流水"为设计理念，通过海浪与云的形象演化，形成了轻盈流动的邻里中心形象。同时，邻里中心圆润贯通的横向线条寓意为水，层层退台的公寓寓意为山，整体形成面向城市展开的高山流水形象，打造具有山水意境的城市建筑。

邻里中心结合层层退台的空间设计了多样化的屋顶绿化，如梯田般重峦叠翠，在打造多层级绿化平台与空中交往空间的同时实现建筑与城市和谐共生。公寓区通过山、水、园的理念形成生态智谷，连续的风雨连廊串联起各栋公寓，形成社区居民交流的活力场所。

With the design concept of "sea tide clouds, high mountains and flowing water", the plan forms a light and flowing neighborhood center image through the evolution of the image of waves and clouds. At the same time, the rounded horizontal lines in the neighborhood center imply water, and the apartments with layers of terraces imply mountains, forming the overall image of mountains and rivers facing the city, creating urban buildings with landscape artistic conception.

The neighborhood center is designed with diversified roof greening in combination with the terraced space, which is stacked like terraced fields. It achieves harmonious coexistence between architecture and city while creating multi-level green platform and air communication space. Through the concept of mountain, water and garden, the apartment area forms an ecological wisdom valley. The continuous corridor connecting wind and rain connects each apartment building, forming a dynamic place for community residents to communicate.

建控·江山赋
JianKong·JiangShanFu

项目地点：河南 南阳
总建筑面积：1 000 000 m²

Location: Nanyang, Henan
Total Building Area: 1 000 000 m²

公园境、滨河家、巷陌围庭院、望闻天地宽。以城市园林的文化肌理作为规划着力点，考虑沿河景观的渗透，建筑与自然相映成趣，创造出与自然融合的建筑群体。

超大玻璃、简洁利落的立面，铸造东方奢隐社区，构筑契合当代精英人士的理想栖居之地。

Park in riverside home lanes around the courtyard Hope WenTian wide In city garden cultural texture as planning focus, considering the penetration of landscape along the river, set each other off becomes an interest, architecture and nature to create architectural group merging with nature.

Large glass, concise and agile facade, casting the east luxury community, contemporary building joint cremation of ideal living place.

柬埔寨西哈努克市 海龙湾度假中心
Hailong Bay Resort Center, Sihanouk, Cambodia

项目地点：柬埔寨 西哈努克
建筑面积：650 000 m²

Location: Sihanouk, Cambodia
Building Area: 650 000 m²

西哈努克港是柬埔寨最繁忙的海岸港口，也是除吴哥窟以外最热门的旅游胜地。项目位于柬埔寨西哈努克港东南方向20 km，项目总体定位为国际旅游度假休闲中心、海洋主题游乐中心、娱乐中心。依托基地优越的细沙碧水、蓝天白云及千米海岸线，从旅游文化角度出发，凸显一山一海，以娱乐综合体和中心商业区为两大核心，以公共娱乐轴线和度假休闲轴线串联整个基地；致力于将旅游氛围沿着核心区和两条轴线延伸，最终与海岸线相连，打造柬埔寨具有国际影响力的度假休闲中心。

Sihanouk port is the busiest coastal port in Cambodia and the hottest tourist attraction except Angkor Wat. The project is located 20 km southeast of Sihanouk port in Cambodia. The overall positioning of the project is international tourism and leisure center, marine theme amusement center and gambling entertainment center. Relying on the excellent fine sand and clear water, blue sky and white clouds and kilometer coastline, from the perspective of tourism culture, it highlights one mountain and one sea, connects the whole base with gambling complex, central business district and two cores, and connects the whole base with public entertainment axis and vacation leisure axis; it is committed to extending the tourism atmosphere along the core area and two axes, and finally connecting with the coastline, so as to create a unique tourism environment in Cambodia A resort and leisure center with international influence.

江苏淮海科技城创智产业园
Creative Industry Park of Jiangsu Huaihai Science and Technology

项目地点：江苏 徐州
用地面积：510 000 m²
获奖情况：第一名中标

Location: Xuzhou, Jiangsu
Site Area: 510 000 m²
Awards: First Place Winner

方案总体布局顺应原有地势，各功能组团沿中轴两侧有序展开，通过广场、道路、绿化等衔接内外空间；同时以建筑的进退、错落和穿插等形式丰富滨河形象和沿街城市界面，最大化发挥基地的自然优势。

建筑采用独栋、多层和高层办公结合的形式，以满足不同规模企业的入驻，贯穿基地的室外连廊形成丰富的空间体验，与开敞空间一起组成了园区内部活泼的交流场所。垂直方向的绿化将园区打造成为生态、绿色的园区。

The overall layout of the scheme conforms to the original topography, and each functional group is arranged orderly along both sides of the central axis, connecting the internal and external space through squares, roads, greening, etc.; at the same time, it enriches the riverfront image and the urban interface along the street in the form of building advance and retreat, scattered and interspersed, so as to maximize the natural advantages of the base.

The building adopts the form of single house, multi-storey and high-rise office to meet the needs of enterprises of different sizes. The outdoor corridor throughout the base forms rich space experience and forms a lively communication space in the park together with the open space. The vertical greening will build the park into an ecological and green park.

南京国际展览中心
Nanjing International Wxhibition Center

项目地点：江苏 南京
用地面积：130 000 m²
获奖情况：中国建筑学会新中国成立60周年建筑创作大奖建设部
　　　　　二等奖、省优秀设计一等奖、国家银奖

Location: Nanjing, Jiangsu
Site Area: 130 000 m²
Awards: Architectural Creation Award for the 60th Anniversary
　　　　of the Founding of the People's Republic of China
　　　　The second prize of the Ministry of Construction, the
　　　　first prize of provincial excellent design, and the
　　　　national silver prize

该项目坐落于风景秀丽的玄武湖和紫金山之间，以其优美的造型、现代的设施、宏伟的体量、完备的功能成为古都南京的一项标志性建筑。南京国际展览中心是按照当代国际展览功能需求建设的大型展览场馆，是集展览、会议、商贸、信息交流、娱乐和餐饮为一体的多功能综合性建筑。

中心的配套设施齐全，建筑造型优美，新颖别致，气势恢宏，已成为古都南京的一道亮丽的风景线。

Located between Xuanwu Lake and Zijin Mountain, it has become a landmark building in the ancient capital of Nanjing with its beautiful shape, modern facilities, magnificent volume and complete functions. Nanjing International Exhibition Center is a large-scale exhibition hall built according to the function of contemporary international exhibition. It is a multi-functional comprehensive building integrating exhibition, conference, commerce, information exchange, entertainment and catering.

The center has complete supporting facilities, beautiful architectural shape, novel and unique, magnificent momentum, which has become a beautiful landscape of the ancient capital of Nanjing.

境外

法国 AREP 设计集团
AREP Group

法国AREP设计集团（以下简称AREP集团）成立于1997年，是一家国际综合性设计公司，隶属于法国国家铁路总公司SNCF。公司总部位于法国巴黎，并在中国、俄罗斯、越南、阿布扎比、卡塔尔、印度等设有子公司。

AREP集团强劲的团队聚集了来自三十多个国家的一千多名员工，其中包括规划师、建筑师、工程师、经济概算师、室内设计师等。业务范围涉及以下领域：城市规划与设计、景观设计、城市公共空间设计、建筑设计、建筑工程设计、室内设计、标识设计、工业设计、城市建筑小品设计和交通仿真模拟。整个团队一直以保证品质为原则，于2001年荣获国际ISO9001资格证书与OPQIBI资格认证，2016年更是创造了1.03亿欧元的营业额。另外，AREP在2017年的世界建筑公司100强建筑设计方向榜单中名列第32位，是法国第一所登上该榜单的建筑设计公司。

Founded in 1997 within SNCF (the French national rail operator), AREP is a an international integrated design company. The company is headquartered in Paris, France and has subsidiaries in China, Russia, Vietnam, Abu Dhabi, Qator and India.

AREP brings together more than 1 000 employees and more than thirty nationalities from diverse disciplines: planners, architects, engineers, budget planners, interior designers and so on. The scope of the design covers the following areas: urban planning design, landscape design, urban public space design, architectural design, engineering design, interior design, signage design, product design, urban furniture design and traffic analogue simulation. The entire team has always adhered to the principle of ensuring the quality and won the international ISO9001 certificate and OPQIBI certification in 2001. The company's turnover for 2016 was 103 million Euros. In addition, AREP ranked 32nd in 2017 in the list of the world's top 100 architectural firms in the world, which is the first French architectural design company to be on the list.

北京-中国总部：
地址：北京市西城区西海南沿48号G座
邮编：100035
电话：010-64636981
传真：010-64636982
邮箱：arepchina@arepchina.com
网址：www.arepgroup.com

Beijing:
Add: Building G, 48# Xihai Nanyan, Xicheng District, Beijing
P.C: 100035
Tel: 010-64636981
Fax: 010-64636982
Email: arepchina@arepchina.com
Web: www.arepgroup.com

上海分公司：
地址：上海市徐汇区泸闵路8075号中区304室
邮编：200233
电话：021-54110912
邮箱：arepchina@arepchina.com
网址：www.arepgroup.com

Shanghai:
Add: Room Middle 304, 8075 Humin Road, Xuhui District, Shanghai
P.C: 200233
Tel: 021-54110912
Email: arepchina@arepchina.com
Web: www.arepgroup.com

成都博览城北站交通枢纽
Chengdu Bolancheng North Station

交通枢纽——城市交通设施/地铁站
Transport — Urban Transport Facilites/Metro Station

项目地点：四川 成都
用地面积：22.00 hm²
启动时间：2015年
竣工时间：2020年

Location: Chengdu, Sichuan
Site Area: 22.00 hm²
Begin Time: 2015
Completion Time: 2020

AREP集团与中国西南建筑设计研究院联合体通过国际竞赛选拔，脱颖而出，以"花园式车站"的设计概念赢得成都博览城北站交通枢纽项目竞赛的第一名。

博览城北站交通枢纽距成都市中心28 km，位于成都天府新区中央公园中，紧邻西部博览城及国际会议中心，距天府新区CBD也仅1 km。在如此重要的地区设置一座拥有地铁、公交、轻轨、停车、商业等复合功能、交通能力可覆盖整个城市的枢纽建筑，是天府新区政府城市规划的重要一步。

博览城北站交通枢纽汇集了即将实施的地铁1号线、18号线和规划中的6号线、眉山线，轻轨和多条公交线路也从此经过。

天府新区的中央公园，总面积约230 hm²，被南北向天府大道、东西向南京路划分为4部分。博览城北站的主体部分即位于南京路的下方。方案以"花园式车站"为概念，通过南北两侧的下沉式花园以及枢纽站的站厅平台，实现中央公园南北向的贯通，以及枢纽站与重要设施的直接联系。

此次设计中AREP集团成功地将交通枢纽建筑与自然景观资源完美地结合为一体，打造了舒适而轻松的绿色城市空间。

AREP Group and China southwest architectural design research institute complex used the concept of "garden station" to get through the international competition and won the design right of Chengdu Bolancheng North Station.

Chengdu Bolancheng North Station is about 28 km away from Chengdu city center, and is located in the central park within Chengdu Tianfu New District, while adjoining to the west expo center and international conference center, with a distance of 1 km away from CBD in Tianfu New district as well. It is a significant step in the tianfu new district regulation plan to set up a central building with the functions of traffic (including, subway, public bus, tram), parking, commercial and other complex functions. Besides, it owns the traffic capacity that could cover the whole city.

This hub gathers tube line 1 and line 18 which is still under planning, and line 6 and line Meishan, which are all upcoming to construction. Trams and many public buses routes pass this area too.

The central park has an acreage of 230 hm², which is divided into four parts by the Tianfu avenue from north to south and Nanjing road from east to west. The main body of hub is just under the Nanjing road. The concept is "Garden station". The connection of central park from north to south is implemented by the two sunken gardens at the south and north sides as well as the hall of the station. Therefore, the station is directly connected with other important facilities as well.

This design connect successfully traffic hub and natural environment, to create a comfortable green urban space.

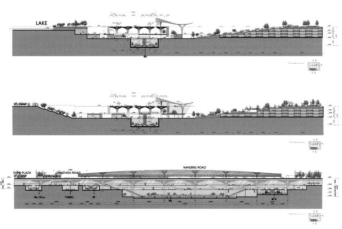

上海金桥车辆段上盖物业综合开发
Shanghai Jinqiao Development

项目地点：上海	Location: Shanghai
建筑面积：976 000 m²	Building Area: 976 000 m²
启动时间：2012 年	Begin Time: 2012
竣工时间：建设中	Completion Time: Under Construction

本项目位于上海浦东新区金桥开发区内，贴邻外环，距上海火车站17.3 km，距浦东机场20 km。项目基地临近金桥出口加工区，同时服务范围覆盖多个产业园区及重要城市节点，是新区重要产业区块的重心之地。土地的复合、集约利用，交通的便捷，宜居的生活环境是贯穿整个项目的几大原则。

本项目南北向长约1 800 m，东西向长约530 m，占地97.48 hm²，总建筑面积约1 000 000 m²。上海地铁9号线、12号线、14号线途经此地，且停车库选址于此，为本项目提供了极大的交通便利。基地东侧隔外环绿化带及运河邻接S20公路，北侧为金海路，西侧为金穗路，南侧为规划中的桂桥路，与周边重要干道巨峰路、申江路联系紧密，交通便捷。

此项目的总图布局经过精心设计：通过高效的交通流线组织、便捷的交通枢纽配套，保证整体项目的可达性与便利性；通过设置中央绿地花园和景观大道形成有效串联，有效地提升整体项目的空间节奏及环境舒适性。明确定位各种不同的城市功能（商业紧邻地铁站、健身体育设施与学校置于街区中心公园地带）；除了住宅之外，项目内的主要公共空间和城市地标，都布置在商业区和交通站点周围。这些充满建筑体量的元素，组成了街区中心地带的城市景观。而纵横交错的人行步道网，造就了饶有自然趣味、便捷而舒适的步行空间与行人通道；街区内道路采用简单便捷的组织方式，两条南北向的主路贯穿整个街区，同时连接各个入口与通道。

通过沿街界面的环境塑造，将功能性与景观性结合，削弱上盖的巨大体量对城市界面的消极影响，结合落地开发的商业办公建筑一体化考虑，把车辆段上盖的消极的城市立面转化成富于变化又富有情趣的功能界面，既为市民创造一站式服务的商业，又使其环境尺度亲切宜人。

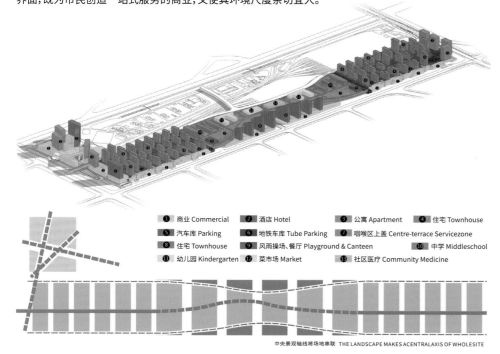

景观轴线图

The project is located in Jinqiao development area of Pudong New District, Shanghai, neighboring the outer-ring highway, 17.3 km to Shanghai Railway Station and 20 km to Pudong Airport. The project site, locating near Jinqiao Export Processing Zone, with services covering multiple industrial parks and major city nodes, is the main area of New Area's important industrial blocks. The principles throughout the whole project include compound and intensive use of land, convenient transportation and livable environment.

This project's south-north length is about 1 800 m, east-west length 530 m, covering 97.48 hm². Shanghai metro line 9, 12 and 14 go by this place, where is also the parking lot, which hugely benefits this project convenient transportation. The site, adjoining Road S20 across the outer-ring green belt and the canal in the east, Jinhai Road in the north, Jinhui Road in the west and Guiqiao Road in planning, is linked with such surrounding major roads as Jufeng Road and Shenjiang Road, providing convenient transportation.

This project's general plot plan is well-designed by AREP: effective traffic streamline and convenient transportation junction facility guarantee the accessibility and convenience of overall project; central green gardens and landscape avenue form effective connection to improve the space cadence and environmental comfort; various urban functions are clearly positioned (the business is close to subway station, sports facilities, and school is placed in the central park); in addition to residential buildings, main public spaces and urban landmarks within the project all locate around shopping area and traffic station, and those features of building volumes constitute the heartland's urban landscape; the crisscross sidewalk web brings natural, convenient and comfortable walk space and walkway; roads within the district are organized in an easy and speedy mode: two main roads in north-south direction run through the whole district and connect with each entrance and channel at the same time.

Conduct environmental interface building along the street and combine functionality with landscape to decrease the negative influences of huge massing of upper covers on the urban interface, and take into account the commercial office building integration of land development to turn the negative urban facades on the car depot into a functional interface featured with variety and entertainment. Businesses of one-stop services shall be created for the public with cordial and pleasant environment.

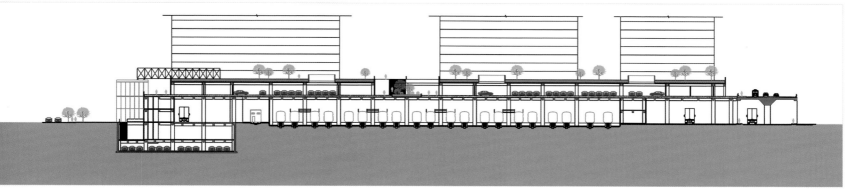

剖面图

美国华贝设计有限公司（中国）
Warton-Pei & Associates(China)

美国华贝设计有限公司总部设在美国纽约，1990年开始进入中国市场，1994年正式设立分支机构。华贝设计集中美设计高手于一体，充分发挥各自优势，互相取长补短，在吸取西方现代设计理念的基础上，传承发扬古老的中国传统文化精华，大胆开拓，不断创造出最优化的设计组合。公司曾先后为美国、南非、瑞典、沙特阿拉伯、阿联酋以及中国等国家和地区进行过设计，其中有25%的工程设计获得世界级奖项。

Warton-Pei & Associates is headquartered in New York, USA. It began to enter China in 1990, and officially established a branch in 1994. Huabei Design is a collection of beautiful design masters, giving full play to their respective advantages, learning from each other's strengths and weaknesses, on the basis of absorbing modern Western design concepts, combining the essence of ancient Chinese traditional culture, boldly exploring, and constantly creating the optimal design combination. The company has successively carried out designs for the United States, South Africa, Sweden, Saudi Arabia, the United Arab Emirates and China and other countries and regions, and 25% of the engineering designs have won world-class awards.

地址：上海市虹桥路2419号雪松阁2D	Add: Tower 2D, Cedar Pavilion, No.2419, Hongqiao Road, Shanghai
电话：021-62695859	Tel: 021-62695859
传真：021-62685266	Fax: 021-62685266
邮箱：wpeilwx@126.com	Email: wpeilwx@126.com
网址：www.wartonpei.com	Web: www.wartonpei.com

上海金山冰雪海养生公园
Ice and Snow Sea Health Park, Shanghai Jinshan

为实现"3亿人上冰雪"的庄严承诺而贡献一份力量，也为增强全民体质、更好地圆梦而实现伟大的民族复兴，我们正在探索、创新和打造一座全新的、亲民的、低造价的、多功能的、低消费的和原生态的中国式的"冰雪海养生公园"。

我们经过近5年的调研和实践，在吸取诸家专长的基础上，摸索、创新了一系列建设室内冰雪场的新理念，命名为"3.1版室内生态养生冰雪场"。其主要的潜在价值是：在同等体量和容量的前提下，造价只是传统室内冰雪场的1/7，建设时间只有1/5，节能提高1/3。1/7、1/5、1/3,3个1简称3.1版，还由于中国室内雪场建设大致可划为三个阶段：2002年上海银七星阶段，2009年绍兴乔波阶段，目前我们研发的生态冰雪场阶段，故统称为"3.1版室内生态冰雪场"。

In order to contribute to the realization of the solemn promise of "300 million people go to the ice and snow", and to enhance the physique of the whole people, better realize their dreams and realize the great national rejuvenation, we are exploring, innovating and building a new, people-friendly, low-cost The cost-effective, multi-functional, low-consumption and original Chinese-style "ice and snow sea health park".

After nearly five years of research and practice, we have explored and innovated a series of new ideas for building indoor ice and snow rinks on the basis of absorbing the expertise of various schools, which we call "Version 3.1 Indoor Ecological Health Ice and Snow Fields". Its main potential value is: under the premise of the same volume and capacity, the construction cost is only 1/7 of the traditional indoor ice and snow field, the construction speed is only 1/5, and the energy saving is increased by 1/3. 1/7, 1/5, 1 /3,3 1 is referred to as version 3.1, and also due to the construction of indoor snow fields in China, it can be roughly divided into three stages: the 2002 Shanghai Yinqixing period, the 2009 Shaoxing Qiaobo period, and the ecological ice and snow field period we developed at present, so Collectively referred to as "Version 3.1 Indoor Ecological Ice and Snow Field".

上海中俄冰雪嘉年华
China and Russia Ice Carnival, Shanghai

主要吸粉亮点 / Main feature

(1) 这里找不到"雾霾"?　　(1) Can't find "smog" here ?
(2) 有"小分子雪"吗?　　　(2) Is there "small molecule snow" ?
(3) 这里有"神奇水"吗?　　(3) Is there "magic water" here ?
(4) 好奇特的体育馆?　　　(4) Curious about the gymnasium ?
(5) 冰雪运动还有"保护神"?　(5) Is there a "protector" in ice and snow sports ?
(6) 断电怎么办?　　　　　(6) What to do when the power is cut off ?
(7) 冰雪场有WI-FI吗?　　　(7) Does the ice and snow field have WI-FI ?
(8) 从冰雪中寻找节能减排?　(8) Looking for energy saving and emission reduction from ice and snow ?
(9) 冰雪能养生?　　　　　(9) Can ice and snow keep you healthy ?
(10) 冰雪能融通文化?　　　(10) Can ice and snow melt culture ?
(11) 影院、餐饮进冰场?　　(11) Cinema and catering into the ice rink ?
(12) "你滑雪、我买单"?　　(12) "You ski, I pay" ?
(13) 小冰雪、大市场?　　　(13) Small ice and snow, big market ?

云澜湾冰雪海嘉年华
Yunlan Bay Ice and Snow Sea Carnival

"莲花"是云澜湾温泉的主旋律,新建的云澜湾冰雪海嘉年华理应融化其中,而且更应为其增添新彩。

故总体设计立意为:四朵并莲四角水,一弯明月一条藕;荷塘得雪更清妍,人经冰霜无忧愁。

"Lotus" is the main theme of "Yunlan Bay Hot Spring", and the newly-built "Yunlan Bay Ice and Snow Sea Carnival" should be melted in it, and it should add new color to it.

Therefore, the overall design concept is: four parallel lotuses and four corners of water, a curved moon and a lotus; the lotus pond is more beautiful when the snow is clear, and people have no worries through frost.

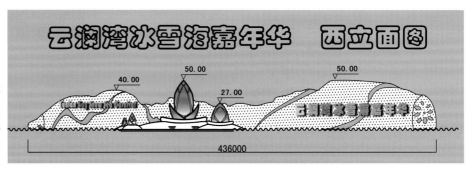

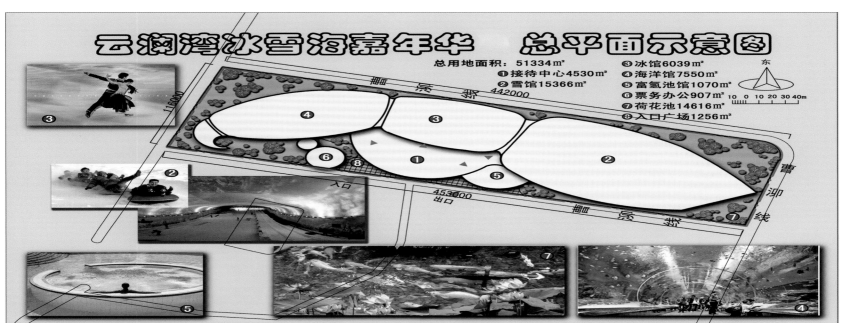

香港道和国际设计有限公司
Hong Kong Daohe International Design Co., Ltd.

香港道和国际设计有限公司1998年创建于香港九龙官塘工业中心，2006年入驻杭州。公司秉承尚道有义、天地人和理念，经过近20年精耕，目前拥有完善的设计系统和操作模式，业务领域包括建筑设计、园林景观设计、室内装饰设计等。

公司由著名设计师、德国柏林大学建筑学博士高正华先生出任院长。香港道和国际设计有限公司是一支富有激情和想象力的综合化设计团队，公司以现代化的设计理念，将规划、建筑、景观、装饰、市政等专业设计集成，达到设计空间美学一体化。对作品崇尚高要求高格局，以人为本，致力于自然、艺术、人性化设计理念的融合，为家园和社会承担更多责任。

多年来，公司负责设计的项目遍及上海、浙江、江西、山东等国内大部分省份，近3年来公司累计承接完成的项目总面积超350万m²，获得客户一致好评，其中龙景·雷迪森庄园酒店项目荣获浙江省建设工程钱江杯奖。

Hong Kong Daohe International Design Co., Ltd. was founded in 1998 in Guantang Industrial Center, Kowloon, Hong Kong, and settled in Hangzhou in 2006. The company adheres to the concept of upholding justice, heaven and earth and people. After nearly 20 years of intensive cultivation, it currently has a complete design system and operation mode. Its business areas include architectural design, garden landscape design, interior decoration design, etc.

The company is headed by Mr. Gao Zhenghua, a well-known designer and a doctor of architecture from the University of Berlin in Germany. Hongkong Daohe International Design Co., Ltd. is a comprehensive design team full of passion and imagination. The company integrates professional design such as planning, architecture, landscape, decoration, and municipal administration with modern design concepts to achieve the integration of design space aesthetics. Reverence for the work, high requirements and high layout, people-oriented, committed to the integration of nature, art, and humanized design concepts, and take more responsibilities for the homeland and society.

Over the years, the company has been responsible for designing projects covering most of the domestic provinces such as Shanghai, Zhejiang, Jiangxi, Shandong, etc. In the past three years, the company has undertaken and completed projects with a total area of 3.5 million m², which has won unanimous praise from customers. Among them, Longjing Reddy The Mori Manor Hotel project won the Qianjiang Cup Award of Zhejiang Province Construction Project.

地址：香港九龙观塘道472-484号观塘工业中心3期6楼A座4室
浙江省杭州市滨江区滨盛路1509号天恒大厦904室
（杭州道和建筑设计有限公司）
邮编：310052
电话：0571-86952575
邮箱：daohejianzhu@163.com

Add: Room 4, Block A, 6th Floor, Phase 3, Kwun Tong Industrial Centre, 472-484 Kwun Tong Road, Kowloon, Hong Kong
Room 904, Tianheng Building, 1509 Binsheng Road, Binjiang District, Hangzhou, Zhejiang
(Hangzhou Daohe Architectural Design Co., Ltd.)
P.C: 310052
Tel: 0571-86952575
Email: daohejianzhu@163.com

安吉吾想园酒店
Wuxiangyuan Hotel in Anji County

项目地点：浙江 湖州
Location: Huzhou, Zhejiang

该项目位于浙江省湖州市安吉县上墅乡罗村村，建筑地下一层，地上以三层为主，局部二层。功能以客房为主，餐饮为辅。客房主要分为40 m²单间、105 m²套房及115 m²复式套房。

The project is located in Luo Village, Shangshu Township, Anji County, Huzhou City, Zhejiang Province. The building has one floor under ground, three floors above ground mainly and two floors above ground locally. The project functions mainly as guest rooms, with catering as a supplement. Guest rooms are mainly divided into 40 m² single rooms, 105 m² suites and 115 m² duplex suites.

高二高姆山度假酒店
Gaomushan Resort Hotel in Gaoer Township

建筑设计师：高正华、刘伟平、邹涛涛
Designers: Zhenghua Gao, Weiping Liu, Taotao Zou

项目位于浙江省磐安县，项目性质为高端度假酒店，规模约8 500 m²，该项目包括一处酒店接待中心和52间度假客房，每排客房前后落差约8 m，可以保证前后视线的无遮挡，并且该项目最大限度地保持原来山体，所有客房采用装配式下部架空的建筑形式。

The project is located in Pan'an County, Zhejiang Province, which functions as a high-end resort hotel with an area of about 8 500 m². The project consists of a hotel reception center and 52 resort rooms, with about 8 m height difference of each row, which can ensure unobstructed view. The project maintains the original mountain to the maximum extent, and all rooms adopt the prefabricated lower overhead building form.

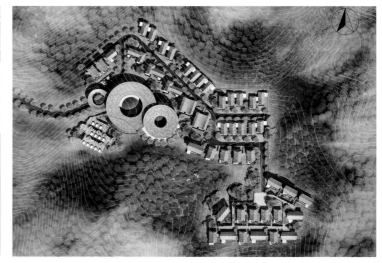

杭州婚庆文化基地
Hangzhou Wedding Culture Base

建筑设计师：高正华、邹涛涛
Designers: Zhenghua Gao, Taotao Zou

项目位于杭州市余杭区江南慢村，项目性质为乡村会客厅，建筑规模约3 500 m²，为一栋主体楼和10栋装配式客房，该项目主要打造艺术公社和婚庆基地，为城市配套提供更多的艺术活动空间。

The project is located in Jiangnanman Village, Yuhang District, Hangzhou City, functioning as a village reception room. With a construction area of 3 500 m², the project consists of one main building and 10 prefabricated guest rooms. The project mainly creates an art commune and a wedding base to provide more space for art activities in the city.

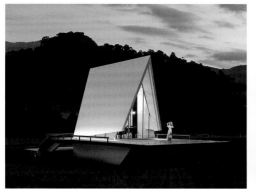

磐安县乌石村乡村度假酒店
Village Resort Hotel in Wushi Village, Pan'an

建筑设计师：高正华、邹涛涛、刘伟平
Designers: Zhenghua Gao, Taotao Zou, Weiping Liu

项目位于浙江省磐安县尖山镇，该地最大的特色就是当地产的乌石，故名乌石村，项目分为两期，总建筑面积约7 500 m²，其中一期为改造区，面积约2 500 m²，改造内容为一间小学和19间老宅，二期为新建部分，总建筑面积约5 000 m²，整个项目以乌石为主要材料，从当地的古民居提取建筑文化符号，打造在地文化属性比较强的高端乡村度假酒店。

The project is located in Jianshan Town, Pan'an County, Zhejiang Province. The biggest feature of the project is the locally produced Wushi, and that's how the name Wushi Village comes. The project is divided into two phases, with a total construction area of about 7 500 m². The first phase is the transformation area of a primary school and 19 old houses, covering about 2 500 m². The second phase is the new building with a total construction area of about 5 000 m². The whole project takes Wushi as the main material and extracts architectural cultural symbols from local ancient dwellings to create a high-end rural resort hotel with strong local cultural attributes.

磐安樱花谷野生酒店
Cherry Blossom Valley Wild Hotel in Pan'an County

建筑设计师：高正华、邹涛涛、刘伟平
Designers: Zhenghua Gao, Taotao Zou, Weiping Liu

项目位于浙江省东阳市樱花园内，该项目为传统的樱花园，樱花花期很短，其他时间都比较萧瑟，利用这一特点，打造一处独具特色的"野奢酒店"，酒店建筑面积约为8 000 m²，55间客房，项目的建设为其他时间段注入了活力。

The project is located in the Cherry Blossom Garden in Dongyang City, Zhejiang Province. This project is a traditional cherry blossom garden. The cherry blossom period is very short. By taking advantage of this feature, a unique wild luxury hotel will be built with a construction area of about 8 000 m², equipped with 55 rooms. The construction of the project will inject vitality during other time periods without cherry blossom.

温州云顶草上世界度假酒店
Genting World Resort Hotel in Wenzhou

建筑设计师：高正华、邹涛涛、刘伟平
Designers: Zhenghua Gao, Taotao Zou, Weiping Liu

项目位于浙江省温州市瓯海区云顶草上世界景区内，项目用地为2.07 hm²，建筑面积约24 000 m²，场地内落差约为50 m，该项目最大限度地利用了坡度，本着少开挖的原则进行设计，项目客房量约为200间，是一处具有规模的会议及度假型酒店。

The project is located in the Genting World Scenic Spot, Ouhai District, Wenzhou City, Zhejiang Province, with a land area of 2.07 hm² and a building area of about 24 000 m². The height difference within the site of the project is about 50 m, which makes maximum use of the slope and is designed in line with the principle of less excavation. The project, equipped with about 200 rooms, is a large-scale conference and resort hotel.

资溪·冠合开元芳草地乡村度假酒店
Zixi · Guanhe New Century Parkview Grass Village Resort Hotel

设计师：高正华
项目地点：江西 抚州
建筑面积：40 502 m²
用地面积：33 430 m²
容积率：1.19

Designer: Zhenghua Gao
Location: Fuzhou, Jiangxi
Building Area: 40 502 m²
Site Area: 33 430 m²
Plot Ratio: 1.19

项目地靠近大觉山景区，是一家基于芳草地品牌调性而打造的运动主题度假酒店。酒店场地地势复杂，采取顺应自然、尊重场地的设计原则。依势采取台地式布局，着重于体现建筑对环境的谦让与共融。酒店设有集中客房48间，低密度客房125间及能容纳200人以上的会议厅。

The project is located near Dajue Mountain Scenic Spot. It is a sports-themed resort hotel based on the style of Parkview Grass brand. The terrain of the hotel site is complex, and the project adopts the design principle of adapting to nature and respecting the site. The platform layout is adopted according to the terrain, striving to reflect the building's humility and harmony to the environment. The hotel has 48 centralized rooms, 125 low-density rooms and a conference hall that can accommodate more than 200 people.

安徽建筑大学·建筑与规划学院
Anhui Jianzhu University School of Architecture and Urban Planning

安徽建筑大学建筑与规划学院现设有建筑系、规划系、景观系以及建筑基础部、实验教学中心，开设建筑学、城乡规划、风景园林3个本科专业，拥有建筑学、建筑学（专业型）、城乡规划学、城市规划（专业型）、风景园林（专业型）等硕士学位授权点。学院现有全职教师近百人，高级职称者近40人，其中博士生导师1人，硕士生导师28人，有安徽省青年皖江学者1人，安徽省教学名师2人。城乡规划学为安徽省高校学科，建筑学为安徽省高峰培育学科，风景园林学为安徽建筑大学重点建设学科。建筑学专业和城乡规划专业分别于2007年和2008年首次通过国家专业教育评估，是国内较早双双通过两大专业本硕评估的院校之一。近5年来，学院获国家优秀教学成果二等奖1项，安徽省优秀教学成果特等奖2项、一等奖2项。

Anhui Jianzhu University School of Architecture and Urban Planning has the Department of Architecture, the Department of Urban-Rural Planning, the Department of Landscape Architecture, the Basic Teaching Department of Architecture, and the Experimental Teaching Center. It has three undergraduate majors: Architecture, Urban & Rural Planning, and Landscape Architecture. It has master's degree authorization points such as Architecture, Architecture (Professional), Urban & Rural Planning, Urban Planning (Professional), and Landscape Architecture (Professional). At present, our School has nearly 100 full-time teachers and nearly 40 senior supervisors, including 1 Doctoral Supervisor, 28 Master Supervisors, 1 Young Wanjiang Scholar of Anhui Province and 2 Teaching Masters of Anhui Province. Urban & Rural Planning is the Peak Discipline of Anhui Province, Architecture is the peak cultivation discipline of Anhui Province, and Landscape Architecture is the key construction discipline of Anhui Jianzhu University. Architecture Major and Urban-Rural Planning Major passed the National Professional Education Assessment for the first time in 2007 and 2008 respectively, which is one of the earliest institutions in China that both passed the undergraduate and postgraduate assessment of the two majors. In recent five years, the School has won one second prize of National Excellent Teaching Achievement, two special prizes and two first prizes of Anhui Excellent Teaching Achievement.

地址：合肥市蜀山区紫云路292号安徽建筑大学南校区实训楼（6号楼）
邮编：230022
传真：0551-63513048

Add: Training building (Building 6) of Anhui University of Architecture South Campus, No.292, Ziyun Road, Shushan District, Hefei
P.C: 230022
Fax: 0551-63513048

安徽建筑大学·城乡规划设计研究院
Anhui Jianzhu University Urban - Rural Planning Design Research Institute

安徽建筑大学城乡规划设计研究院隶属于安徽建筑大学，主要成员与技术骨干为安徽建筑大学建筑与规划学院教师，是集设计与科研于一体的国家甲级规划设计单位，技术实力雄厚，专业配置齐全合理，配备城市规划、道路交通、城市设计、市政工程、园林绿化、环境工程等专业人员。共有专业技术人员42人，其中高级技术职称人员16人。

Anhui Jianzhu University Urban - Rural Planning Design Research Institute is attached to Anhui Jianzhu University. Its main members and technical backbone are teachers from the School of Architecture and Urban Planning of Anhui Jianzhu University. It is a national Class-A planning and design institution integrating design and scientific research with strong technical strength and complete and reasonable professional staff. It is equipped with professionals in urban planning, road traffic, urban design, municipal engineering, landscaping and environmental engineering. There are 42 professional and technical personnel, including 16 personnel with senior technical title.

地址：合肥市金寨南路856号安徽建筑大学北区
电话：0551-63513105
传真：0551-63513115
邮箱：jy-ah@126.com

Add: North Section of Anhui Jianzhu University, No.856, South Jinzhai Road, Hefei
Tel: 0551-63513105
Fax: 0551-63513115
Email: jy-ah@126.com

绍兴上虞·东关古运河滨水地区城市设计研究
Study on Urban Design of Shaoxing Shangyu · Dongguan Ancient Canal Waterfront Area

项目主持与负责：吴强、顾康康
项目创意总设计师：吴强
主要设计人员：张祥钰、杜勇、陈婧淑、陈亚伟、储成伟、孔文志、曹雪峰、王智玮、徐昕、胡中昱、蒋园
项目地点：浙江 绍兴
设计单位：安徽建筑大学城乡规划设计研究院

Project Leader and Responsibility: Qiang Wu, Kangkang Gu
Chief Creative Designer of the Project: Qiang Wu
Designers: Xiangyu Zhang, Yong Du, Jingshu Chen, Yawei Chen, Chengwei Chu, Wenzhi Kong, Xuefeng Cao, Zhiwei Wang, Xin Xu, Zhongyu Hu, Yuan Jiang
Location: Shaoxing, Zhejiang
Design Unit: Anhui Jianzhu University Urban - Rural Planning Design Research Institute

本设计源于浙江绍兴"全国高等院校浙东古运河（上虞段）概念设计竞赛"的研究性课题之选题。设计地块位于浙江省绍兴市上虞区东关街道，总用地面积约270 hm²。设计地块包含一、二、共三个部分，用地面积分别为92.2 hm²、87.6 hm²、90 hm²。

This design comes from the research topic of "Conceptual Design Competition of the Ancient Canal of East Zhejiang (Shangyu Section) among National Universities" in Shaoxing, Zhejiang. The design plot is located in Dongguan Street, Shangyu District, Shaoxing City, Zhejiang Province, with a total land area of about 270 hm². The design plot consists of three parts, with land area of 92.2 hm², 87.6 hm² and 90 hm² respectively.

基础研究

1、区位空间关系分析

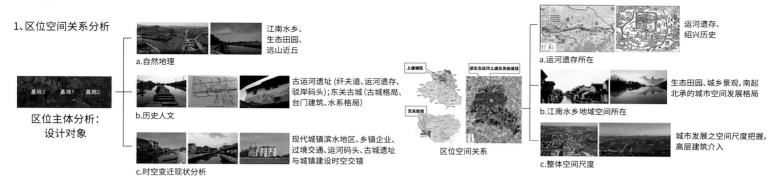

区位主体分析：设计对象

a.自然地理 —— 江南水乡、生态田园、远山近丘

b.历史人文 —— 古运河遗址（纤夫道、运河遗存、驳岸码头）；东关古城（古城格局、台门建筑、水系格局）

c.时空变迁现状分析 —— 现代城镇滨水地区、乡镇企业、过境交通、运河码头、古城遗址与城镇建设时空交错

区位空间关系

a.运河遗存所在 —— 运河遗存、绍兴历史

b.江南水乡地域空间所在 —— 生态田园、城乡景观，南起北承的城市空间发展格局

c.整体空间尺度 —— 城市发展之空间尺度把握，高层建筑介入

2、上位规划控引

a《绍兴市城市总体规划(2018-2035)方案》
——"保护老城、中心集聚、生态维护、协调发展"

b《东关街道战略规划》
——生态田园与城市高质量发展的指向

c《2021年上虞区美丽乡村示范带建设三年行动计划》
——江南水乡带、古运河、民俗体验建设工程

3.3《2021年上虞区美丽乡村示范带建设三年行动计划》

1.建设目标——2021年—2023年，巩固美丽乡村建设成果，充分挖掘区域特色与优势，围绕"串珠成链、共同富裕、引领示范"，建造主题突出、特色明显、宜居宜业的山水田园雅聚地，农民富裕富足生活样板地，先行培育打造共同富裕美丽乡村示范带3条。在示范带上重点建设10个左右有上虞辨识度、有引领带动力的特色精品片，推动城乡共同富裕，高标准实施乡村振兴战略。在此基础上，全区每年培育30个左右拥有美丽景观、彰显乡土风情、蕴涵乡愁韵味、展现上虞特色的"锦绣villages"。

2.建设任务——孝德文化核心、锦绣虞南示范带、江南水乡示范带、都市田园示范带

3.建设内容
(一) 共性化建设内容
环境整治建设工程、农村公厕提升工程、美丽建设工程、场景建设工程
(二) 特色化建设内容
锦绣虞南示范带：林相改造建设工程、美丽管溪建设工程、数字农业工程、业态培育工程
江南水乡示范带："江南威尼斯"建设工程、古运河建设工程、民俗体验建设工程
都市田园示范带：数字田园工程、仙果长廊建设工程、雅聚空间建设工程

——对于民生建设的关注，引发对街区内部基础设施改善与环境整治的审视与思考

d《大运河（绍兴段）遗产保护规划》、《浙江省大运河核心监控区国土空间管控通则》
——文化遗产价值、历史街区、环境整治、生态环境、历史景观特征

e《绍兴市上虞区东关单元(Z11-1)控制性详细规划》—— 上虞城区西部门户型城市组团

整体空间建构概念设计

设计中的核心问题：如何把握基地所在城市发展战略、上位规划控引、区位空间审视的特质性，及其一、二、三基地内在空间的关联来进行城市设计的功能与意象，在历史见证、文化传承、城市特色与旅游文化形象传媒的"象、形、境、意"形态空间战略、战术的整体空间建构与设计。

总的指导思想：把握基地所在区域、城市、地段环境区位空间关系审视的特质性，突出古运河历史遗存，见证保护与文化传承，融贯整合历史文化保护传承、城市设计、山水城市、城市更新的思想方法，努力建构江南水乡、山水写意、文脉传承、独具上虞门户景观标志、城市旅游文化形象。

设计原则：建筑、规划、景观三位一体，区位意象整合并行，理性基础、感性强化，宏观感知、整体把握，纵引横跨、问道特质的概念设计总原则。

Conceptual design of overall space construction

Core issues in the design: How to grasp the characteristics of the urban development strategy of the base, the control of the upper planning, the location space review, and the correlation of the inner space of the first, second and third bases, to carry out the function and image of urban design, as well as the overall space construction and design of the form space strategy and tactics of "image, form, environment and meaning" in the historical witness, cultural inheritance, urban characteristics and tourism cultural image media.

General guiding ideology: It aims to grasp the characteristics of examining the spatial relationship of the region, city and environmental location, highlight the protection of historical relics of the ancient canal, witness the inheritance of culture, and integrate the thoughts and methods of historical and cultural protection and inheritance, urban design, landscape city and urban renewal, striving to construct an overall image form space which coincides with the centennial birthday of the Communist Party of China to witness the future history based on characteristics of urban waterfront area with Jiangnan water town, landscape brushwork, and cultural heritage, Shangyu gateway landscape logo, urban tourism culture image media and ink.

Design principle: architecture, planning, landscape all in one, location image integration, rational basis, perceptual strengthening, macro perception, overall grasp.

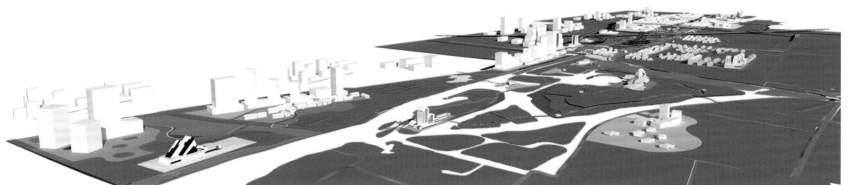

——整体鸟瞰图（由基地一荷花荡生态湿地遗址公园遥望三个地块）

整体空间概念设计建构特色

1. 突出基地资源的特质性保护与利用及整体空间的相对性、连续性和时空性。

2. 突出古运河历史遗存见证与时空文化的演绎，尤以纤夫文化精神的品位与问道，致敬中国共产党百年华诞。

3. 突出城市更新的感性与理性的城市设计整体空间建构的价值力寻——城市生态文明、城乡融合发展、乡村振兴与美丽乡村建设，文化自信、历史文化传承的当代中国城乡建设的使命。

4. 突出中国意象——山水写意、江南水乡、文脉传承的上虞门景地域文化，自然、历史、人文时空特质的标志性空间的建构。

Characteristics of overall spatial conceptual design construction

1. Highlight the unique protection and utilization of resources, and the relativity, continuity and space-time of the overall space.

2. Highlight the witness of the historical relics of the ancient canal and the interpretation of time and space culture, especially to take the taste of the spirit of the culture of the slender man to salute to the centennial birthday of the Communist Party of China.

3. Highlight the value search of the perceptual and rational overall space construction of urban renewal - in line with the contemporary Chinese urban construction mission of urban ecological civilization, urban-rural integrated development, rural revitalization and beautiful countryside construction, cultural confidence and historical and cultural inheritance.

4. Highlight the Chinese image: The construction of symbol space with the spatial and temporal characteristics of Shangyu Menjing regional culture, nature, history and humanity of Jiangnan water town, landscape brushwork and cultural heritage.

寒塘霁月凝荷香——荷花荡生态遗址公园概念设计（基地三）
The Moon in the Pond with Lotus Fragrance - Conceptual Design of Hehuadang Ecological Relics Park (Base 3)

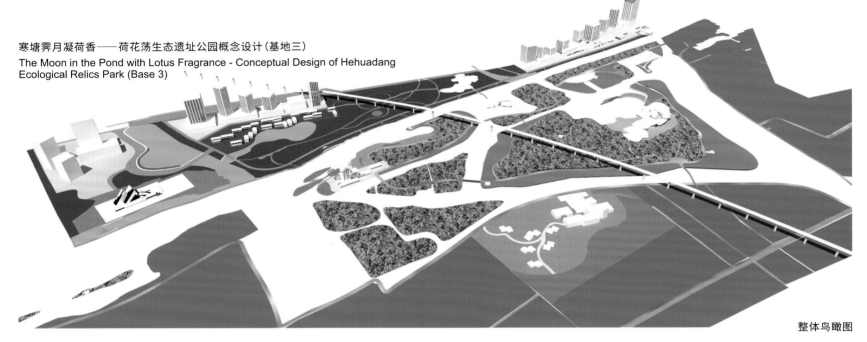

整体鸟瞰图

关于荷花荡生态遗址公园的命题解读，就一二三基地的历史遗存空间来看，可谓缘西东进，横贯三个基地，但却存在资源特质上的相对识异性，诚如本基地的区位空间关系特质——上虞东关古运河历史遗存残境、禅意文化，最具特质山水写意的独特自然地理空间所在。这是不同于远在荒郊野外、空旷之所的一般遗址公园之境，恰是古今人居场所之证之承之今日泼墨寄情之所，另有高铁穿越的破宁之困，但大隐于朝、中隐于市、小隐于野之禅境，禅修、禅悟的入世之道，岂不平添了荷花荡遗址古今天上人间的玩味，这便是荷花荡生态遗址公园的命题由来。

As for the proposition interpretation of Hehuadang Ecological Relics Park, in terms of the historical relic space of the first, second and third bases, it can be said that the three bases are along the west and east, but there are relative differences in resource characteristics. Just like the characteristics of the location and spatial relationship of this base - the historical relic of the ancient canal in Dongguan of Shangyu and the Zen culture, they are the most unique natural geographical spaces. This is different from the general relics park in the wilderness and empty place. It is just the place where the ancient and modern living places bear the witness of today's ink pouring and expressing emotions. There are also the high-speed rail crossing by to break the peace. As a common saying goes, while the "lesser hermit" lives in seclusion in the country, the "greater hermit" does so in the city. This is the origin of the proposition of Hehuadang Ecological Relics Park.

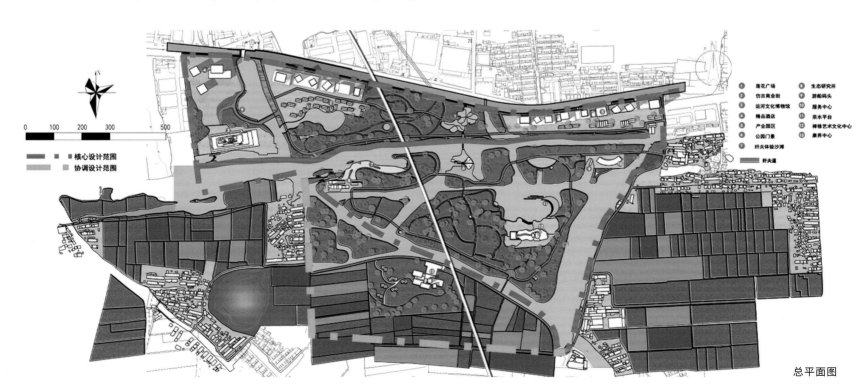

总平面图

1. 禅修文化艺术中心

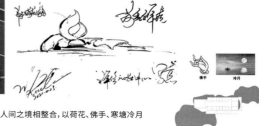

将荷花荡之自然资源特质,与荷香凝月的天上人间之境相整合,以荷花、佛手、寒塘冷月之写意来突显禅修文化中心建筑创作的禅之品位。

The natural resource characteristics of Hehuadang is integrated with the image of fragrant lotus and the moon. The freehand brushwork of lotus, Finger Citron, cold pond and cold moon is used to highlight the taste of Zen in the architectural creation of meditation culture center.

禅修文化艺术中心效果图

2. 荷花荡生态湿地遗址公园门景建筑

荷花

帆船

荷花荡生态湿地遗址公园门景建筑以荷之秀叶,莲之绽放,帆之写意,整合建构,突显其生态历史遗存资源的特质,更显其禅境文化的入世之道,将北之莲花广场作为入园的"前奏",一桥之飞虹渡,寒塘冷月凝荷香。

The integrated construction of the gate landscape architecture of Hehuadang Ecological Wetland Relics Park highlights the characteristics of its ecological and historical resources, and also shows the way of its Zen culture's entry into the world. "Lotus" in Buddha means the life of Buddha's wisdom, and the lotus rises from the silt without staining. The Lotus Square in the north is used as the "prelude" to the garden.

荷花荡生态湿地遗址公园门景建筑效果图

3. 生态科研科普中心

骏马

山水

生态科研科普中心以"山石"的整体空间意象,融合"马"的汉字艺术形式。空间上,两山对峙,挟纤石道纤夫艰难前行,意象上,一为脚踹乱石,入劲山(运河博物馆),一为一仞孤山奔马起,快马飞驿入东关,突显古运河历史遗存之苍雄,纤夫文化之民族魂魄,其中的纤夫艰行之体验,前置于一仞孤山的风帆,令背负纤绳行走在祖先纤石一道的游人倍感历史的苍劲雄浑。

Ecological Scientific Research and Popularization Center uses the overall spatial image of "mountains and rocks" to integrate the Chinese character art form of "horse". In terms of space, the two mountains confront each other, carrying the tracker stone path forward with difficulty. In terms of image, one is trekking into Jinshan (Canal Museum), the other is flying to Dongguan, highlighting the vicissitude of ancient canal historical relics and the national soul of the tracker culture. The arduous journey experience of the tracker is set on a solid sail on an isolated mountain, making the tourists feel the vigorous history when carrying the rope to walk in the ancestor stone road.

生态科研科普中心效果图

4. 纤石文化艺术博物馆

山石

纤夫

在建筑形态空间创作中,突显建筑与环境及纤石文化一项之沧海浑厚、原始旷野之意境,以石的肌理、山的形势建构纤夫山重水复之雄浑劲道,引发纤夫文化之民族精神的问道。在功能建构中以运河历史见证、文脉文化传承演绎与学术研究交流为主要职能。

The creation of architectural form space highlights the artistic conception of the profound sea and primitive wilderness of architecture, environment and Qianshi culture. It constructs the vigorous way of restoring mountains and rivers with the texture of stones and the terrain of mountains, and arouses the national spirit of tracker culture. In the functional construction, the main functions are canal historical witness, cultural inheritance and deduction, and academic research and exchange.

纤石文化艺术博物馆效果图

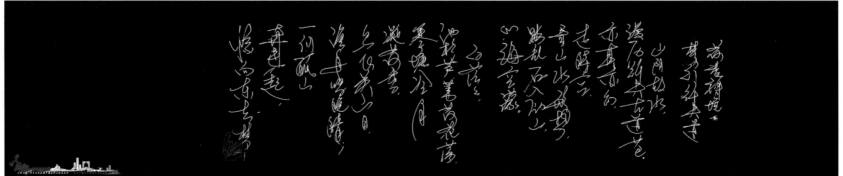

城市设计意境:荷香禅境,梦行纤夫道

山阴故水,漫历纤夫古道苍。亦真亦幻。走时空,看山水形势,踏乱石入劲山,心海空濛。白茫茫,池杉芦苇荷花荡,寒塘冷月凝荷香。鸟飞关山日,渔舟唱晚晴;一仞孤山奔马起,惊向东去。梦……
——吴强 2021.5

红帆踏浪·向前、向前、向前——纤夫时空文化广场城市设计（基地一）
Red Sail Treading Waves · Forward, Forward, Forward -- Urban Design of the Tracker Space-time Cultural Square (Base 1)

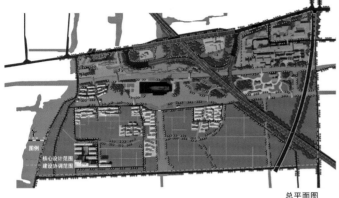

总平面图

1、上虞城市建设文化艺术中心

艺术中心建筑创作，以"门"字艺术、台门建筑文化传承元素、"绍兴桥"艺术形式、"水际线"的形式语汇、"山水城市"意象要素的写意整合，来凸显艺术中心所在地域文化的识异与标志特质建构，将门景建筑的功能巧妙地组织到形态空间意象表达之中；"桥"的筑基空间沟通了南北水域，串联水上游线，恰是"桥"的功能与形态空间融贯整合的城市设计品位与品质的表达。

1. Shangyu Urban Construction of Culture and Art Center

In the creation of art center architecture, the character art of "gate", the cultural inheritance element of terrace architecture, the art form of "Shaoxing Bridge", the formal vocabulary of "water line", and the image elements of "landscape city" are integrated to highlight the distinguishing and marking characteristics of the regional culture where the art center is located. The function of gate landscape architecture are cleverly organized into the expression of the morphological space image. The foundation space of the "bridge" communicates the north and south waters and connects the line of water, which is exactly the expression of the taste and quality of urban design integrated with the function and form space of "bridge".

"门"字　　台门建筑　　江南水乡-桥文化

上虞城市建设文化艺术中心

2、红帆踏浪文化艺术中心

红帆踏浪文化艺术中心建筑创作深刻把握古运河与古纤道的历史文化内涵与精神内核，将运河文化与纤夫精神之时空写意、中国共产党的百年征程，落墨于红帆踏浪·纤夫时空文化艺术中心的建筑写意空间；文化中心建筑形态的建构，以"船"形、"帆"态，结合"水际线"的形式语言，隐喻了中国共产党嘉兴南湖红船之源与百年征程的劈波斩浪、不断前行。运用写意、国画的表现手法，将文化中心建筑的景观与场所有机整合。整个造型极富厚重、张力、简笔的国画意境与雕塑之美，呈现出文化要素与形态空间融贯整合的城市设计意境与品位的力寻与力作。

船　　帆

2. Red Sail Treading Waves Culture and Art Center

The architectural creation of Red Sail Treading Waves Culture and Art Center deeply grasps the historical and cultural connotation and spiritual core of the ancient canal river and the ancient track road, and puts the brushwork space of canal culture and the spirit of the tracker - the century-long journey of the Communist Party of China into the architectural brushwork space of Red Sail Treading Waves Culture and Art Center. The construction of the architectural form of the cultural center, in the form of "boat" and "sail", combined with the formal language of "water line", implicitly refers to the original source of the Communist Party of China having meeting on the red boat on the South Lake in Jiaxing and the continuous progress of the century-long journey. The expression techniques of freehand brushwork and traditional Chinese painting are used to organically integrate the landscape and site of the cultural center architecture. The artistic conception of traditional Chinese painting and the beauty of sculpture in the whole shape are very rich, strong and simple, presenting the force search and masterpiece of the artistic conception and taste of urban design integrated with the cultural elements and form space.

红帆踏浪文化艺术中心效果图

3、室外桥文化博物馆

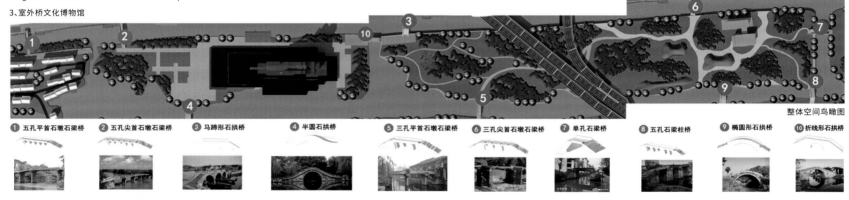

整体空间鸟瞰图

① 五孔平首石墩石梁桥　② 五孔尖首石墩石梁桥　③ 马蹄形石拱桥　④ 半圆石拱桥　⑤ 三孔平首石墩石梁桥　⑥ 三孔尖首石墩石梁桥　⑦ 单孔石拱桥　⑧ 五孔石梁柱桥　⑨ 椭圆形石拱桥　⑩ 折线形石拱桥

室外"桥文化博物馆"设计立足于古纤道的保护与利用，兼顾旅游文化职能，通过紧邻纤道设置木栈道，以及将基地范围内运河两岸通过传统的绍兴桥文化的演绎，于不同节点处设置不同形式的桥，利用慢行系统串联不同的人文场所空间，以此来展现江南水乡在桥文化这一笔上的落墨，通过这一设计手法在整体空间上形成室外的"桥文化博物馆"。

The design of the outdoor "Bridge Culture Museum" is based on the protection and utilization of the ancient fiber road, taking into account the functions of tourism and culture. By setting up a wooden plank road next to the fiber road, and through the interpretation of the traditional Shaoxing bridge culture on both sides of the canal within the base, it is located at different nodes. Set up different forms of bridges, and use the slow-moving system to connect different cultural spaces, so as to show the bridge culture of Jiangnan water town. The falling ink on the top, through this design method, forms an outdoor "Bridge Culture Museum" in the overall space.

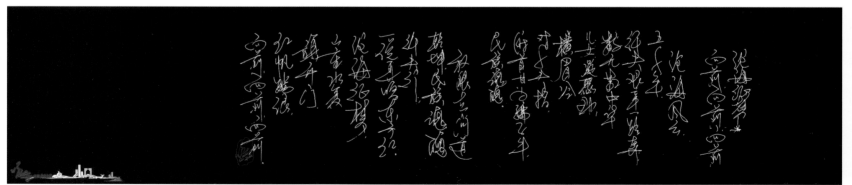

城市设计意境：沧海征梦，向前、向前、向前

沧海风云。五千年，纤夫艰辛一路奔，数无数中华儿女。凝聚那，横眉冷对千夫指，俯首甘为孺子牛，民族魂魄。放眼去。问道乾坤。民族魂魄纤夫行，一说再唱东方红。沧海征梦，山重水复旗开门，红帆踏浪，向前向前、向前。

——吴强 2021.5

乡愁一樽女儿红——东关历史街区保护与更新城市设计(地块二)
Nostalgia with A Bottle of Nverhong - Urban Design for Preservation and Renewal of Dongguan Historic Block (Plot 2)

整体空间鸟瞰图

1、女儿红酒文化中心

女儿红酒文化中心意象要素选取

女儿红酒文化中心效果图

女儿红酒文化中心建筑创作深刻把握女儿红酒文化的历史渊源即女儿红黄酒与婚宴嫁娶活动的密切关系,以酒器"爵"、篆体"贺",亭亭玉立待出嫁的少女"荷花"共同建构女儿红酒文化博物馆的建筑空间表达。既是曲水流觞女儿情的流溢,亦是女儿红酒文化本身的形象展示。

The architectural creation of Nverhong Spirits Culture Center deeply grasped the historical origin of Nverhong spirits culture, that is, the close relationship between Nverhong yellow rice spirits and wedding ceremony activities. The architectural space expression of Nverhong spirits culture museum is jointly constructed by the spirits vessel "Jue", seal character of "He", and the girl "lotus" who is to be married. It is not only the overflow of the daughter's love, but also the image of Nverhong spirits culture itself.

女儿红酒文化中心效果图

整体空间效果图

① 廊桥 ② 老台门建筑 ③ 水街 ④ 章裕泰台门 ⑤ 竺可桢故居 ⑥ 老孙家台门 ⑦ 范家台门 ⑧ 东关小学 ⑨ 竺可桢中学 ⑩ 东关历史街区门景 ⑪ 徐恒兴台门 ⑫ 东关中学 ⑬ 驳船码头 ⑭ 万象春台门 ⑮ 莲桥 ⑯ 五猖庙戏台 ⑰ 东关农贸市场 ⑱ 乡愁文化博物馆 ⑲ 戏台 ⑳ 女儿红酒文化创意街区 ㉑ 女儿红酒厂 ㉒ 女儿红酒文化中心 ㉓ 大师工作坊 ㉔ 吉祥庵 ㉕ 天华禅寺 ㉖ 亭桥

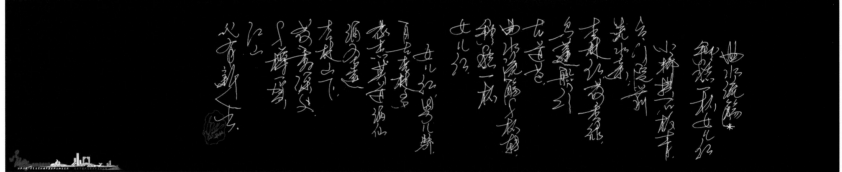

城市设计意境:曲水流觞,乡愁一樽女儿红

小桥拱,石板青,台门院前流水亲;杏林红,荷香绿,乌篷船行古道苍。曲水流觞千杯转,乡愁一樽女儿红。女儿红,男儿骄,自古杏林多豪杰。莫道酒仙酒圣远,杏林山下,荷香深处。千樽一贺,江山代有新人出。

——吴强 2021.5

2、水街节点设计

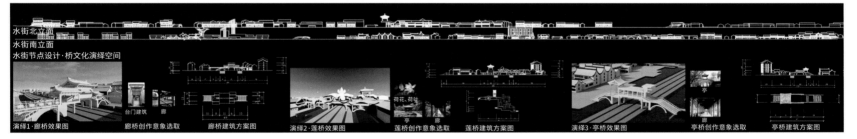

水街北立面
水街南立面
水街节点设计·桥文化演绎空间

演绎1·廊桥效果图　廊桥创作意象选取　台门建筑　廊　廊桥建筑方案图　演绎2·莲桥效果图　莲桥创作意象选取　荷花、荷叶　莲桥建筑方案图　演绎3·亭桥效果图　亭桥创作意象选取　亭　廊　亭桥建筑方案图

水街桥文化之时空演绎,以绍兴传统拱桥文化艺术的特质为蓝本,分别构筑"亭廊之桥"、"台门廊桥"和"莲香之桥",其中"莲香之桥"的创意实是东关古运河历史遗存荷花荡空间的观照,更是一、二、三基地城市设计整体空间的建构。

The interpretation of Water Street bridge culture is based on the characteristics of Shaoxing traditional arch bridge culture and art. "Pavilion Gallery Bridge", "Terrace Corridor Bridge" and "Lotus Bridge" are respectively constructed. Among them, the "Lotus Bridge" is actually originated from the spatial correlation of the historical relics of Dongguan Ancient Canal, and it is also the overall space construction of the urban design of Base 1, Base 2 and Base 3.

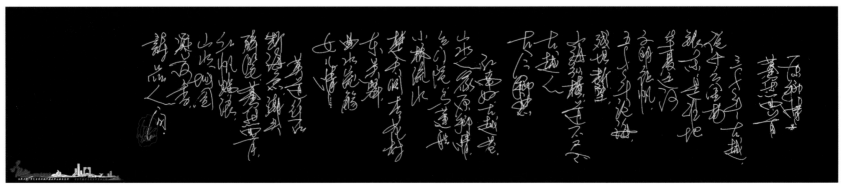

城市设计意境:原乡情,蓦然回首

三千年古越,说什么金柯银东;是谁把华夏运河文明征帆;五千年沧海,残垣断壁,山海纵横,道不尽古越今愁。江南好,古越苍。山水人家原乡情。台门院、乌篷船,小桥流水楚天阔;杏花村、东关驿,曲水流觞女儿情……莫道纤石断海急,潮头醉笔,蓦然回首:红帆踏浪、山水城园凝荷香。诗品人间。

——吴强 2021.5

清华大学建筑设计研究院有限公司
ARCHITECTURAL DESIGN & RESEARCH INSTITUTE OF TSINGHUA UNIVERSITY CO., LTD.

官方微信公众号　官方微信视频号　官方抖音号　官方微博号　官方bilibili号

清华大学建筑设计研究院有限公司(THAD)成立于1958年，为国内知名建筑设计机构。依托于清华大学深厚广博的学术、科研和教学资源，作为建筑学院、土水学院等院系教学、科研和实践相结合的基地，秉承"精心设计、创作精品、超越自我、创建一流"的奋斗目标，设计水平在国内名列前茅。现有工程设计人员1 300余人，其中中国科学院、中国工程院院士6人，勘察设计大师4人。设计并建成的工程已获得国家级、省部级优秀设计奖达600余项，位居全国建筑设计机构前列。

Founded in 1958, the Architectural Design & Research Institute of Tsinghua University Co.,Ltd. (THAD) is a leading architectural design institute based in Beijing, China. Leveraging on Tsinghua University's profound and extensive academic achievement and competitiveness in the interdisciplinary field, THAD, as a comprehensive platform that combines academic study, professional research, and project implementation of Tsinghua University, is adhering to the striving goal of "Designing Meticulously, Creating the finest, Transcending Ourselves, Striving for the best". Its design ranks among the best in China. Currently, THAD has more than 1 300 design professionals, including 6 academicians of the Chinese Academy of Sciences and the Chinese Academy of Engineering, and 4 National Design Masters. Our projects have won over 600 outstanding design awards at the national level and provincial level, ranking top among China's architectural design institutes.

地址：北京海淀区清华大学建筑设计中心楼　　企划部　　　　　　　　　　　　　　　Add: Architectural Design Center, Tsinghua　　Planning Department
邮编：100084　　　　　　　　　　　　　　　电话：010-62789996　　　　　　　　　　　　University, Beijing, China　　　　　　　　　Tel: 010-62789996
网址：www.thad.com.cn　　　　　　　　　　邮箱：thad_branding@thad.com.cn　　　　P.C: 100084　　　　　　　　　　　　　　　　Email: thad_branding@thad.com.cn
　　Web: www.thad.com.cn

国家跳台滑雪中心
National Ski Jumping Center

项目地点：河北 张家口
建筑面积：24 200 m²
用地面积：204.92 hm²
赛道剖面设计：汉斯马丁

Location: Zhangjiakou, Hebei
Building Area: 24 200 m²
Site Area: 204.92 hm²
Track Profile Design: Hans-Martin Renn

国家跳台滑雪中心的跳台剖面因与中国传统吉祥饰物"如意"的S形曲线契合,因此被形象地称为"雪如意",是我国首座符合国际标准的跳台滑雪场地,工程量大、建设难度高。"雪如意"的选址充分利用自然地貌,拥有两条赛道,落差分别为136.2 m和114.7 m,与所在山谷的落差、形状及比赛需求高度契合。区别于以往的国内外跳台设计,"雪如意"首次设立了位于运动员出发区上部的顶峰空间,运动员从跳台眺望,远处依稀可见明代长城遗迹,顶部"飞碟"式的实心空间中被挖开一处圆形空间,使顶峰形成了3 900 m²的环形场地。

The National Ski jumping Center is figuratively called "Xueruyi" because its jumping profile matches the S-shaped curve of the Traditional Chinese auspicious decoration "Ruyi". It is the first ski jumping site in China that meets international standards, with the largest amount of engineering and the highest difficulty in construction. Xue Ruyi is located to take full advantage of the natural landscape, with two courses, with a drop of 136.2 m and 114.7 m respectively, and the drop and shape of the valley are highly suited to the needs of the race. Different from the previous platform designs at home and abroad, "Xueruyi" for the first time set up a peak space located at the upper part of the athletes' starting area. From the platform, the athletes can see the remains of the Ming Dynasty Great Wall in the distance, The "flying saucer-like" solid space at the top is cut into a circular space, resulting in a 3 900 m² circular site at the summit.

九寨沟景区沟口立体式游客服务中心
Visitor Center of Jiuzhai Valley National Park

项目地点：四川阿坝藏族羌族自治州
建筑面积：30 649.84 m²
用地面积：8.996 hm²

Location: Aba Tibetan and Qiang Autonomous Prefecture, Sichuan
Building Area: 30 649.84 m²
Site Area: 8.996 hm²

　　九寨沟景区沟口立体式游客服务中心建设项目位于九寨沟风景名胜区沟口。项目总建筑面积约30 000 m²，总用地面积8.996 hm²。建设内容包含游客服务中心、国际交流中心、荷叶宾馆改造、林卡景观、白水河和翡翠河驳岸加固、立交桥及引道建设等。作为景区标志性建筑及门户，项目建成后，将为每天最多4.1万人次游客提供交通接驳及保障性服务。

Visitor Center of Jiuzhai Valley National Park project is located in Jiuzhaigou scenic spot Goukou. The total construction area of the project is about 30 000 m², and the site area is 8.996 hm². The construction includes tourist service center, international exchange center, transformation of Heye Hotel, Linka landscape, reinforcement of baishui river and Emerald River revetment, overpass and approach construction, etc. As the landmark building and gateway of the scenic spot, the project will provide transportation and guarantee services for up to 41 000 tourists per day upon completion.

雄安新区容东片区C组团
Group C Rongdong Area, Xiong'an New District

项目地点：河北 雄安
建筑面积：1 578 000 m²
用地面积：1 270 hm²

Location: Xiong'an, Hebei
Building Area: 1 578 000 m²
Site Area: 1 270 hm²

设立河北雄安新区，是党中央做出的一项重大历史性战略选择，是千年大计、国家大事。容东片区作为新区开发建设的先行区，承担着首期居民拆迁安置任务，肩负着探索建设经验、创新开发模式的重要使命。一栋栋建筑拔地而起，由图纸变为现实，这离不开建设者们的努力。如今建设项目已陆续交付，居民不但住上了新居，在教育、商业、医疗等方面也都享受到了实实在在的便利。

The establishment of Xiong'an New District in Hebei province is a major historic and strategic choice made by the CPC Central Committee. It is a major project of the millennium and a major national event. Rongdong Area undertakes the task of residents' demolition and resettlement in the first phase, and shoulders the important mission of exploring the construction experience and innovating the development mode. A pillar of building, from the drawing into reality, which is inseparable from the efforts of the builders. Now construction projects have been delivered, residents not only live in new homes, education, business, health care and other tangible convenience.

中国第一历史档案馆新馆
The First Historical Archives of China

项目地点：北京
建筑面积：99 470 m²

Location: Beijing
Building Area: 99 470 m²

中国第一历史档案馆1925年建馆，是我国历史上第一个现代意义的专业档案机构，是我国近现代档案事业的开始。原馆在北京故宫西华门内，现有条件已不能满足档案保存、展示需要。"文化是一个国家、一个民族的灵魂"。新馆建设工程，有利于抢救和保护明清档案文化资源，有利于弘扬祖国优秀传统文化，有利于提升国家文化软实力。新馆位于北京市东城区祈年大街，总建筑面积99 470 m²，高45 m。

Founded in 1925, The First Historical Archives of China is the first modern professional archival institution in Chinese history and the beginning of China's modern archival career. China's First Historical Archives is located in the Xihua Gate of the Imperial Palace in Beijing, but the existing conditions can no longer meet the needs of archival preservation and display. Culture is the soul of a country and a nation. The construction of the new museum is conducive to saving and protecting the cultural resources of the Ming and Qing Dynasties, promoting the excellent traditional culture of the motherland, and promoting the soft power of national culture. The new building is located on Qinian Street, Dongcheng District, Beijing, with a total floor area of 99 470 m² and a height of 45 m.

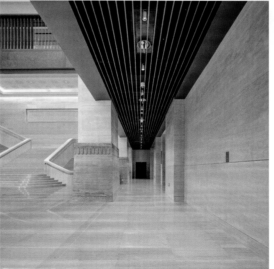

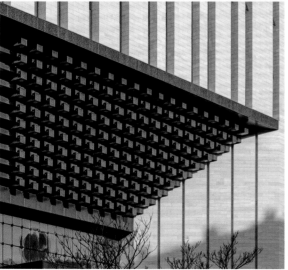

清华大学图书馆北馆
Tsinghua University Library North Wing

项目地点：北京
建筑面积：14 959 m²

Location: Beijing
Building Area: 14 959 m²

清华大学图书馆是该校园的重要标志性建筑，始建于1919年，由美国建筑师亨利·墨菲（Henry Murphy）设计。其后历经两次扩建：20世纪30年代由杨廷宝设计的二期扩建和20世纪80年代由关肇邺设计的三期扩建。北馆暨李文正馆，是对图书馆建筑的四期扩建，地处清华大学校园核心区。设计力求在延续校园历史文脉的同时，满足当代教学与学科发展需要。

The Tsinghua University Library is a landmark building on the Tsinghua University campus. It was initially designed by American architect Henry Murphy, and the construction was completed in 1919, followed by two extensions: the second phase of construction designed by Yang Tingbao in the 1930s, and the third phase of construction designed by Guan Zhaoye in the 1980s. This project is the fourth phase of extension of the library complex, located in the core area of the campus. The library embodies the continuance of the historical context of the university, and at the same time addresses the need for development in regards to teaching and academic disciplines.

亚洲金融大厦暨亚洲基础设施投资银行总部永久办公场所
Asia Financial Centre and AIIB Headquarters

项目地点：北京	Location: Beijing
建筑面积：389 972 m²	Building Area: 389 972 m²
用地面积：6.12 hm²	Site Area: 6.12 hm²
合作单位：德国gmp国际建筑设计有限公司	Cooperative Design: German GMP International Architectural Design Co., Ltd.

亚洲金融大厦暨亚洲基础设施投资银行总部永久办公场所是一个高标准国际金融机构总部办公场所，庄重、简约、绿色、包容，是实现了国际一流生态、节能技术水准的绿色建筑。建筑整体格局严整有序、方正内敛，内部空间穿插交融、开放共享。建筑以营造高品质办公场所和室内外交流空间为目标，通过在室内空间引入系列化开放的共享交流空间，赋予建筑全新的公共交流体验，营造融合绿色、交往、共享的内外空间环境。建筑在室内环境、空气品质、生态智能及绿色节能等方面创新集成运用先进的设计、建造、运维技术，在高品质建筑、先进建造、智能运维等多层面进行了创新探索性实践。

Asia Financial Centre and AIIB Headquarters is a high-standard international financial institution headquarters office space. It is solemn, simple, green and inclusive. It is a green building that has achieved world-class ecological and energy-saving technology standards. The overall structure of the building is orderly, square and restrained, and the internal space is interspersed, open and shared. The goal of the building is to create high-quality office space and indoor and outdoor communication space. By introducing a series of open shared communication space into the interior space, the building is endowed with a new public communication experience and creates an internal and external space environment integrating green, communication and sharing. In terms of indoor environment, air quality, ecological intelligence, green energy saving and other aspects, advanced design, construction, operation and maintenance technologies have been innovatively integrated into the building, and innovative and exploratory practices have been carried out in the aspects of high-quality construction, advanced construction, and intelligent operation and maintenance.

中国建筑科学研究院有限公司 建筑设计院
China Academy of Building Research

扫描查看更多信息

中国建筑科学研究院有限公司建筑设计院（以下简称"建筑设计院"）于1985年2月7日经城乡建设环境保护部批准成立，初始名称为"中国建筑科学研究院综合设计研究部"，2002年改制为"建研建筑设计研究院有限公司"，2007年又由国有控股独立法人企业整体并入到中国建筑科学研究院，沿用名称"中国建筑科学研究院建筑设计院"，2017年根据公司制改制的需要，更名为"中国建筑科学研究院有限公司建筑设计院"。

建筑设计院是中国建筑科学研究院有限公司（以下简称"公司"）直属事业部，使用并维护管理公司所持有的建筑工程设计甲级、风景园林甲级、建筑专业资信甲级、工程造价咨询乙级、人防专业乙级、军工涉密业务咨询服务单位安全保密条件备案等资质，开展建筑工程综合设计、城乡规划以及设计咨询等业务，是国内甲级设计院中第一批通过ISO 9001质量管理体系认证的单位。

建筑设计院的业务范围包括：建筑工程设计及前期策划；建筑智能化、城乡规划及景观设计、建筑室内外装饰装修的设计；绿色建筑、海绵城市、健康建筑、建筑工业化、人防与地下空间工程、BIM工程等的研究、设计与咨询，以及建筑工程检测和人防工程检测等业务。

建筑设计院一贯坚持"与大师同行、与品牌同行"的发展思路，借助公司整体技术实力和人才优势，从国内走向国际，广泛开展国际合作，先后与贝聿铭、矶崎新、马里奥·博塔、冯·格康、里卡多·波菲、斯蒂芬·霍尔等世界级大师合作设计，打造了一批标志性建筑。

建筑设计院的作品涵盖总部办公、医疗康养、文化博览、商业综合体、会展、住宅、城市更新和既有建筑改造等建筑设计领域，其中较为知名的精品建筑包括中国银行总部大厦、中国疾病预防控制中心一期、中国国家博物馆改扩建工程、成都来福士广场、世界机器人大会永久会址（亦创国际会展中心）、北京清河橡树湾、深圳市慢性病防治院改扩建工程、实创医谷产业园、国家会展中心（天津）、中国红岛国际会展中心、黄河国家博物馆、雄安城市计算（超算云）中心、深圳市前海综合交通枢纽上盖物业、海口国际免税城、杭州大会展等。

China Academy of Building Research Co., Ltd. (abbreviation "Architectural Design Institute") was established on February 7, 1985 with the approval of the Ministry of Urban and Rural Construction and Environmental Protection. The initial name was "Comprehensive Design and Research Department of China Academy of Building Sciences". In 2002, it was restructured into "Architectural Design and Research Institute Co., Ltd.". In 2007, it was integrated into the China Academy of Building Sciences by a state-owned holding independent legal entity, The name of "Architectural Design Institute of China Academy of Building Sciences" was followed. In 2017, it was renamed as "Architectural Design Institute of China Academy of Building Sciences Co., Ltd." according to the needs of the company's system reform.

The Architectural Design Institute is a business unit directly under China Academy of Building Sciences Co., Ltd. (hereinafter referred to as the "Company"). It uses and maintains the qualifications held by the management company, such as Class A in architectural engineering design, Class A in landscape architecture, Class A in building professional credit, Class B in engineering cost consulting, Class B in civil air defense, and the filing of security and confidentiality conditions for military secret related business consulting service units, to carry out comprehensive design of architectural projects Urban and rural planning, design consulting and other businesses are the first batch of domestic first-class design institutes to pass the ISO 9001 quality management system certification.The business scope of the architectural design institute includes: architectural engineering design and preliminary planning; Intelligent building, urban and rural planning and landscape design, interior and exterior decoration design of buildings; Research, design and consultation of green buildings, sponge cities, healthy buildings, building industrialization, civil air defense and underground space projects, BIM projects, as well as building engineering testing and civil air defense engineering testing.

The Architectural Design Institute has always adhered to the development idea of "walking with masters and brands". With the help of the company's overall technical strength and talent advantages, it has carried out extensive international cooperation from domestic to international. It has successively cooperated with world-class masters such as I.M. Pei, Jisaki Xin, Mario Bota, Feng Gekang, Ricardo Boffin and Stephen Hall to design and build a number of landmark buildings.

The works of the Architectural Design Institute cover such architectural design fields as headquarters office, medical care, cultural exhibition, commercial complex, exhibition, residence, urban renewal and existing building reconstruction, among which the more famous boutique buildings include Bank of China Headquarters Building, China Disease Prevention and Control Center Phase I, China National Museum Reconstruction and Expansion Project, Chengdu Raffles Square, the permanent site of the World Robotics Congress (also the International Convention and Exhibition Center) Beijing Qinghe Oak Bay, Reconstruction and Expansion Project of Shenzhen Institute for Chronic Disease Prevention and Treatment, Shichuang Medical Valley Industrial Park, National Convention and Exhibition Center (Tianjin), China Red Island International Convention and Exhibition Center, Yellow River National Museum, Xiong'an Urban Computing (Chaosuan Cloud) Center, properties on Qianhai Comprehensive Transportation Hub in Shenzhen, Haikou International Duty Free City, Hangzhou Convention and Exhibition, etc.

照片提供：gmp，摄影师：CreatAR Images

中国工艺美术馆、中国非物质文化遗产馆
CTCM

项目地点：北京
主要功能：文化博览
建筑面积：91 126 m²
合作设计单位：德国gmp国际建筑设计有限公司
项目时间：2013年—2021年

Location: Beijing
Main Functions: Cultural Expo
Building Area: 91 126 m²
Cooperative Design: German GMP International Architectural Design Co., Ltd.
Project Time: 2013-2021

照片提供：gmp，摄影师：CreatAR Images

照片提供：gmp，摄影师：CreatAR Images

该项目是党中央、国务院在"十三五"期间决定建设的国家重点文化设施，是展示中国优秀传统文化、弘扬中华民族审美观与社会主义核心价值观、树立文化自信与民族自信的国家级文化殿堂。该项目圆满竣工投入使用，标志着我国又新增一处代表国家和首都文化形象、彰显新时代文化繁荣发展气象的重要文化地标，其建成填补了我国工艺美术和非物质遗产国家级博物馆的空白。

As a national key cultural establishment approved by the CPC Central Committee and the State Council during the 13th Five-Year Plan period, this project is a national cultural palace where we can display Chinese excellent traditional culture, promote the Chinese national aesthetic view and socialist core values, as well as establish cultural confidence and national confidence. The project has been successfully completed and put into use, making it another new important cultural landmark representing the cultural image of the country and the capital and highlighting the cultural prosperity of the new era. Its completion fills the gap in the field of arts and crafts and intangible heritage museum.

国家会展中心（天津）
National Convention & Exhibition Center (Tianjin)

项目地点：天津
主要功能：会展
建筑面积：1 352 000 m²
方案单位：德国gmp国际建筑设计有限公司
项目时间：2013年－2021年（一期完成），
　　　　　2019年至今（二期在建）

Location: Tianjin
Main Functions: Exhibition
Building Area: 1 352 000 m²
Cooperative Design: German GMP International Architectural Design Co., Ltd.
Project Time: 2013-2021 (Phase I Completed),
　　　　　　　2019 to Now (Phase II Under Construction)

该项目作为商务部和天津市政府的合作项目，是商务部继广州、上海之后，在全国布局的第三个国家级现代化展馆。其功能完备、配套齐全、交通顺畅，展厅全部单层无柱、超强承载能力，同时采用地源热泵、屋面光伏、智慧管理平台技术，并取得绿色建筑三星级、A级装配式建筑等认证，令其成为会展建筑的翘楚。项目分两期建设，其中一期790 000 m²，二期562 000 m²。共设置32个现代化12 500 m²单层展厅，室内净布展面积400 000 m²，室外布展面积150 000 m²，建筑高度24 m，中央大厅高度34 m。

The National Convention & Exhibition Center (Tianjin) is jointly built by the Ministry of Commerce of the People's Republic of China and Tianjin Municipal People's Government. As the third national modern exhibition hall in the country following Guangzhou and Shanghai, it has complete functions, complete facilities and smooth traffic. The exhibition hall is all single-storey without column, with super bearing capacity. At the same time, it adopts ground source heat pump, roof photovoltaic, intelligent management platform technology, and has obtained the green building three-star certification and A-level prefabricated building certification, making it a leader in exhibition building field. The project is divided into two phases-the first phase is 790 000 m² and the second phase is 562 000 m². A total of 32 modern single-storey exhibition halls of 12 500 m² are set up, with a net indoor exhibition area of 400 000 m² and an outdoor exhibition area of 150 000 m². The building is 24 m high, and the central hall is 34 m high.

中国美术出版大厦
（人民美术出版社新址）
China Fine Arts Publishing Building
(New Site for People's Fine Arts Publishing House)

项目地点：北京	Location: Beijing
主要功能：办公	Main Functions: Office
建筑面积：14 000 m²	Building Area: 14 000 m²
项目时间：2010年－2019年	Project Time: 2010-2019

该项目为人民美术出版社新址办公楼，主要功能包括办公和小型美术馆。建筑项目地上七层，地下两层，其中地上建筑面积8 000 m²，地下建筑面积6 000 m²，建筑高度32 m。

The project is the new office building of the People's Fine Arts Publishing House, whose main functions include office building and small art gallery. The project has seven floors above ground and two floors underground, including an above-ground building area of 8 000 m², an underground building area of 6 000 m², and a building height of 32 m.

北京亦创国际会展中心
（世界机器人大会永久会址）
Beijing Etrong International Exhibition & Convention Center (Permanent Venue of the World Robot Conference)

项目地点：北京	Location: Beijing
主要功能：会展	Main Functions: Exhibition
建筑面积：86 919 m²	Building Area: 86 919 m²
项目时间：2016年－2017年	Project Time: 2016-2017

该项目位于北京市大兴区荣昌东街6号，是北京亦庄经济开发区传统产业转型示范和城市更新的重要项目。项目自竣工起，标志着21世纪初建设的北人印刷车间厂房以世界机器人大会为契机，正式成功转型为可以容纳20万人同时使用的集产、研、展一体的国际会议及会展中心。

Located at No.6 Rongchang East Street, Daxing District, Beijing, it is an important project of traditional industrial transformation demonstration and urban renewal in Beijing Yizhuang Economic Development Zone. Its completion marks that the Beiren Printing Workshop built in the early 21st century has been successfully transformed into an international conference and exhibition center integrating production, research and exhibition that can accommodate 200 000 people.

中国·红岛国际会议展览中心
China· Hongdao International Exhibition & Convention Center

项目地点：山东 青岛
主要功能：会展
建筑面积：488 000 m²
方案及初设（建筑）单位：德国gmp国际建筑设计有限公司
项目时间：2016年－2019年

Location: Qingdao, Shandong
Main Functions: Exhibition
Building Area: 488 000 m²
Cooperative Design: German GMP International Architectural Design Co., Ltd.
Project Time: 2016-2019

该项目定位为青岛新"窗"，是山东省最大的会展经济综合体，配有酒店、商业、会议和办公功能，是一座综合性会展建筑。室内展览面积15万m²，室外展览面积20万m²。建筑群呈现"H"形的对称布局，将北部城区与南部胶州湾海岸景观连为一体，具有逻辑明晰、和谐统一的整体感。

Positioned as the new "window" of Qingdao, it is a comprehensive exhibition building and the largest exhibition economic complex in Shandong Province, equipped with hotels, businesses, conferences and offices. The indoor exhibition area is 150 000 m², and the outdoor exhibition area is 200 000 m². The building groups adopt an H-shaped symmetrical layout, connecting the northern urban area with the coastal landscape of Jiaozhou Bay in the south, with an integrated sense of clear logic and harmony.

天津生态城图书档案馆
Tianjin Eco-City Library and Archives

项目地点：天津
主要功能：图书档案馆
建筑面积：67 000 m²
方案及初设（建筑）单位：德国gmp国际建筑设计有限公司
项目时间：2012年－2018年

Location: Tianjin
Main Functions: Library and Archives
Building Area: 67 000 m²
Cooperative Design: German GMP International Architectural Design Co., Ltd.
Project Time: 2012-2018

该项目主要功能为图书馆、档案馆、附属办公、餐饮及图书销售。建筑地上5层，无地下室，建筑高度35 m。建筑沿东、北两侧紧临的故道河展开布置，与周围自然景观和城市景观融为一体，富有文化内涵和时代气息。图书馆部分藏书量为100万册。

The main functions of the project are library, archives, ancillary offices, catering and book sales. The project has five floors above ground without basements. It has a building height of 35 m. Arranged along the ancient river on the east and north sides, the building is integrated with the surrounding natural landscape and urban landscape, which is rich in cultural connotation and the flavor of the times. The collection of the library is 1 million volumes.

嘉铭东枫产业园
Jiaming Dongfeng Industrial Park

项目地点：北京	Location: Beijing
主要功能：办公	Main Functions: Office
建筑面积：39 220 m²	Building Area: 39 220 m²
方案单位：德国gmp国际建筑设计有限公司	Cooperative Design: German GMP International Architectural Design Co., Ltd.
项目时间：2012年—2018年	Project Time: 2012-2018

  该项目是一组院落式总部基地办公楼,地上5层、地下2层,总建筑高度29.7 m,由东塔、西塔、北塔3座主建筑构成,每栋建筑首层都拥有5.8 m层高的企业展厅。中心设有下沉式中央广场,每栋设有2层平台花园和屋顶花园,形成了立体花园式办公空间,办公区自然光覆盖率达到87%,是北京首家获得LEED-CS铂金认证的城市商务花园项目,是驻京国际化公司设立企业总部的最佳选择。

  The project is a group of courtyard-style headquarters office buildings with a total building height of 29.7 m, including five floors above ground and two floors underground. It consists of three main buildings-the East Tower, the West Tower and the North Tower. There is a 5.8 m high enterprise exhibition hall on the first floor of each building. The center is equipped with a sunken central square, and each building has a two-floor platform garden and roof garden, forming a three-dimensional garden-style office space. The coverage rate of natural light in the office area reaches 87%. It is rated as the first urban business garden project in Beijing to obtain LEED-CS platinum certification, and it is the best choice for international companies in Beijing to set up corporate headquarters.

华润清河橡树湾（海淀区清河镇住宅及配套工程）
China Resource Qinghe Oak Bay (House and Supporting Project in Qinghe Town, Haidian District)

项目地点：北京	Location: Beijing
主要功能：住宅及配套	Main Functions: Residence and Supporting Facilities
建筑面积：813 259 m²	Building Area: 813 259 m²
方案单位：北京中联环建文建筑设计有限公司、澳大利亚柏涛墨尔本建筑设计有限公司深圳代表处	Cooperative Design: Beijing Zhonglian Huanjianwen Architectural Design Co., Ltd., Shenzhen Representative Office of Australia Botao Melbourne Architectural Design Co., Ltd.
项目时间：2006年—2018年	Project Time: 2006-2018

该项目位于北京市海淀区清河镇,为住宅小区及住宅配套项目,是华润置地在全国橡树湾系列产品的开山之作。项目以南北向板式高层住宅建筑为主,结合部分东西向短板单元,通过平面上的巧妙布局以及高度上的错落变化,营造出形态各异的半开敞式院落空间;通过整体协调充分发挥各组团形态丰富、个性迥异的特征,展示出建筑群的韵律之美,项目已成为京北海淀具有很强品牌知名度与影响力的明星楼盘。

Located in Qinghe Town, Haidian District, Beijing, it is a residential community and residential supporting project, which is the first work of Oak Bay series products of China Resources in the country. The project mainly consists of north-south slab-type high-rise residential buildings, combined with some east-west short slab units, creating a semi-open courtyard space with different forms through the clever layout on the plane and the scattered changes in height. The project has become a star property with strong brand awareness and influence in Haidian District of Beijing by giving full play to the diverse forms and different personalities of each group and showing the beauty of rhythm of the building.

中国人民大学附属中学丰台学校
Fentai School of the High School Affiliated to Renmin University of China

项目地点：北京	Location: Beijing
主要功能：教育	Main Function: Education
建筑面积：133 300 m²	Building Area: 133 300 m²
项目时间：2015年—2021年	Project Time: 2015-2021

该项目位于北京市丰台区王佐镇，主要功能为中小学教学楼及配套设施，是人大附中规模最大的分校。2015年我院凭借优秀的原创方案在竞标评选中一举中标。设计内容包含整体规划、单体建筑、室内精装、专项建筑、园林景观、校园文化概念规划、夜景照明概念规划。通过对全过程设计的把控，最终实现了建筑方案的高完成度，也实现了建成一座植入人大附中先进教育理念并引领京西南中小学整体教育水平的旗舰型校园建筑的项目定位。

As the largest branch school affiliated to Renmin University of China, the project is located in Wangzuo Town, Fengtai District, Beijing. Its main functions are teaching buildings and supporting facilities for middle and primary schools. In 2015, we won the bid with our excellent original program. The design content includes overall planning, single building, interior decoration, special architecture, garden landscape, campus culture concept, night scene lighting concept. Through the control of the whole design process, the high degree of completion of the architectural scheme was finally achieved, and the project positioning of building a flagship campus building implanted with advanced educational concepts of the High School affiliated to Renmin University of China and leading the overall education level of middle and primary schools in southwest Beijing was also realized.

中坤广场改造项目
Zhongkun Plaza Renovation Project

项目地点：北京	Location: Beijing
主要功能：办公	Main Functions: Office
建筑面积：400 000 m²	Building Area: 400 000 m²
项目时间：在建	Project Time: Under Construction

该项目是北京三环上的重要地标建筑，不仅承载着周边发展的历史记忆，更代表着区域的城市形象。改造设计从整合城市形态入手，充分利用现有300 m长的城市界面，对建筑形象进行保留、更新和提升处理，使其更具科技感与现代感。对区域原有业态进行梳理，结合新兴需求，在保留生活记忆的同时植入新亮点和新业态，办公、商业、交通、生态场所等多功能复合型的城市空间相辅相成，实现了以新的姿态带动周边发展、提振区域活力的项目目标定位。

As an important landmark building on Beijing's Third Ring Road, the project not only carries the historical memory of the surrounding development, but also represents the urban image of the region. Starting from the integration of urban forms, the renovation design makes full use of the existing 300 m long urban interface, and carries on the preservation, updating and upgrading of the architectural image to make it more scientific and modern. By combing the original business forms of the region and combining with the emerging needs, the project implants new highlights and new business forms while preserving the living memory. The multi-functional urban spaces such as office, business, transportation and ecological places complement each other, realizing the project positioning of driving the development of the surrounding areas and boosting the regional vitality with a new attitude.

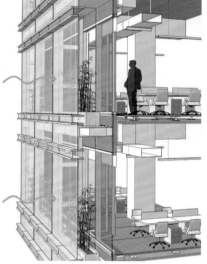

航天规划设计集团有限公司
Aerospace Planning and Design Group Co., Ltd.

航天规划设计集团有限公司（简称航天设计集团）是中国航天建设集团有限公司（即中国航天科工七院，简称七院）所属从事工程服务业务板块的新平台，是七院下属二级法人单位，注册资金26 500万元，总部位于北京。

航天设计集团拥有工程设计综合甲级、城乡规划编制甲级、工程勘察综合甲级、测绘甲级、工程咨询单位资信甲级、工程监理甲级、工程造价咨询甲级、地质灾害危险评估甲级、压力管道设计等资质；拥有建筑工程施工总承包、电子与智能化工程专业承包、环保工程专业承包、建筑机电安装工程专业承包、建筑装修装饰工程专业承包等建筑业企业资质；具有对外工程总承包经营权。

航天设计集团设有总体部、创新院、综合设计院、勘察设计院、专项设计院、事业部、区域分公司、全资子公司、控股公司等50余家下属单位。以"高端策划咨询、工程咨询设计和EPC、工程信息化、装备设计与集成、综合科技服务"为五大发展方向。经营范围涵盖工程设计；工程勘察；测绘服务；岩土工程；工程监理；技术开发、技术咨询、技术服务；压力管道设计；产品设计；电子元器件产品研发；技术装备制造；保温隔热材料；保温隔热材料技术开发；规划管理；环境评价及检测；经济信息咨询；货物进出口、代理进出口、技术进出口；招投标代理；地质勘察服务；施工总承包；专业承包；技术检测；销售建筑材料；信息系统集成服务；工业设计服务；工程管理服务；工业机器人制造；园林绿化工程施工等。

航天设计集团现有员工1 700余人，取得各专业国家最高级别注册资格者500余人。具有高级专业技术职称的360余人，其中研究员（教授级高级工程师）百余人。先后有近300项咨询设计荣获中国土木工程詹天佑奖及其他国家级、省部级以上优秀设计奖项。

航天设计集团站在新的历史起点上，承载着50多年深厚文化积淀的航天建设使命，将坚持以"融入航天、建设航天、创新驱动、转型重生"为方针，紧跟国家产业政策，聚焦七院发展战略，实现经济规模高质量增长，以一流的作品、一流的质量、一流的作风、一流的诚信，精心打造航天优质品牌，为国防事业、航天事业和社会各界真诚服务。

Aerospace Planning and Design Group Co., Ltd. is a new platform for engineering service business belonging to China Aerospace Construction Group Co., Ltd. (the Seventh Academy of Aerospace Science and Engineering of China, referred to as the Seventh Academy for short), a secondary legal entity under the Seventh Academy, with a registered capital of 265 million yuan and headquartered in Beijing.

Aerospace Design Group has the following qualifications: comprehensive engineering design grade A, urban and rural planning grade A, comprehensive engineering survey grade A, surveying and mapping grade A, engineering consulting unit credit grade A, engineering supervision grade A, engineering cost consulting grade A, geological disaster risk assessment grade A, pressure pipeline design, etc; It has the qualifications of construction enterprises such as general contracting of construction projects, specialized contracting of electronic and intelligent projects, specialized contracting of environmental protection projects, specialized contracting of building mechanical and electrical installation projects, specialized contracting of building decoration projects, etc; It has the right of general contracting for foreign projects.

Aerospace Design Group has more than 50 subordinate units, including the General Department, Innovation Institute, Comprehensive Design Institute, Survey and Design Institute, Special Design Institute, Business Division, regional branches, wholly-owned subsidiaries and holding companies. The five major development directions are "high-end planning consulting, engineering consulting design and EPC, engineering informatization, equipment design and integration, and comprehensive scientific and technological services". The business scope covers engineering design; Engineering survey; Surveying and mapping services; Geotechnical engineering; Project supervision; Technical development, technical consultation and technical services; Penstock design; Product design; R&D of electronic components; Technical equipment manufacturing; Thermal insulation materials; Technical development of thermal insulation materials; Planning management; Environmental assessment and detection; Economic information consultation; Import and export of goods, agency import and export, technology import and export; Bidding agency; Geological survey service; General construction contracting; Professional contracting; Technical inspection; Sales of building materials; Information system integration service; Industrial design services; Engineering management services; Industrial robot manufacturing; Landscaping project construction, etc.

The Aerospace Design Group currently has more than 1 700 employees, and more than 500 people have obtained the highest level registration qualification in various professional countries. There are more than 360 people with senior professional and technical titles, including more than 100 researchers (professor level senior engineers). Nearly 300 consulting designs have won the Zhan Tianyou Award of China Civil Engineering and other national, provincial and ministerial excellent design awards.

Standing at a new historical starting point and carrying the mission of aerospace construction with more than 50 years of profound cultural accumulation, Aerospace Design Group will adhere to the principle of "integration into aerospace, construction of aerospace, innovation driven, transformation and rebirth", closely follow the national industrial policy, focus on the development strategy of the Seventh Academy, and achieve high-quality economic growth. With first-class works, first-class quality, first-class style and first-class integrity, we will carefully build a high-quality aerospace brand and sincerely serve the national defense, aerospace and all sectors of society.

地址：北京市大兴区西红门镇春和路39号院3号楼A座
邮编：100162
电话：010-89060999
传真：010-89060995

Add: Building A, Tower 3, No.39 Courtyard, Chunhe Road, Xihongmen Town, Daxing District, Beijing
P.C: 100162
Tel: 010-89060999
Fax: 010-89060995

北京新机场东航机务维修区项目
Maintenance Area Project of China Eastern Airlines in Beijing New Airport

项目地点：北京
主要功能：飞机维修基地
建筑面积：154 561 m²
项目时间：2017年－2019年

Location: Beijing
Main Function: Airlines Maintenance Area
Biluding Area: 154 561 m²
Project Time: 2017-2019

该项目主要功能为飞机维修库、航材库、特种车库、工装设备厂房、化工品库、附属办公及配套等。飞机维修库包括机库大厅及3层附楼，机库大厅为单层，开间为138 m，进深90 m，建筑高度为38 m。建筑沿东侧机场飞行区展开布置，机库紧邻飞行区滑行道。整个基地规划设计以飞机维修的工艺流程为基础，打造东航品质的员工热爱的家。

The main functions of the project are aircraft maintenance warehouse, aviation material warehouse, special garage, tooling and equipment plant, chemical material warehouse, auxiliary office and supporting facilities, etc. The aircraft maintenance warehouse includes the hangar hall and the three-storey annex building. The hangar hall is single-storey, with a space of 138 m, a depth of 90 m and a building height of 38 m. The building is arranged along the east flight area of the airport, and the hangar is adjacent to the flight area taxiway. The planning and design of the whole base is based on the aircraft maintenance process to create a quality home that China Eastern Airlines employees love.

上海微小卫星工程中心卫星研制项目
Satellite Development Project of Shanghai Engineering Center for Microsatellites

项目地点：上海	Location: Shanghai
主要功能：卫星总装厂房	Main Founction: General Assembly
总建筑面积：94 794.14 m²	Total Building Area: 94 794.14 m²
项目时间：2015年—2019年	Project Time: 2015-2019

本项目主要用于卫星的总装、集成与测试、部件装配等，包含1号—6号厂房共6栋主要单体建筑。主要建筑地上5层，地下1层，建筑高度23.8 m。1号厂房布置在园区北侧，2#厂房布置在园区中央，3号、4号厂房在2号厂房两侧呈对称布置，5号、6号厂房布置在南侧园区主入口两侧，整体布局功能合理，满足现代化生产需要。

This project is mainly used for satellite final assembly, integration and testing, component assembly, etc., including six main single buildings from Plant 1# to Plant 6#. The main building has five floors above ground and one floor underground, with a height of 23.8 m. Plant # 1 is arranged on the north side of the park. Plant # 2 is arranged in the center of the park. Plant # 3 and Plant # 4 are arranged symmetrically on both sides of Plant # 2. Plant # 5 and Plant # 6 are arranged on both sides of the main entrance to the park on the south side. The overall layout has reasonable function to meet the needs of modern production plants.

长阳航天城科技园项目
Project of Changyang Aerospace City Technology Park

项目地点：北京
主要功能：研发办公
建筑面积：121 000 m²
项目时间：2013年—2018年

Location: Beijing
Main functions: R&D office
Biluding Area: 121 000 m²
Project time: 2013-2018

该项目主要功能为研发办公，建筑地上9层，地下2层，建筑高度45 m，在方案创作过程中通过"母题化"的壁柱设计元素，对应着超越时空的探索精神，体现航天建筑的文化内涵和时代气息。

The main function of the project is research and development office. The building has nine floors above ground and two floors underground, with a building height of 45 m. The design element of "motif" pilaster is adopted in the process of scheme creation, corresponding to a spirit of exploration beyond the time and space and reflecting the cultural connotation and the spirit of the times of space architecture.

海德园
Hyde Park

项目地点：深圳
建筑面积：233 400 m²
用地面积：39 000 m²

Location: Shenzhen
Building Area: 233 400 m²
Site Area: 39 000 m²

海德园项目致力于打造极具活力的城市空间和高品质居住社区；以最佳景观角度和超大视距，追求庭院空间的最大化与流动性，打造多样化、多元化的居住产品类型。项目中创建了一个由公共空间、公共绿化、架空层花园、屋顶花园等组成的全方位的立体绿化体系，将庭院作为生态体系的一部分，与周边自然生态相互渗透，极大地丰富了小区环境和城市环境，形成闹与静、开放与封闭、公共与私密以及地面与空中的全系列绿化体系。营造出一种个性、休闲、时尚的生活方式。

The project is committed to creating a very dynamic urban space and a high-quality residential community with the best landscape angle and large visual distance, in pursuit of the maximization and mobility of the courtyard space, to create diversified types of residential products. The project creates an all-round three-dimensional greening system consisting of public space, public greening, overhead garden, roof garden, etc., making the courtyard a part of the ecological body and interpenetrate with the surrounding natural ecology, which greatly enriches the residential environment and urban environment, achieving a full series of greening system of noise and quietness, opening and closeness, public space and private space, ground and air, and creating a personalized, casual and fashionable lifestyle.

礼贤小学
Lixian Primary School

项目地点：浙江 衢州
主要功能：48班小学
建筑面积：27 858 m²
项目时间：2015年—2017年

Location: Quzhou, Zhejiang
Main Functions: Class 48 Primary School
Building Area: 27 858 m²
Project Time: 2015-2017

该项目位于"南孔圣地"浙江衢州，是一座国家历史文化名城。本着功能、经济、社会、环境四个效益兼顾的原则，在设计上力求塑造具有时代特点、鲜明个性、丰富文化内涵的校园建筑形象。建筑群体无论是总体布局还是建筑的形态、色彩上都保持与当地文化的协调统一，规整的布局、鲜明的建筑色彩、以庭院来组织建筑空间的手法和坡屋顶的运用等都充分体现对衢州当地建筑传统和特点的尊重与发扬。

This project is located in Quzhou City, Zhejiang Province, a national historical and cultural city known as the "holy land of South Confucius". In line with the principle of taking into account the four benefits of function, economy, society and environment, the design strives to shape the image of campus construction with the characteristics of the times, distinct personality and rich cultural connotation. The architectural groups maintain the coordination and unity of local culture in terms of overall layout, architectural form and color. Regular layout, bright architectural color, the technique of organizing architectural space by courtyards and the application of sloping roofs all fully reflect the respect and promotion of local architecture in Quzhou City.

星空标准厂房项目
Star Standard Plant Project

项目地点：四川
建筑面积：140 318.84 m²
用地面积：111 653.33 m²

Location: Sichuan
Building Area: 140 318.84 m²
Site Area: 111 653.33 m²

项目拟在简阳投资建设"一带一路"高通量宽带卫星产业基地，以高通量宽带卫星为核心，建设高通量宽带卫星运营基地、应用研发基地为核心，带动相关产业聚集，建成国际高通量宽带卫星网络运营中心，服务"一带一路"。通过国际卫星互联网，带动互联网文化创意产业的发展、航天科技文化旅游体验、建立空间信息前沿技术研发基地，形成产、学、研一体的高科技产业集群。

The project plans to invest in the construction of "The Belt and Road" high-throughput broadband satellite industrial base in Jianyang, with high-throughput broadband satellite as the core. It aims to build an international high-throughput broadband satellite network operation center with high-throughput broadband satellite operation base and application research and development base as the core to drive related industries to gather, serving the implementation of the "The Belt and Road". Through the international satellite Internet, it will promote the development of Internet cultural and creative industries, the experience of space science and technology cultural tourism, establish research and development bases for frontier space information technologies, and form high-tech industrial clusters integrating production, learning and research.

智能制造基础件产业集群建设项目(一期)
Intelligent Manufacturing Basic Parts Industrial Cluster Construction Project (Phase I)

项目地点：贵州 贵阳 Location: Guiyang, Guizhou
建筑面积：477 015.33 m² Building Area: 477 015.33 m²

本项目建设用地位于贵阳市经济技术开发区小孟工业园区，西临开发大道，北临东西走向的主干道花孟路，南临规划中的横二路，距离地块东650 m处与贵阳南北走向的城市快速路花冠路相连，东面比邻规划中的纵一路；建设用地形状呈较为规整的矩形，南北长946 m，东西长705 m，建设用地面积47.85 hm²(不含林地)；一期总建筑面积240 074 m²，新建25个单体。

The construction land is located in Xiaomeng Industrial Park, Guiyang Economic and Technological Development Zone. The construction land of this project is adjacent to Kaifa Avenue in the west, east-west Huameng Road in the north, Henger Road in the plan in the south. It is connected to Huaguan Road, a north-south urban express road of Guiyang, 650 m away from the east of the plot, and Zhongyi Road in the plan in the east. The shape of the construction land is a fairly regular rectangle, about 946 m long from north to south and about 705 m long from east to west. The construction land area is about 47.85 hm², excluding forest land. The total construction area of Phase I (including 25 new single buildings) is 240 074 m².

中国航发沈阳发动机研究所JG12项目
Project JG12 of AECC Shenyang Engine Research Institute

项目地点： 辽宁
建筑面积： 21 446 m²
用地面积： 134 hm²

Location: Liaoning
Building Area: 21 446 m²
Site Area: 134 hm²

JG12项目致力于建设全国为数不多的航空发动机露天试验基地,对我国航空发动机的研制定型有重要战略意义,建设规模21 446 m²。项目实施后,功能和性能指标比肩国际水平,具备航空发动机多种露天试验能力,初步形成较为完整的航空发动机试验体系,保障发动机试验任务。

Project JG12, with a construction scale of 21 446 m², is committed to building one of the few aviation engine open-air test bases in China, which has important strategic significance for the development of Chinese aviation engine. After the implementation of the project, the function and performance indexes are basically equal to the international level. It has the ability of various aero-engine open-air tests, and has initially formed a relatively complete aero-engine test system to ensure the completion of engine test task.

北京

中国中轻国际工程有限公司
China Light Industry International Engineering Co., Ltd.

中国中轻国际工程有限公司(CLIEC)即原中国轻工业北京设计院,成立于1953年1月。2000年10月成为交中央管理的大型科技型设计企业,现为国务院国资委监管的中央企业"中国保利集团有限公司"的下属企业。

经过70年的创业、发展、壮大,中国中轻国际已成为国内外知名的大型科技型企业,中国中轻国际坚持以市场为先导,以项目为中心,以创新为动力,以服务为宗旨,以质量为保证,为顾客提供全过程、多方位、专业化、质量高、效果好的服务。

China Light Industry International Engineering Co., Ltd. (CLIEC), i.e. the former "China National Bejing Contracting & Engineering Institute for Light Industry" was founded in January 1953. In October 2000, the management of the company was transferred to the central government and became an engineering enterprise of science and technology. It is now a subsidy of Poly Group which is supervised and administered by State-owned Assets Supervision and Administration Commission of the State Council.

Through pioneering, development and expansion of more than 70 years, CLIEC has become a large-scale high-tech type enterprise well known both at home and abroad. CLIEC insists on taking the market as the guidance, the projects as working center, innovation as the driving force, service as the aim and quality as assurance to provide the customers with whole procedure, multi-orientation, professional, high quality, better benefit and satisfactory services.

地址:北京市朝阳区白家庄东里42号	Add: No.42, Baijiazhuangdongli, Chaoyang District, Beijing
邮编:100026	P.C: 100026
电话:010-65826061　13811480249	Tel: 010-65826061　13811480249
邮箱:cliec@cliec.cn	Email: cliec@cliec.cn
网址:www.cliec.cn	Web: www.cliec.cn

宝应新城吾悦广场
Wuyue Square of Baoying New City

设计师:闫金栋、王甫、施玮、刘志存、范妍君	Designers: Jindong Yan, Pu Wang, Wei Shi, Zhicun Liu, Yanjun Fan
项目地点:江苏 扬州	Location: Yangzhou, Jiangsu
建筑面积:123 140.05 m²	Building Area: 123 140.05 m²
用地面积:6.043 hm²	Site Area: 6.043 hm²
建设单位:宝应亿盛房地产开发有限公司	Construction Unit: Baoying Yisheng Real Estate Development Co., Ltd.

北京工商大学良乡校区二期
Liangxiang Campus of Beijing Technology and Business University Phase II

设计师：何嘉欣、杨振宇、苏匹夫、红格尔
项目地点：北京
建筑面积：286 875.41 m²
用地面积：376 750 m²
建设单位：北京工商大学

Designers: Jiaxin He, Zhenyu Yang, Pifu Su, Geer Hong
Location: Beijing
Building Area: 286 875.41 m²
Site Area: 376 750 m²
Construction Unit: Beijing Technology and Business University

北京美中宜和妇儿医院
Beijing Amcare Women's & Children's Hospital

设计师：何嘉欣、施玮、刘志存、红格尔、范妍君、苏匹夫	Designers: Jiaxin He, Wei Shi, Zhicun Liu, Geer Hong, Yanjun Fan, Pifu Su
项目地点：北京	Location: Beijing
建筑面积：38 744 m²	Building Area: 38 744 m²
用地面积：21 774 m²	Site Area: 21 774 m²
建设单位：北京美中宜和妇儿医院有限公司	Construction Unit: Beijing Amcare Women's & Children's Hospital Co., Ltd.

北京市顺义区域医疗中心
Beijing Shunyi Medical Center

设计师：何嘉欣、解志军、谷丽娜、红格尔、范妍君
项目地点：北京
建筑面积：242 087.8 m²
用地面积：80 851 m²
建设单位：北京顺义区医院

Designers: Jiaxin He, Zhijun Xie, Lina Gu, Geer Hong, Yanjun Fan
Location: Beijing
Building Area: 242 087.8 m²
Site Area: 80 851 m²
Construction Unit: Beijing Shunyi District Hospital

中关村科技园区丰台园产业基地东区三期
Zhongguancun Science and Technology District Fengtaiyuan Industrial Base East Zone Phase III

设计师：程春雪、牛欣宇、肇洋、何梓璇、王溪　　Designers: Chunxue Cheng, Xinyu Niu, Yang Zhao, Zixuan He, Xi Wang
项目地点：北京　　Location: Beijing
建筑面积：587 445.18 m²　　Building Area: 587 445.18 m²
用地面积：138 224.255 m²　　Site Area: 138 224.255 m²
建设单位：北京丰科新元科技有限公司　　Construction Unit: Beijing Fengke Xinyuan Technology Co., Ltd.

迈百瑞生物医药(苏州)有限责任公司生物医药创新中心及运营总部
Biopharmaceutical Innovation Center and Operation Headquarters Construction Project of Maverick Biopharmaceutical (Suzhou) Co., Ltd.

设计师：程春雪、张晓、曹伟、李宁	Designers: Chunxue Cheng, Xiao Zhang, Wei Cao, Ning Li
项目地点：江苏 苏州	Location: Suzhou, Jiangsu
建筑面积：138 733.08 m²	Building Area: 138 733.08 m²
建设单位：迈百瑞生物医药(苏州)有限责任公司	Construction Unit: Maverick Biopharmaceutical (Suzhou) Co., Ltd.

北京大兴国际机场临空经济区（廊坊）物流港
Beijing Daxing International Airport Airport Economic Zone (Langfang) Logistics Port Comprehensive Infrastructure Construction Project

设计师：奚艳、牛欣宇、陈丽丽、孙冬雪
项目地点：大兴国际机场临空经济区
建筑面积：85 994.38 m²
用地面积：144 063.78 m²
建设单位：北京大兴国际机场临空经济区（廊坊）管理委员会

Designers: Yan Xi, Xinyu Niu, Lili Chen, Dongxue Sun
Location: Daxing International Airport Airport Economic Zon
Building Area: 85 994.38 m²
Site Area: 144 063.78 m²
Construction Unit: Management Committee of Beijing Daxing International Airport Airport Economic Zone (Langfang)

海口综合保税区高端食（药）材加工标准厂房项目
Haikou Comprehensive Bonded Area High-end Food (Drug) Materials Processing Standard Factory

设计师：解志军、赵光耀、关鹏、牛欣宇、王文悦、石家成
项目地点：海南 海口
建筑面积：117 745.02 m²
用地面积：49 720.93 m²
建设单位：海口保税建设发展有限公司

Desingners: Zhijun Xie, Guangyao Zhao, Peng Guan, Xinyu Niu, Wenyue Wang, Jiacheng Shi
Location: Haikou, Hainan
Building Area: 117 745.02 m²
Site Area: 49 720.93 m²
Construction Unit: Haikou Bonded Construction Development Co., Ltd.

北京行者匠意建筑设计咨询有限责任公司
Beijing Pilgrim Architecture Design Co., Ltd.

北京行者匠意建筑设计咨询有限责任公司(行者工坊)成立于中国北京。在此之前，分布于法国和美国等地的几位合伙人经过长时间的思考和沟通，一致同意只有在中国的实践才是他们将来的建筑方向，于是齐聚北京共同创建了行者工坊。踏实立足于中国，以国际化的视角来做设计。在十多年的耕耘中，行者工坊将他们的建筑热情，投入到中国各地的建筑实践中，先后参与辽宁、安徽、广东和北京等地的多个建筑项目。公司代表作赤山湖游客服务中心、2019北京世界园艺博览会央视动画馆等小型公建取得了业主及同人的一致好评。今后，行者工坊仍然坚持深入挖掘传统文化的精髓，并结合现代主义的主旨努力创造属于"吾土、吾民、吾生活"的建筑。

Beijing Pilgrim Architecture Design Co., Ltd. (Pilgrim) was established in Beijing, China. After a long period of reflection and communication, several partners from France and the United States agreed that only in China was the future direction of their architecture, so they got together in Beijing to create the Walker Workshop. Firmly based in China, with an international perspective to do design. In more than 10 years of cultivation, Walker Workshop has put their passion into architectural practice in various parts of China, and has participated in several architectural projects in Liaoning, Anhui, Guangdong and Beijing. The representative works of the company, such as the Visitor Service Center of Chishan Lake and the CCTV Animation Hall of 2019 Beijing International Horticultural Exhibition, have won the praise of the owners and colleagues. In the future, the workshop will continue to dig deep into the essence of traditional culture and strive to create buildings belonging to "our land, our people and our life" combined with the theme of modernism.

地址：北京市西城区车公庄大街9号五栋大楼6层
电话：010-88365055
传真：010-88365055
邮箱：1439022792@qq.com
网址：www.padesign.cn

Add: 6th Floor, Building 5, No.9, Chegongzhuang Street, Xicheng District, Beijing
Tel: 010-88365055
Fax: 010-88365055
Email: 1439022792@qq.com
Web: www.padesign.cn

东营市河口第一中学
No.1 Middle School of Hekou in Dongying City

设计师：刘磊、高天鹏　　Designers: Lei Liu, Tianpeng Gao
项目地点：山东 东营　　Location: Dongying, Shandong
建筑面积：86 000 m²　　Building Area: 86 000 m²
用地面积：17.7 hm²　　Site Area: 17.7 hm²
容积率：0.48　　Plot Ratio: 0.48
竣工时间：2016年　　Completion Time: 2016

  东营市河口第一中学位于山东省东营市河口区。学校位于城市边缘，用地平坦而开阔，自然条件较好。校园的设计吸收了中国传统古建筑群和园林的经营策略，探讨了边界内与外、场地疏与密、空间开与合、单体连与离这四对范畴，从而确定了南侧以教学和行政为主的"前区"和北侧以宿舍、食堂、体育锻炼为主的"后区"。在此基础上"前区"强调礼制的严谨和秩序，采取了古典的中轴对称、一主两辅的拱卫之势；"后区"则以现代主义的功能优先原则，灵活布局，联系方便，开合自如。这样的方案有效地同时解决了功能的划分、空间的构架和景观交通体系的骨架问题，最终成为一个主次分明、疏密相生、开合有度的教学园区。

  No.1 Middle School of Hekou is located in Hekou District, Dongying City, Shandong Province. The school is located at the edge of the city. The land is flat and open, enjoying excellent natural conditions. The design of the campus has absorbed the management strategy of traditional Chinese ancient buildings and gardens, and discussed the four pairs of categories: inside and outside the boundary, sparse and dense site, open and close space, and single unit connection and separation. Thus, the "front area" in the south is determined to focus on teaching and administration, and the "back area" in the north is determined to focus on dormitory, canteen and physical exercise. On this basis, the "front area" emphasizes the rigor and order of the ritual system, adopting the classical middle axis symmetry with one main building and two auxiliary buildings. The "back area" is based on functional priority principle of modernism style with flexible layout and convenient contact. Such a scheme effectively solves the problems of functional division, spatial framework and skeleton of landscape transportation system at the same time, making the project a teaching park with distinct priorities and proper degree of opening and closing.

2019北京世界园艺博览会－央视动画馆设计
Design of 2019 Beijing Expo Park-CCTV Animation Pavilion

设计师：刘磊、郝悦	Designers: Lei Liu, Yue Hao
项目地点：北京	Location: Beijing
建筑面积：368 m²	Building Area: 368 m²
用地面积：1 000 m²	Site Area: 1 000 m²
容积率：0.37	Plot Ratio: 0.37
竣工时间：2019	Completion Time: 2019

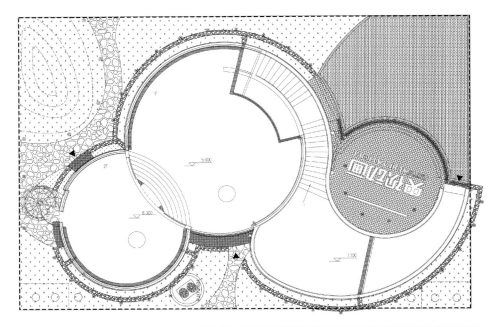

中央电视广播总台央视动画馆是2019北京世界园艺博览会的企业馆之一，它位于展园西部教育与未来展园，与万科投资的植物展园一路之隔。建筑物由三部分组成：位于中央的圆形动画放映厅、位于放映厅西面的一层纪念品售卖区和二层的办公室以及位于东侧的大坡道。这些功能由两道连续的曲面墙体婉转包绕。曲面的外墙由红色的铝制风动片（风铃片）覆盖。这些14 cm直径、1 mm厚的圆形铝片通过小孔安装在水平伸出的螺杆上，保证它能够随着风的吹动上下摇曳。当风吹拂时，整面弧形墙体上的风动片随风摇曳，变幻莫测，远观仿佛正面红色的墙体都被吹起了涟漪，正好像哪吒挥舞的红绫一般。央视动画馆诠释了建筑不仅是凝固的音乐也可能是流动的音乐这一理念。

CCTV Animation Pavilion is one of the corporate pavilions of 2019 Beijing Expo Park. It is located in the Education and Future Exhibition Park in the western Expo Park, across the road from the Plant Pavilion invested by Vanke. The building consists of three parts: a circular animation screening hall in the center, a souvenir selling area on the first floor and offices on the second floor on the west side of the screening hall, and a large ramp on the east side. These functions are circumscribed by two continuous curved walls. The curved exterior walls are covered by red aluminum pneumatic plates (wind chime plates). The round aluminium sheets, 14 cm in diameter and 1 mm thick, are mounted through small holes on horizontal screw rods, allowing them to move up and down as the wind blows. When the wind blows, the pneumatic plates on the entire curved wall sway in the wind and are unpredictable. From a distance, it seems that the red wall has been blown into ripples, just like the red satin wielded by Nezha. CCTV Animation Pavilion creates an interpretation that architecture is not only frozen music, but also flowing music.

北京东方华脉工程设计有限公司
China Humax Engineering Design Co., Ltd.

北京东方华脉工程设计有限公司成立于1999年，于2007年与北京数家知名建筑设计公司合并，并重新扩充组建，具有国家建设部颁发的甲级建筑工程设计资质、甲级城乡规划资质、甲级风景园林工程设计专项资质和发改委颁发的工程咨询资质，已成为国内建筑界有影响力的综合性建筑设计公司。

公司的业务范围涵盖城市规划与设计、居住区规划（住宅、别墅、高档公寓等）、公共建筑（办公、商业、教育、医疗、酒店等）、古建筑仿古建筑等中式传统建筑、园林景观规划设计、室内设计、项目前期策划及可行性研究等方面，至今已完成了数百万平方米的工程设计任务，公司在发展中强调创新与经验的融合，技术与科研的结合，客户范围涉及国内20多个省市，既有政府部门、大型国企，也有大型房地产开发企业，都与公司建立了长期的业务关系。公司也非常注重国际间的交流与合作，在与国际知名建筑设计公司和设计所的合作过程中，注重学习国外的先进设计理念和设计流程，不但在创作思路上有了新的认识，在设计管理工作上也得到了很大的启发，开阔了眼界，积累了经验。

China Humax Engineering Design Co., Ltd. was established in 1999, and merged with several well-known architectural design firms of Beijing in 2007. After re-expansion and reformation, Huamx presently is qualified with Class A Design Qualification of Construction & Engineering, Class A Planning Qualification, Class B Specific Landscape & Engineering Design Qualification issued by National Ministry of Construction, and Engineering Consultation Qualification issued by National Development and Reform Commission. Humax has become an influential and comprehensive architectural design company in China.

HUMAX's business scope includes: urban planning and design, residential community planning (residence, villa, high grade apartment, etc.), public facilities (office, commercial building, education building, medical building, hotel, etc.), Chinese classical building (ancient style), landscape design, interior design, preliminary planning of the construction project and feasibility research, and its total design area now is up to millions of square meters. HUMAX Has always been emphasizing on the integration of innovation and experience, technology and research, and it Has established stable cooperative relationships with government departments, state-owned enterprises and real estate developers in over 20 provinces and cities. It also focuses on international communication and cooperation, and its ideas and design management were inspired during cooperation with well-known foreign construction design companies.

地址：北京市海淀区白石桥东方圆大厦14层　　Add: 14th, Floor, Dongfangyuan Building, Baishiqiao, Haidian District, Beijing
电话：010-88029902　　010-88026181　　Tel: 010-88029902　　010-88026181

博湖博物馆及游客中心项目
Bohu Museum and Tourist Center Project

设计师：孙久强	Designer: Jiuqiang Sun
项目地点：新疆 博湖	Location: Bohu, Xinjiang
建筑面积：9 676.97 m²	Building Area: 9 676.97 m²
用地面积：23 002.14 m²	Site Area: 23 002.14 m²
容积率：0.42	Plot Ratio: 0.42

项目位于巴音郭楞蒙古自治州博湖县，开都河下游，博斯腾湖西岸。主要功能为县博物馆及博斯腾湖景区游客接待集散中心两部分。博斯腾湖是中国最大的内陆淡水吞吐湖，博斯腾湖景区是国家5A级旅游景区。博物馆是县域未来最大的博物馆。整体建筑是新疆巴音郭楞蒙古自治州重点项目，是博湖县新地标性建筑和新名片。

The project is located in Bohu County, Bayingol Mongolian Autonomous Prefecture, downstream of Kaidu River and on the west bank of Bositeng Lake. The main function consists of the county museum and Bositeng Lake Tourist Reception Center. Bositeng Lake is the largest inland freshwater lake in China. Bositeng Lake Scenic Spot is a national 5A tourist attraction. The museum is the largest museum in the county in the future. The overall building is a key project of Xinjiang Bayingol Mongolian Autonomous Prefecture, a new landmark building and a new name card of Bohu County.

绛县涑水河新区（一期）建设项目总承包（EPC）项目

EPC Project of Sushui River New District, Jiang County (Phase I)

设计师：杨通顺
项目地点：山西 绛县
建筑面积：260 000 m²
用地面积：25 hm²

Designer: Tongshun Yang
Location: Jiangxian, Shanxi
Building Area: 260 000 m²
Site Area: 25 hm²

绛县涑水新区规划采取开放式、外向型、多元化的发展思路，一期建设项目包括商业中心、涑水大酒店、人民广场、体育中心和城市会客厅等，总规划用地面积25 hm²，总建筑面积260 000 m²。其中，城市会客厅功能包括政务服务中心、文化馆、图书馆、科技馆、博物馆、档案馆、人民剧院和游客服务中心，建筑面积80 000 m²。

Sushui New District in Jiang County is designed to be open, export-oriented and diversified. The first phase of the construction projects include the commercial center, Sushui Hotel, the People's Square, the Sports Center and the city meeting room. The total planning area is 25 hm² and the total construction area is 260 000 m². The functions of the urban meeting room include government affairs service center, cultural center, library, science and technology museum, museum, archives, people's theater and tourist service center, with a construction area of 80 000 m².

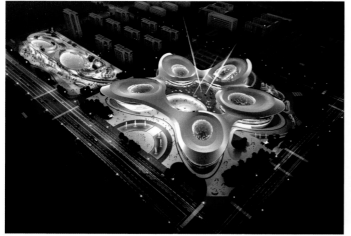

龙湖锦璘原著
Glorious Palace

设计师：魏彪、邓东坡、贾浩	Designers: Biao Wei, Dongpo Deng, Hao Jia
项目地点：辽宁 沈阳	Location: Shenyang, Liaoning
建筑面积：190 711 m²	Building Area: 190 711 m²
用地面积：7.35 hm²	Site Area: 7.35 hm²
容积率：2.0	Plot Ratio: 2.0

该项目为大型住宅小区，业态为洋房、高层、公寓、商业，立面为现代风格，采用极简的线条、配合局部铝板的立面材质，勾勒出极具现代感的立面意境。

The project is a large residential community, including foreign-style houses, high-rises, apartments, and commercial buildings, with modern style facade. The use of minimalist lines with partial aluminum facade material outlines an artistic conception of the facade with a modern sense.

太原万达沙河澜山销售物业大区景观设计
Landscape Design of Taiyuan Wanda Shahe Lanshan Sales Property Region

设计师：张亮、白志媛、王璇、杨瑞
项目地点：山西 太原
建筑面积：18 917 m²
用地面积：8 hm²
容积率：3.5

Designers: Liang Zhang, Zhiyuan Bai, Xuan Wang, Rui Yang
Location: Taiyuan, Shanxi
Building Area: 18 917 m²
Site Area: 8 hm²
Plot Ratio: 3.5

项目位于山西省太原市，景观设计以太原本土生态环境及人们对于社区景观的诉求为切入点，整体概念为让运动公园住进社区，社区中庭分别由森语会客厅、海洋奇趣乐园、律动高尔夫组成了20 000 m²超大天然氧吧，并且利用宅间空间打造健身起航区、国民训练场、进击橄榄球等各种运动模块，将社区与运动真正融合起来。

The project is located in Taiyuan City, Shanxi Province. The landscape design takes the local ecological environment of Taiyuan and people's demands for community landscape as the starting point. The overall concept is to let the sports park live in the community. The community atrium consists of a 20 000 m² natural oxygen bar, which is composed of Senyu Meeting Room, Ocean Fun Park and the dynamic golf. In addition, various sports modules such as fitness departure area, national training ground, attacking rugby and so on are built in the space to truly integrate our community with sports.

唐山·水山樾城一期、二期
Tangshan · Shuishan Yuecheng (Phase I and Phase II)

设计师：魏民、王勇、赵罕、闫冬
项目地点：河北 唐山
建筑面积：225 178.41 m²
用地面积：68 241.87 m²
容积率：2.5

Designers: Min Wei, Yong Wang, Han Zhao, Dong Yan
Location: Tangshan, Hebei
Building Area: 225 178.41 m²
Site Area: 68 241.87 m²
Plot Ratio: 2.5

本项目位于唐山市开平区，北临南新东道，集合商业、学校和住宅等多功能，共同构成了这一地区颇具活力的新型生活社区。商业与住宅和谐共存，商业为住宅提供了生活的便利和优雅的生活氛围，住宅以"对外封闭、对内开放"的布局为住户提供了安静的居住空间，并给商业带来了活力和商机。

The project is located in Kaiping District, Tangshan City, adjacent to Nanxindong Road in the north, forming a dynamic new living community in this area integrating businesses, school and residential buildings. Businesses and residential buildings coexist in harmony. Businesses provide living convenience and elegant living atmosphere. The residential buildings, with the layout of "closed to the outside, open to the inside", provide quiet living space for residents, and brings vitality and business opportunities to the businesses.

冬奥会环境建设项目(二期)
Environmental Construction Project of Winter Olympics (Phase II)

设计师：韩震、文志明、阮艾悦、葛晓梅、彭佳媛、戴晟、周兵
项目地点：北京
总设计面积：15万m²

Designers: Zhen Han, Zhiming Wen, Aiyue Ruan, Xiaomei Ge, Jiayuan Peng, Sheng Dai, Bing Zhou
Location: Beijing
Total Site Area: 150 000 m²

冬奥会是中国向世界呈交的一张彩色名片，其中延庆小海坨赛区是北京赛区的重要节点。本次设计区域集中在小海坨赛区周边的京礼高速、京银路、松闫路道路两侧，重点提升道路两侧的周边环境及景观。通过景观提升展示延庆山林自然风光和生态基底，设计中充分考虑冬季景观、冰雪资源、提炼雪花造型与肌理，结合乡土植物和乡土材料，从大绿化、大景观、看得见远山、记得住乡愁的角度出发，充分展示冰雪元素、冬奥风情。具体设计采用了梳理、整合、塑形等手法。

Winter Olympics is a colorful business card presented by China to the world and Xiaohaituo competition area in Yanqing is an important node of Beijing competition area. The design area is concentrated improving the surrounding environment and landscape on both sides of Beijing-Chongli Expressway, Jingyin Road and Songyan Road around Xiaohaituo competition area. The natural scenery and ecological base of Yanqing mountains and forests are displayed through landscape enhancement. The design gives full consideration to winter landscape, snow and ice resources while snowflake modeling and texture are refined, with combination of local plants and materials. Snow and ice elements and flavours of Winter Olympics are fully displayed from the perspective of big green, big landscape, seeing distant mountains and remembering homesickness. The specific design adopts combing, integration, and shaping methods.

沈阳远洋上河风景项目
Shenyang Sino-Ocean Shanghe Scenics Project

设计师：袁晖、石庆方、马本君、孙毅	Designers: Hui Yuan, Qingfang Shi, Benjun Ma, Yi Sun
项目地点：辽宁 沈阳	Location: Shenyang, Liaoning
建筑面积：433 951 m²	Building Area: 433 951 m²
用地面积：17.44 hm²	Site Area: 17.44 hm²
容积率：1.8	Plot Ratio: 1.8

该项目为大型居住区，业态为多层及高层住宅、门市、集中商业以及综合配套，立面为新亚洲风格，公建采用石材配合金属线脚，住宅采用石材与真石漆搭配，勾勒出既现代又凸显国风的立面意境。

The project is a large residential area, with multi-storey and high-rise residential buildings, nretail sales, concentrated businesses and comprehensive supporting facilities. The facade adopts new Asian style. The public buildings are made of stone with metal lines, and the residential buildings are made of stone with real stone paint, which outlines the artistic conception of the facade that is both modern and highlights the national style.

920街坊项目
920 Neighborhood Project

设计师：孙久强
项目地点：江苏 盐城
建筑面积：50 000 m²
用地面积：20 hm²

Designer: Jiuqiang Sun
Location: Yancheng, Jiangsu
Building Area: 50 000 m²
Site Area: 20 hm²

项目处于连接新老镇区的关键位置，毗邻AAAA级景区荷兰花海，有良好的产业、客群基础。

定位融入斗龙港生态组团，凸显老街人文历史积淀、条田特色格局，打造盐城创新文旅新名片。

以"生活着的人文老街，全方位的文旅体验"为主题，以垦管局遗址更新、酱油工坊、船屋民宿、河畔剧场等项目为引领，是集餐饮、住宿、酒吧、文创等多业态文旅体验于一体的项目。

It is located in a key position connecting the new towns and the old ones, adjacent to the AAAA scenic spot Holland Flower Sea, and has a good industry and customer base. Doulong Port ecological group is integrated to highlight the cultural history of the old street, the characteristics of the strip field pattern, creating a new name card of innovative cultural tourism in Yancheng.

With the theme of "Living Cultural Old Street, All-round Cultural Tourism Experience", it is a cultural tourism experience project integrating catering, accommodation, bar, cultural innovation and other business forms, led by the renovation of the ruins of Reclamation Administration Bureau, soy sauce workshop, houseboat hostel, riverside theater and other projects.

重庆

重庆市设计院有限公司
Chongqing Architectural Design Institute CO.,LTD.

扫描查看更多信息

重庆市设计院有限公司发轫之始可追溯至20世纪20年代，随重庆首次建市(1929)及抗战陪都建设等重大历史事件，在重庆城市建设的早期现代化进程中承担了大量勘察、规划与设计(市政、建筑、园林)工作。1950年，以原民国时期重庆工务局、怡信工程司、重庆建筑公司设计部、重庆市下水道工程处等部分技术力量整合组建了重庆市人民政府建设局设计部(即重庆市设计院前身)，在百废待兴的重庆大地上创造出多项重要文化、体育、办公及市政基础设施。经数十年持续耕耘，重庆市设计院已发展成为实力雄厚、特色鲜明的现代化综合甲级勘察设计和工程咨询企业，在山地建筑与城市设计、超高层建筑、医疗建筑、文教建筑、商业文旅、城市更新、乡村振兴等领域具有较为深厚的积淀与优势，曾荣获全国建设系统文明建设先进单位、全国"政府放心、用户满意"先进单位、全国建筑设计行业先进单位以及"重庆市五一劳动奖状先进单位""全国工程勘察设计百强企业""当代中国建筑设计百家名院"等荣誉称号。

The beginning of Chongqing Architectural Design Institute Co., Ltd. can be traced back to the 1920s. With the first construction of Chongqing (1929) and the construction of the capital during the War of Resistance against Japan and other major historical events, it undertook a lot of survey, planning and design (municipal, architectural, garden) work in the early modernization process of Chongqing's urban construction. In 1950, the Design Department of the Construction Bureau of the People's Government of Chongqing (the predecessor of Chongqing Design Institute) was formed by integrating some technical forces such as the Chongqing Public Works Bureau, Yixin Engineering Department, the Design Department of Chongqing Construction Company, and the Chongqing Sewer Engineering Department in the former Republic of China period, creating a number of important cultural, sports, office and municipal infrastructure on the land of celebration. After decades of continuous efforts, Chongqing Design Institute has developed into a modern comprehensive Class A survey and design and engineering consulting enterprise with strong strength and distinctive characteristics. It has a profound accumulation and advantages in mountain building and urban design, super high-rise buildings, medical buildings, cultural and educational buildings, business and cultural tourism, urban renewal, rural revitalization and other fields. It has won the National Advanced Unit for Building a Civilized Construction System Honorary titles such as the national advanced unit of "government reassurance and customer satisfaction", the national advanced unit in the architectural design industry, the "Chongqing May Day Labor Award Advanced Unit", the "National Top 100 Engineering Survey and Design Enterprises", and the "Top 100 Contemporary Chinese Architectural Design Institutes".

地址：重庆市渝中区人和街31号	Add: No.31 Renhe Street, Yuzhong District, Chongqing, China
邮编：400015	P.C: 400015
电话：023-63854124　023-63619826	Tel: 023-63854124　023-63619826
传真：023-63856935	Fax: 023-63856935
邮箱：CQADI@cqadi.com.cn	Email: CQADI@cqadi.com.cn
网址：www.cqadi.com.cn	Web: www.cqadi.com.cn

南江华润希望乡村规划及建筑设计项目
China Resources Hope Town

项目总负责：褚冬竹
方案设计：喻焰、邓宇文、秦文智、曾昱玮、朱维嘉、周凌志、兰慧琳、王雨寒、宋登高、董志凌、刘欣雨、张乃千、苏红、杨欣、黎柔含、阳蕊、朱羽翼(重庆大学建筑城规学院)
核心区历史建筑测绘：陈蔚、邹一玮、庞春勇、张江涛(重庆大学建筑城规学院)
景观概念设计：王中德、余林冰、吴有鹏(重庆大学建筑城规学院)
施工图设计：杨洋、查理文、罗建兵、郑志斌、田俊永、任炼、李丹、李建、姜骁、张德馨、周松、张丁、殳馨雨、蒲国华、王紫旭、罗天池(重庆市设计院有限公司)
项目地点：四川 南江
用地面积：40.9 hm²

Project Leader: Dongzhu Chu
Project Design: Yan Yu, Yuwen Deng, Wenzhi Qin, Yuwei Zeng, Weijia Zhu, Lingzhi Zhou, Huilin Lan, Yuhan Wang, Denggao Song, Zhiling Dong, Xinyu Liu, Naiqian Zhang, Hong Su, Xin Yang, Rouhan Li, Rui Yang, Yuyi Zhu (School of Architecture and Urban Planning Chongqing University)
Core Historic Building Mappingn: Wei Chen, Yiwei Zou, Chunyong Pang, Jiangtao Zhang (School of Architecture and Urban Planning Chongqing University)
Landscape Concept Design: Zhongde Wang, Linbing Yu, Youpeng Wu (School of Architecture and Urban Planning Chongqing University)
Construction Drawing Design: Yang Yang, Wen Zhali, Jianbing Luo, Zhibin Zheng, Junyong Tian, Lian Ren, Dan Li, Jian Li, Xiao Jiang, Dexin Zhang, Song Zhou, Ding Zhang, Xinyu Shu, Guohua Pu, Zixu Wang, Tianchi Luo (Chongqing Architectural Design Institute CO.,LTD.)
Location: Nanjiang, Sichuan
Site Area: 40.9 hm²

　　华润希望小镇是华润集团践行央企社会责任，紧密围绕红色文化、乡村振兴战略的公益性探索实践。该项目为华润的第十二个小镇，选址于四川南江。规划和建筑方案设计基于传统村落的自然禀赋与历史传承，以农业生产为承托载体，综合优化土地利用、产业发展、民居改造、福祉提升、环境整治、生态保育和文化复兴，重塑古村印象，延续红色记忆，发扬耕读文化，在结构、空间、形态、风貌、技术等多个层次上完成村落的全面升级，塑造出一个精神火种不断、地域特色鲜明、适应时代发展的乡村振兴规划与设计作品。

　　China Resources Hope Town is a public welfare exploration practice of China Resources Group to practice social responsibility as a central enterprise, closely centering on red culture and rural revitalization strategy. The 12th town of China Resources is located in Nanjiang, Sichuan province. The planning and architectural design is based on the natural endowment and historical inheritance of traditional villages, and takes agricultural production as the carrier to comprehensively optimize land use, industrial development, residential renovation, welfare improvement, environmental remediation, ecological conservation and cultural revival, so as to reshape the impression of ancient villages, continue the red memory and carry forward the farming culture. Complete the comprehensive upgrading of the village from multiple levels of structure, space, form, style and technology, shaping a rural revitalization planning and design work with constant spiritual fire, distinctive regional characteristics and adapting to the development of The Times.

长嘉汇弹子石老街
Changjiahui Danzishi Old Street

项目地点：重庆
竣工时间：2018年6月
建筑面积：50 906.75 m²
用地面积：21 463.15 m²
设计团队：汤启明、张引、何晓、杨博、李兵、包行健、熊洁、陈鸿翔、何畅、赖韬、张朝云、刘文君、樊明玉、杨薇、陈进、徐向茜、李昌平、高方明、李远智、陈国伟

Location: Chongqing
Completion Time: June 2018
Building Area: 50 906.75 m²
Site Area: 21 463.15 m²
Design Team: Qiming Tang, Yin Zhang, Xiao He, Bo Yang, Bing Li, Xingjian Bao, Jie Xiong, Hongxiang Chen, Chang He, Tao Lai, Chaoyun Zhang, Wenjun Liu, Mingyu Fan, Wei Yang, Jin Chen, Xiangqian Xu, Changping Li, Fangming Gao, Yuanzhi Li, Guowei Chen

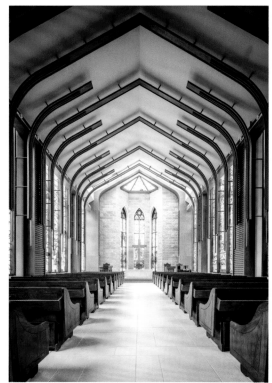

"长嘉汇弹子石老街"（G304/02地块D组团）项目位于重庆南岸区，地处长江东岸，毗邻长江、嘉陵江两江交汇处，用地高踞长江之滨，鸟瞰长江，俯首两江交汇，远眺重庆大剧院，遥望朝天门码头，自然和人文景观资源独特。在尊重原有保护性规划原则的前提下，呈现了山地滨水商业街区特别的风貌。本项目分为两个组团，南区滨江商业连接重庆南滨路全国重点文物保护单位"法国水师兵营"，延伸至嘉陵江江边，北区弹子石老街向上延伸连接北侧弹子石老城区。项目含多层商业20栋及相应的7层地下车库，多层商业建筑及对应的地下车库沿山势从半山路入口位置（1956年黄海高程系海拔254.200 m）向下延伸至泰昌路段（1956年黄海高程系海拔215 m），整个老街按标高分为7个台地，各台地由踏步和扶梯竖向连接。项目以体验式商业的空间组织理念为指导，通过重庆传统街巷空间的经典元素梯坎及坝子的组合，通过平面和垂直方向上的空间肌理变化，呼应着重庆人记忆中老街的模样。

"Changjia Hui Danzi Stone Old Street" (G304/02 block D Group) project is located in the southern district of Chongqing, located on the east bank of the Yangtze River, adjacent to the intersection of the Yangtze River and Jialing River. The land is high on the bank of the Yangtze River, overlooking the Yangtze River, overlooking the intersection of the two rivers, overlooking the Chongqing Grand Theater, and overlooking the Chaotianmen, with unique natural and cultural landscape resources. On the premise of respecting the original protective planning principles, it presents the special features of the waterfront business district in the mountains. The project is divided into two groups. The business of Binjiang in the southern district is connected to the "French Navy Barracks", a national key cultural relic protection unit in Nanbin Road, Chongqing, and extends to the Jialing River. The old marble stone street in the northern district extends upward to connect the old marble stone city in the north. he project includes 20 multi-storey commercial buildings and corresponding 7-storey underground garages. The multi-storey commercial buildings and corresponding underground garages extend down from the entrance of Banshan Road (254.200 m above sea level of Huanghai elevation system in 1956) to the Taichang Road section (215 m above sea level of Huanghai elevation system in 1956) along the mountain. The whole Old Street is divided into 7 platforms according to elevation. Each platform is connected vertically by steps and escalators.Guided by the spatial organization concept of experiential commerce, the project echoes the appearance of old streets in Chongqing people's memory through the combination of the classic elements of traditional street Spaces in Chongqing: terraces and bazi, and the spatial texture changes in the plane and vertical directions.

解放碑至朝天门步行大道品质提升
Quality Improvement from Jiefangbei to Chaotianmen Pedestrian Avenue

项目位置：重庆	Location: Chongqing
设计师：徐千里、余水	Designers: Qianli Xu, Shui Yu
方案设计：许书、章玲、王瑞、陈世林、吴静琪、付逸、王倩倩	Schematic Design: Shu Xu, Ling Zhang, Rui Wang, Shilin Chen, Jingqi Wu, Yi Fu, Qianqian Wang
设计经理：肖虎	Design Manager: Hu Xiao
施工图设计：刘扬、孙礼平、胡海蒂、杨频驷、杨翔宇、陈吉林、黄皇	Construction Documents: Yang Liu, Liping Sun, Haidi Hu, Pinsi Yang, Xiangyu Yang, Jilin Chen, Huang Huang
市政设计：饶友平、卢光明、任家林、陈亚平、漆建伟	Municipal Design: Youping Rao, Guangming Lu, Jialin Ren, Yaping Chen, Jianwei Qi
广告设计：陈婧、熊甲艳、张程	Advertising Design: Jing Chen, Jiayan Xiong, Cheng Zhang
建筑面积：45 000 m²(立面)	Building Area: 45 000 m² (Elevation)

改造后的下新华路
Renovated Lower Xinhua Road

改造前的罗汉寺广场
The square in front of the Luohan Temple before renovation

改造前的罗汉寺广场
The square in front of the Luohan Temple before renovation

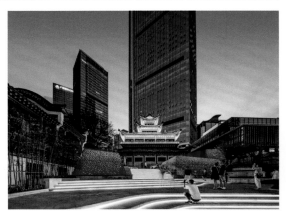

改造后的罗汉寺前广场
The square in front of the Luohan Temple after renovation

左一：改造前的新华路80号
80th Xinhua Road before renovation

左二：改造前的小商品市场
Small Commodity Market before renovation

右一：改造后的新华路80号
Renovated 80th Xinhua Road

右二：改造后的小商品市场
Renovated Small Commodity Market

右三：改造后的圣名国际服装城
Renovated Shengming Masion

项目旨在完善重庆母城核心区的步行空间，推进现代人文的都市社区，采取EPC模式，以5大类19个小类来综合构建一条功能完善的世界级文旅步行大道，内容涉及建筑景观、广告店招、导向标识、智慧灯杆、箱体迁改、夜景灯饰等9大类工作，以道路U形界面为基础来提升整体城市的形象，创造多个城市公共空间，从使用人群的全龄化角度来提升城市的功能，带动城市的发展，是"城市核心商业街区"式的城市更新项目。

This project aims to improve the walking space of the core city area in Chongqing, and promote modern humanistic urban community. It takes the EPC mode with five categories and nineteen sub-categories to comprehensive build a fully functional world-classcultural and tourism walking avenue, content involves architecture, landscape, store design, way finding, wisdom light pole, night view lighting and so on. It is an "urban core commercial block" type urban renewal project to improve the image of the whole city based on the U-shaped interface of the road, create multiple urban public spaces, improve the city's function from the perspective of the full age of users and drive the development of the city.

重庆环球金融中心
Chongqing World Financial Center

项目地点：重庆
竣工时间：2015年2月
建筑面积：204 611 m²
用地面积：5 800 m²
设计单位：重庆市设计院有限公司、大原建筑设计咨询（上海）有限公司
项目团队：邓小华、余波、汤启明、李正春、李祖原、黎明、周强、钟文泉、胡珏、张蕾、
杨航超、徐小林、谭平、周爱农、黄显奎

Location: Chongqing
Completion Time: February 2015
Building Area: 204 611 m²
Site Area: 5 800 m²
Design Organization: Chongqing Architectural Design Institute CO.,LTD.
Dayuan Architectural Design Consulting (Shanghai) CO., LTD.
Design Team: Xiaohua Deng, Bo Yu, Qiming Tang, Zhengchun Li, Zuyuan Li, Ming Li, Qiang Zhou, Wenquan Zhong, Jue Hu, Lei Zhang, Hangchao Yang, Xiaolin Xu, Ping Tan, Ainong Zhou, Xiankui Huang

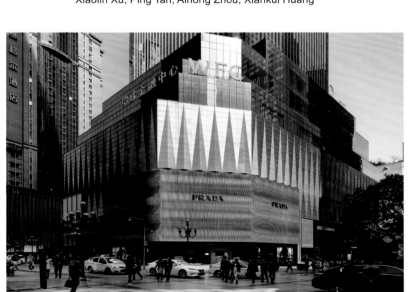

重庆环球金融中心位于重庆市渝中区解放碑核心商圈,高度339 m,地上70层,地下6层,是集甲级写字楼、高档商业、酒店、观光等于一体的城市地标性超高层综合体。项目建筑形象独特,以简洁、现代的建筑处理手法强调新的美学体验,并融入东方文化元素,追求精神理念的崇高与丰富。宛如一颗闪闪发亮的宝石,镶嵌在城市的云端,展现着城市的品位与特质,书写着城市奋发向上的精神追求,成为渝中半岛的制高点和璀璨的城市之冠,形成重庆市乃至长江中上游的标志性建筑。

Chongqing World Financial Center is located in the core business District of Jiefangbei, Yuzhong District, Chongqing, with a height of 339 m, 70 floors above ground and 6 floors underground. It is a landmark super-high-rise complex integrating grade A office buildings, high-end commercial hotels and sightseeing.The architectural image of the project is unique, emphasizing new aesthetic experience with simple and modern architectural treatment techniques, and integrating eastern cultural elements to pursue the sublimity and richness of spiritual concepts.t is like a shining gem embedded in the clouds of the city, showing the taste and characteristics of the city and writing the spiritual pursuit of the city. It has become the commanding height of the Yuzhong Peninsula and the crown of the shining city, and has become a landmark building of Chongqing city and even the middle and upper reaches of the Yangtze River.

丹寨旅游小镇
Danzhai County

项目地点：贵州	Location: Guizhou
竣工时间：2017年8月	Completion Time: August 2017
建筑面积：56 118.43 m²	Building Area: 56 118.43 m²
用地面积：143 886.6 m²	Site Area: 143 886.6 m²
设计团队：陶立、周含川、王子驹、王志学、卢仕银、张晓鸥、阴忠欣、夏虹、高建华、袁振华、张文茂、吴赅、杨鹏、陈维果、朱敏、王奎、汤海涛、胡海蒂、邹光陶、谭竹荃	Design Team: Li Tao, Hanchuan Zhou, Ziju Wang, Zhixue Wang, Shiyin Lu, Xiaoou Zhang, Zhongxin Yin, Hong Xia, Jianhua Gao, Zhenhua Yuan, Wenmao Zhang, Gai Wu, Peng Yang, Weiguo Chen, Min Zhu, Kui Wang, Haitao Tang, Hattie Hu, Guangtao Zou, Zhuquan Tan

项目位于贵州省黔东南州丹寨县，用地呈南北向狭长状，南北长约1 060 m，东西宽约77至100 m不等。为保持最初对城市的美好印象，反映城市特征和峰、谷概念的再现，一是与自然环境紧密结合，东湖水库旁独特的自然景观和场地变化是本设计最重要的灵感来源，亦是本案最重要的环境和资源依托，形成独特的背景，是本案最大的优势所在；二是历史文脉延续，县境内多民族聚居，有苗族、汉族、水族、布依族等21个少数民族，丹寨文化，星河璀璨。采用苗族元素融入建筑，体现传统古建筑的传承，丰富城市界面，提供一个顺应历史文脉的良好景区形象，该项目的建成已成为新丹寨城市概念的典范。

The project is located in Danzhai County, Southeast Guizhou Province, guizhou Province. The land is long and narrow from south to north, with a length of about 1 060 m from north to south and a width of 77 to 100 m from east to west. To maintain the initial response to the goodness of city image, city characteristics and the concept of peak and valley, it is closely integrated with the natural environment, near the east lake reservoir unique natural landscape and change of this design is the most important sources of inspiration, also is the most important environmental and resource based on this case, form a unique background, is the biggest advantage of this case. Second, the continuation of the historical context, the county is inhabited by many ethnic groups, miao, Han, shui, Buyi and other 21 ethnic minorities. Danzhai culture is brilliant. Miao elements are integrated into the architecture, reflecting the inheritance of traditional ancient architecture, enriching the urban interface, and providing a good image of the scenic spot that conforms to the historical context. The completion of the project has become a model of the concept of new Danzhai city.

重庆

中煤科工重庆设计研究院（集团）有限公司
CCTEG CHONGQING ENGINEERING (GROUP) CO., LTD.

中煤科工重庆设计研究院（集团）有限公司组建于1953年，原隶属燃料工业部。2020年成立集团，被认定为"国家高新技术企业"，现为国务院国资委管辖的中国煤炭科工集团有限公司全资二级子企业。公司下设40多个勘察设计院所、2家全资子公司、1家控股子公司、12个驻外分支机构，从业人员2 900余人，涵盖40余个专业、硕士学历以上、高级职称及各类注册师1 300余人。经过近70年的转型改革，公司从一个部属专业院发展成为一家技术力量雄厚、技术装备先进、人才荟萃、成果丰硕、文化优秀的大型综合性设计集团。拥有各类甲乙级资质40余项，业务范围遍及国内30个省、市、自治区。在产业上以工程勘察、设计、咨询、监理及工程总承包为主体，在行业上以建筑、市政、环保及煤炭为主导，在科研成果上以绿色建筑、装配式建筑、智慧城市、BIM技术等新技术为主的格局，逐步发展为全过程的工程技术咨询、设计集成和投融建一体化工程建设为主的工程建设全领域、全专业技术服务产品集群。公司先后获得120余项荣誉称号，连续6年入选RCC瑞达恒评选的中国十大建筑设计院榜单，勘察设计企业全国排名前100强，具有较强的影响力和较高的品牌度。

CCTEG Chongqing Engineering (Group) Co., Ltd. was established in 1953 and was originally under the Ministry of Fuel Industry. The group was established in 2020 and was recognized as a "national high-tech enterprise". It is now a wholly-owned secondary subsidiary of China National Coal Science and Industry Corporation under the jurisdiction of the State-owned Assets Supervision and Administration Commission of the State Council. The company has more than 40 survey and design institutes, 2 wholly-owned subsidiaries, 1 holding subsidiary, 12 overseas branches, more than 2 900 employees, covering more than 40 majors, master's degree or above, senior professional titles and various There are more than 1 300 registered teachers. After nearly 70 years of transformation and reform, the company has developed from a subordinate professional institute into a large-scale comprehensive design group with strong technical force, advanced technical equipment, talented people, fruitful achievements and excellent culture. It has more than 40 grades A and B qualifications, and its business scope covers 30 provinces, municipalities and autonomous regions in China. In the industry, the main body is engineering survey, design, consulting, supervision and general project contracting; in the industry, it is dominated by construction, municipal, environmental protection and coal; and other new technologies, and gradually develop into a full-field, full-professional technical service product cluster based on the whole process of engineering technical consultation, design integration, and integration of investment, financing and construction. The company has won more than 120 honorary titles successively, and has been selected for the list of China's top ten architectural design institutes selected by RCC Ruidaheng for six consecutive years.

地址：（马家堡院区）重庆市渝中区长江二路179号／（虎头岩院区）经纬大道780号2#、3#楼
邮编：400016 / 400042
电话：023-68725010（公司办公室）　　023-68725035（生产运营中心）
传真：023-68811613　　邮箱：2504960779@qq.com
网址：www.cqmsy.com　　官微公众号：中煤科工重庆设计集团（微信号：ccteg-ccec）

Add: (Majiapu Campus) No.179, Changjiang 2nd Road, Yuzhong District, Chongqing / (Hutouyan Campus) 2# and 3# Buildings, No.780, Jingwei Avenue
P.C: 400016 / 400042
Tel: 023-68725010 (Company Office)　　023-68725035 (Produc Operation Center)
Fax: 023-68811613　　Email: 2504960779@qq.com　　Web: www.cqmsy.com
WeChat Account: CCTEG Chongqing Engineering Group (WeChat: ccteg-ccec)

金山国际商务中心一期改造升级项目（城市更新）
Phase I Renovation and Upgrading Project of Jinshan International Business Center (Urban Renewal)

设计师：李英军、周光均	Designers: Yingjun Li, Guangjun Zhou
项目地点：重庆	Location: Chongqing
建筑面积：67 843.73 m²	Building Area: 67 843.73 m²
用地面积：4.64 hm²	Site Area: 4.64 hm²

本项目现更名为东原—奥天地，属于城市更新重点项目。11年间，该片区广场已处于设施陈旧损坏、运动场地荒废的状态，存在巨大的商业配套、运动健康、商务休闲的需求缺口，亟待提升改造。本项目为纯地下大型商业改造项目，地下共3层，更新改造范围为除营业的射击馆外的区域共计67 843.73 m²。创造了地下商业与公园强烈的竖向联系，同时也把公园的景观主题带入地下商业空间，实现地上地下、室内室外的紧密联系。建筑表皮大面积采用铝板、不锈钢结合玻璃的材质，质感轻盈耐久性强。结合中央下沉广场的边缘，面向中央下沉广场和轻轨设置的韵律十足的"指环"构架，形成怀抱整个项目的形象展示面。"指环"由钢结构结合彩釉玻璃构成，竖梃中央预留LED灯光点位，玻璃上藏着白天能完全"隐形"的LED屏幕。"指环"构架不仅拥有日间通透的公园质感，在夜间色彩斑斓的灯光和LED渲染下，也能展现绚烂光影，吸引人群，汇聚活力和高端商业气质，成为新区重要的城市展示面。

This project is now renamed Dongyuan The Oval, which belongs to the key project of urban renewal. In the past 11 years, the square in this area has been in a state of obsolete and damaged facilities and abandoned sports venues, and there is a huge gap in demand for commercial supporting facilities, sports health, business and leisure, which is in urgent need of upgrading and renovation. This project is a pure underground large-scale commercial reconstruction project, with three floors underground, covering a total area of 67 843.73 m² except the operating shooting gallery. It creates a strong vertical connection between the underground commerce and the park, and also brings the landscape theme of the park into the underground commercial space, realizing the close connection between the space above ground and the space underground, indoor space and outdoor space. The large surface of the building is made of aluminum plate, rusty steel and glass, which has strong durability and light shape. Combined with the edge of the central sunken square, the rhythmic "ring" frame facing the central sunken square and the light rail forms an image display surface embracing the whole project. The "ring" is composed of steel structure combined with colored glazed glass. LED lights are reserved in the center of vertical spike, and the LED screen can be completely "invisible" in the daytime. The "ring" structure not only has a transparent park texture in the daytime, but also displays gorgeous light and shadow under the colorful lights and LED rendering at night, attracting crowds, gathering vitality and high-end commercial temperament, and becoming an important urban display surface in the new district.

重庆市青少年活动中心（重庆首个全钢结构装配式公共建筑）
Chongqing Youth Activity Center (The First All-steel Structure Prefabricated Public Building in Chongqing)

设计师：杨第、谢明焕	Designers: Di Yang, Minghuan Xie
项目地点：重庆	Location: Chongqing
建筑面积：47 572 m²	Building Area: 47 572 m²
用地面积：97 904 m²	Site Area: 97 904 m²

重庆市青少年活动中心是重庆"十三五"期间市级公共文化基础设施建设重点工程，是重庆城市文化名片与面向国际的新时代地标建筑，是多功能、数字化、低碳节能、绿色环保的公共化服务综合体。建设内容包括青少年宫、少儿图书馆两部分，主要结构类型为钢框架结构，是重庆首个全钢结构装配式公共建筑。设计利用园内地形，取重庆大山大水之势，依山筑台，形成一座立体园林，如同周边绿山层峦叠嶂，水声山色，竞来相娱；流畅的外檐平台融山、融水、融绿，将葱郁自然与建筑、与城市充分衔接。形象统一而不失灵动，造型如蝴蝶飞舞于山地之间，为青少年提供了一个充满乐趣的空间。

Chongqing Youth Activity Center is a key project of municipal public cultural infrastructure construction during the "13th Five-Year Plan" period of Chongqing. It is the culture name card of Chongqing city and the landmark building of the new era facing the world. It is also a multi-functional, digital, low-carbon, energy-saving, green and environment-friendly public service complex. The construction consists of two parts, the youth palace and the children's library. The main structure type is steel frame structure, which is the first all-steel structure prefabricated public building in Chongqing. The design takes advantage of the topography of the garden, and builds a platform against the mountains to form a three-dimensional garden, just like the surrounding green mountains with the sound of the water and the scenery, competing to entertain each other. The smooth eaves platform merges mountains, water and green, fully connecting the lush nature with architecture and the city. And the image is unified without lack of agility and the shape is like a butterfly flying between the mountains, providing fun space for teenagers.

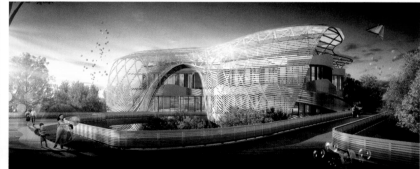

重庆市人民医院
Chongqing People's Hospital

项目地点：重庆	Location: Chongqing
建筑面积：167 499 m²	Building Area: 167 499 m²
用地面积：5.578 hm²	Site Area: 5.578 hm²

重庆市人民医院（两江院区）是位于重庆两江新区照母山片区的一家三甲综合性医院，是重庆市政府重点工程建设项目。一期工程共建设1 000张床位，二期建成后将达1 480张床位。一期总建筑面积16.7万m²，包含医疗综合楼、高压氧舱、全科医师临床培养基地、地下车库等，停车位共1 150个。

hongqing People's Hospital (Liangjiang Hospital) is located in Zhaumushan District, Liangjiang New District, Chongqing. It is a 3A comprehensive hospital and a key construction project of Chongqing Government. A total of 1 000 beds will be built in the first phase and 1 480 beds will be built in the second phase. The first phase covers a total construction area of 167 000 m², including a medical complex building, hyperbaric oxygen chamber, clinical training base for general practitioners, underground garage, etc., with a total of 1 150 parking lots.

重庆市规划展览馆迁建EPC项目
Relocation EPC Project of Chongqing Planning Exhibition Hall

项目地点：重庆	Location: Chongqing
建筑面积：17 000 m²	Building Area: 17 000 m²
用地面积：6 168 m²	Site Area: 6 168 m²

重庆市规划展览馆原址位于重庆朝天门广场，现迁建落子重庆弹子石南滨路，是展示宣传重庆的重要窗口之一，是全国科普教育基地、全国青少年教育基地、国家AAAA级旅游景区。可观最美渝中夜景，周边拥有历史悠久的巴渝文化、开埠文化、大禹文化、码头文化、抗战遗址文化等珍贵文化遗存，更兼得素有"重庆外滩"之美誉的南滨路如玉带拱卫。建筑方案以"山路起风景"为设计概念，将规划展览馆定义为两江交汇观景点、南滨路城市会客厅新起点。通过四个维度阐释"山路起风景"的理念，诠释了对规划展览馆成为重庆新的标志性建筑及公共活动场所的必然性和迫切性，深度展示了重庆历史文脉和"行千里、致广大"的重庆气度，进一步彰显出重庆"山城"与"江城"的双重魅力。

Chongqing Planning Exhibition Hall, originally located in Chaotianmen Square, Chongqing, is now relocated to Nanbin Road, Daozishi, Chongqing. It is one of the important windows to display Chongqing, and is also the national science popularization education base, the national youth education base, and the national AAAA level scenic spot. You can appreciate the most beautiful night view of Chongqing. The project is surrounded by precious cultural relics such as Bayu culture, religious culture, port culture, Dayu culture, dock culture, Anti-Japanese War site culture, etc. Moreover, Nanbin Road, known as the "Chongqing Bund", is like a jade belt. The architectural scheme takes "scenery from mountain road" as the design concept, and defines the planning exhibition hall as the new starting point of the observation spot of the intersection of two rivers and the urban meeting room of Nanbin Road. The concept of "scenery from mountain road" is explained through four dimensions, interpreting the necessity and urgency of the planning exhibition hall becoming a new landmark building and public activity place of Chongqing, deeply demonstrating the historical context of Chongqing and the spirit of "a thousand mile leads to a great distance", and further highlighting the dual charm of "Mountain city" and "River city" of Chongqing.

贵州义龙新区海庄、花月光伏电站EPC项目（新能源）
Haizhuang and Huayue Photovoltaic Power Station EPC Project (New Energy) in Yilong New District, Guizhou

项目地点：贵州
项目总容量：70 MWp

Location: Guizhou
Total Capacity: 70 MWp

贵州义龙新区海庄、花月光伏电站，是公司承接的首个新能源项目。项目总容量70 MWp，包括光伏发电场本体、厂区内集电线路、110 kV升压站及16 km的110 kV送出线路等，是贵州可再生能源发展"十四五"规划重点项目，为当地创造了较好的社会经济与生态环境效益，更为公司探索新业务、开拓新市场打下了坚实的基础。

Haizhuang and Huayue Photovoltaic Power Station in Yilong New District, Guizhou, is the first new energy project undertaken by the company. The project has a total capacity of 70 MWp, including the photovoltaic power station, the current collection line within the plant, the 110 kV booster station and the 110 kV transmission line of about 16 km, etc. It is the key project of Guizhou's renewable energy development of the "14th Five-Year Plan" and has created good social, economic and ecological environmental benefits for the local area. Moreover, the company has laid a solid foundation for exploring new business and opening up new markets.

陕西正通煤业矿井水净化处理工程
Sewage Treatment Project of Shaanxi Zhengtong Coal Mining

项目地点：陕西 咸阳　　Location: Xianyang, Shaanxi
设计处理能力：96 000 m³/d　　Design Processing Power: 96 000 m³/d

本工程位于陕西省咸阳市长武县地掌镇，设计处理能力96 000 m³/d，是目前我国规模最大的矿井水净化处理工程。项目高起点、高定位引进先进环保技术，创新管理方式，高标准做好环保工作，采用"两级预沉、机械加速澄清池、V型滤池"的核心技术工艺，处理后水质达到了《地表水环境质量标准》（GB3838－2002）三类水质要求，出水主要作为煤矿生产和生活用水，多余部分达标外排至泾河。

The project is located in Dizhang Town, Changwu County, Xianyang City, Shaanxi province, with a designed treatment capacity of 96 000 m³/d, which is currently the largest mine sewage treatment project in our country. The project has a high starting point and high positioning, with the introduction of advanced environmental protection technology, innovative management mode, and high standard environmental protection work. The project adopts the core technology of "two-stage pre-sedimentation, mechanical accelerated clarifying tank, V-shaped filter tank". The water quality after treatment meets the class III water quality standard of "Environmental Quality Standards for Surface Water" (GB3838-2002). The effluent is mainly used as coal mine production and domestic water. The excess part which reaches the standard is discharged to Jing River.

重庆

高驰国际设计有限公司
GAOCHI INTERNATIONAL DESIGN

高驰国际设计有限公司(简称高驰国际)成立于2003年,现有经验丰富的各专业设计人才近500人,专为城市建设投资方提供咨询、设计、管理的全面解决方案,以及全专业、全过程、全产业链的优质服务。经过近20年发展沉淀,在挖掘产品价值、提升产品品质、增强产品创新力和竞争力等方面积累大量实战经验。高驰国际作为国家高新技术企业,致力于人居环境的创新研究,坚持创新、技术和服务的企业设计理念,在经营业绩、技术实力、团队建设、品牌打造及行业交流等方面取得优秀的成绩,正向着西部领先、具有全国影响力的大型综合建筑设计企业集团这一宏伟目标而坚定迈进。

Founded in 2003, Gaochi International Design has nearly 500 experienced professional design talents. It provides comprehensive solutions for consulting, design, and management for urban construction investors, as well as high-quality professional, whole-process, and whole-industry chain solutions. Serve. After nearly 20 years of development and precipitation, it has accumulated a lot of practical experience in mining product value, improving product quality, and enhancing product innovation and competitiveness. As a national high-tech enterprise, Gaochi International is committed to innovative research on the living environment, adheres to the corporate design concept of innovation, technology and service, and has achieved excellent results in business performance, technical strength, team building, brand building and industry exchanges. The achievements are firmly moving towards the grand goal of being a large-scale comprehensive architectural design enterprise group that is leading in the west and has national influence.

地址:重庆市重庆市北部新区金开大道1号天湖公园内　邮编:401121	Add: Tianhu Park, Jinkai Avenue, North New Area, Chongqing　P.C: 401121	
电话:023-63942469　传真:023-63942469	Tel: 023-63942469　Fax: 023-63942469	
邮箱:sjy@jinke.com　网址:www.gc-id.com	Email: sjy@jinke.com　Web: www.gc-id.com	

成都文殊坊二期
Chengdu Wenshufun Phase II

设计师:钟洛克、何凌峰、江怡鸥　　Designers: Luoke Zhong, Lingfeng He, Yiou Jiang
项目地点:四川 成都　　Location: Chengdu, Sichuan
建筑面积:100 000 m²　　Building Area: 100 000 m²
用地面积:5.8 hm²　　Site Area: 5.8 hm²
容积率:1.2　　Plot Ratio: 1.2

项目位于成都市一环内,文殊院南侧,距离天府广场约2 km,处于天府文化中心板块,历史文化源远流长,商业氛围景观资源丰富。项目北区为历史文化保护严格控制区,项目中部为较大尺度建筑,通过文堂、书院恢宏气势展示其影响力,辅以水池延伸全局。项目南部以街巷及新中式建筑构成,利于多元业态和现代零售商业组织布局。

The project is located in the first ring road of Chengdu, on the south side of Wenshufun, about 2 km away from Tianfu Square. It is located in the central plate of Tianfu culture, with a long history and culture and rich commercial atmosphere and landscape resources. The north area of the project is a strictly controlled area for historical and cultural protection. The central part of the project is a large-scale building. The magnificent momentum of the Wentang and the Academy shows its influence, and the pool extends the overall situation. The southern part of the project is composed of streets and new Chinese buildings, which is conducive to the organization and layout of diversified business forms and modern retail business.

   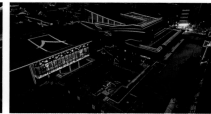

成都杜甫草堂片区规划
Chengdu Du Fu Thatched District Planning

设计师：万欣、刘昌林、程建、李坤
项目地点：四川 成都
规划面积：118.25 hm²

Designers: Xin Wan, Changlin Liu, Jian Cheng, Kun Li
Location: Chengdu, Sichuan
Site Area: 118.25 hm²

项目位于青羊区文化核心草堂片区，属于成都市的文化教育核心区域。项目以"文学、美术、手工、音乐"4大板块分区，形成以杜甫草堂为中心，多条生态廊道配合7大核心节点的多元中轴的格局，建立文旅赋能、商业增效、形象提质、风貌优化的城市空间，打造一站式文旅体验、功能复合的天府文化艺术大区。

The project is located in the cultural core Du fu thatched area of Qingyang District, which is the core area of culture and education in Chengdu. The project is divided into four sections of "literature, Art, handicrafts, music", forming a multi axis pattern with Du fu thatched as the center, multiple ecological corridors and seven core nodes, establishing an urban space with cultural and tourism empowerment, business efficiency, image improvement and style optimization, and creating a one-stop cultural and tourism experience and multi-functional Tianfu culture and art area.

重庆江北嘴紫金大厦
Chongqing Jiangbeizui Zijin Building

设计师：钟洛克、李恒、谭力铭、王军军
项目地点：重庆
建筑面积：60 831.43 m²
用地面积：0.9 hm²
容积率：7.2

Designers: Luoke Zhong, Heng Li, Liming Tan, Junjun Wang
Location: Chongqing
Building Area: 60 831.43 m²
Site Area: 0.9 hm²
Plot Ratio: 7.2

项目位于重庆江北嘴核心地段，城市配套完善。规划上采用"双塔"的布局方式，业态涵盖5A级写字楼及高端公寓。建筑设计采用倾泻而下的曲线线条赋予建筑生机与活力。流动的曲线预示建筑蓬勃向上生长。线性元素与规则线条的交叉融合让建筑与周边环境融为一体，也使建筑立面呈现出强烈的生态性、未来感。

The project is located in the core area of jiangbeizui, Chongqing, with complete and favorable urban supporting facilities. In the planning, the planning layout of "double towers" is adopted, and the business type covers 5A grade office buildings and high-end apartments. The architectural design uses the downward curve lines to give the building vitality and vigor. The flowing curve indicates that the building is growing vigorously. The intersection and integration of linear elements and regular lines make the building integrate with the surrounding environment, and also make the building facade show a strong sense of ecology and future.

成都金牛区商业综合体
Chengdu Jinniu District Commercial Complex

设计师：江怡鸥、刘雪松、陈珈乐、王淑华
项目地点：四川 成都
建筑面积：56 420 m²
用地面积：1.5 hm²
容积率：3.0

Designers: Yiou Jiang, Xuesong Liu, Jiale Chen, Shuhua Wang
Location: Chengdu, Sichuan
Building Area: 56 420 m²
Site Area: 1.5 hm²
Plot Ratio: 3.0

项目位于成都金牛区国宾板块，城市配套完善。规划上采用3栋塔楼，兼顾城市形象营造丰富天际线。在第五立面——公园屋顶，创造多维的城市公共活动空间。业态布局定位精准，增加了人才社区需求的定制业态。建筑外形延续现代川韵，对川西传统建筑韵味进行现代诠释。

The project is located in the state guest section of Jinniu District, Chengdu, with complete and generous urban supporting facilities. Three towers are planned. Give consideration to the urban image and create a rich skyline. On the roof of the fifth facade Park, create a multidimensional urban public activity space. Precise positioning of business layout and customized business formats that increase the demand of talent community. The shape of the building continues the modern Sichuan charm. Modern interpretation of the charm of Western Sichuan Traditional Architecture.

重庆荣昌棠悦府
Chongqing Rongchang Tang Yue Palace

设计师：何凌峰、肖正江
项目地点：重庆
建筑面积：481 877 m²
用地面积：12.8 hm²
容积率：2.76

Designers: Lingfeng He, Zhengjiang Xiao
Location: Chongqing
Building Area: 481 877 m²
Site Area: 12.8 hm²
Plot Ratio: 2.76

西永翰粼天辰
Xiyong Hanlin Tianchen

设计师：贺东海、刘雪松、廖国华
项目地点：重庆
建筑面积：185 356 m²
用地面积：9 hm²
容积率：2.0

Designers: Donghai He, Xuesong Liu, Guohua Liao
Location: Chongqing
Building Area: 185 356 m²
Site Area: 9 hm²
Plot Ratio: 2.0

铜梁原乡溪岸
Tongliang Hometown Creek Bank

设计师：何凌峰、冉娟
项目地点：重庆
建筑面积：441 250 m²
用地面积：18.86 hm²
容积率：1.8/1.6

Designers: Lingfeng He, Juan Ran
Location: Chongqing
Building Area: 441 250 m²
Site Area: 18.86 hm²
Plot Ratio: 1.8/1.6

重庆博翠宸章
Chongqing Imperial Mansion

设计师：王梓西
项目地点：重庆
建筑面积：113 484 m²
用地面积：5.4 hm²
容积率：1.5

Designer: Zixi Wang
Location: Chongqing
Building Area: 113 484 m²
Site Area: 5.4 hm²
Plot Ratio: 1.5

广东

广东省建筑设计研究院有限公司
GuangDong Architectural Design & Research Institute Co., Ltd.

广东省建筑设计研究院有限公司(GDAD)创建于1952年,是新中国第一批大型综合勘察设计单位之一,改革开放后第一批推行工程总承包业务的现代科技服务型企业,全球低碳城市和建筑发展倡议单位、国家高新技术企业、全国科技先进集体、全国优秀勘察设计企业、当代中国建筑设计百家名院、全国企业文化建设示范单位、广东省文明单位、广东省抗震救灾先进集体、广东省重点项目建设先进集体、广东省守合同重信用企业、广东省勘察设计行业领军企业、广州市总部企业、综合性城市建设技术服务企业。

GDAD现有全国工程勘察设计大师2名、广东省工程勘察设计大师5名、享受政府津贴专家13名、教授级高工逾100名,具有素质优良、结构合理、专业齐备、效能显著的人才梯队。

Guangdong Architectural Design and Research Institute Co., Ltd. (GDAD), founded in 1952, is one of the first large-scale comprehensive survey and design units in New China. Carbon City and Building Development Initiative Unit, National High-tech Enterprise, National Advanced Collective of Science and Technology, National Excellent Survey and Design Enterprise, Top 100 Contemporary Chinese Architectural Design Institutes, National Demonstration Unit of Enterprise Culture Construction, Guangdong Province Civilized Unit, Guangdong Province Advanced Earthquake Relief and Disaster Relief Collective, advanced collective for key project construction in Guangdong Province, contract-abiding and trustworthy enterprise in Guangdong Province, leading enterprise in the survey and design industry in Guangdong Province, headquarters enterprise in Guangzhou, and comprehensive urban construction technology service enterprise.

GDAD currently has 2 national engineering survey and design masters, 5 Guangdong engineering survey and design masters, 13 experts enjoying government allowances, and more than 100 professor-level senior engineers. It has a talent echelon with excellent quality, reasonable structure, complete professionalism and remarkable efficiency.

地址：广州市荔湾区流花路97号
电话：177 0206 9202
邮箱：gdad_uad@163.com
网址：www.gdadri.com

Add: No.97, Liuhua Road, Liwan, Guangzhou
Tel: 17702069202
Email: gdad_uad@163.com
Web: www.gdadri.com

广州110KV猎桥变电站
110kV Lieqiao Substation

项目设计及完成年份：2018年7月31日－2021年5月30日
主创：陈雄
设计团队：黄俊华、高原、陈俊明、陈仁杰、陈细明、胡冰清、杨竣凯、郑培鑫
项目地点：广东 广州
建筑面积：4 992 m²
合作方：广州电力设计院有限公司
客户：广东电网有限责任公司广州供电局
摄影版权：凯剑视觉 / 李开建、李开庆

Design & Completion Time: July 31, 2018 to May 30, 2021
Leader Designer: Xiong Chen
Design Team: Junhua Huang, Yuan Gao, Junming Chen, Renjie Chen, Ximing Chen, Bingqing Hu, Junkai Yang, Peixing Zhen
Location: Guangzhou, Guangdong
Building Area: 4 992 m²
Partners: Guangzhou Electric Power Design Institute Co., Ltd.
Clients: Guangzhou Power Supply Bureau, Guangdong Power Grid Co., Ltd.
Photo Credits: KJ_VISION / Kaijian Li, Kaiqing Li

该项目为国际首创"站馆合一"变电站。开创性提出"变电站、公共性、科普性"兼具的多功能组合体,打破传统变电站功能结构单一问题,解决市民与"厌恶型"建筑间的社会矛盾,外立面延续周边建筑圆润优雅的形态特征,打造对外开放的电力科普体验中心,拓展市民活动空间,建造珠江边的"月光宝盒"和城市时尚新地标,让一座典型的"厌恶型"工业建筑,华丽转身为高品质的工业建筑和公共建筑的复合体。

The world's first "station and hall integration" substation. The groundbreaking multi-functional combination of "substation, publicity, popular science" is proposed to break the single problem of the traditional substation function and structure, and solve the social contradiction between citizens and disgusting buildings. The open electric power science experience center expands the space for citizens' activities, and builds a "moonlight treasure box" by the Pearl River, a new landmark of urban fashion. Turn a typical disgusting industrial building into a complex of high-quality industrial buildings and public buildings.

南粤古驿道梅岭驿站
Nanyue Ancient Meiling Post Pavilion

项目设计及完成年份：2018年3月1日－2019年1月31日	Design & Completion Time: March 1, 2018 to January 31, 2019
主创：陈雄	Leader Designer: Xiong Chen
设计团队：黄俊华、郭其轶、李珊珊、许尧强、龚锦鸿、陈进于、金少雄、曾祥、戴力、钟伟华、林全攀、倪俍、李沛华	Design Team: Junhua Huang, Qiyi Guo, Shanshan Li, Yaoqiang Xu, Jinhong Gong, Jinyu Chen, Shaoxiong Jin, Xiang Zeng, Li Dai, Weihua Zhong, Qaunpan Lin, Liang Ni, Peihua Li
项目地点：广东 韶关	Location: Shaoguan, Guangdong
建筑面积：515.4 m²	Building Area: 515.4 m²
客户：广东省住房和城乡建设厅	Clients: Guangdong Provincial Department of Housing and Urban-Rural Development
摄影版权：凯剑视觉 / 李开建、李开庆	Photo Credits: KJ_VISION / Kaijian Li, Kaiqing Li

    南粤古驿道是广东历史上交通运输、经济交流和文化传播的重要通道，是线性文化遗产的主体。梅岭驿站选址于广东韶关南雄梅关景区旁的梅岭古村，既助力南粤古驿道修复活化，也是对乡土建筑进行现代转译的探索。

    设计从3个维度出发：对村民日常生活有利、对南粤古驿道的价值发掘有益、乡土环境里的在地性与传统性相结合。选址上整合古村落资源，让梅关古道旅游价值更加丰富。通过对乡土建筑语境与传统建筑的转译，采用现代单元式的设计手法，以等差模数制为组合逻辑，结合当地传统的青砖、灰瓦、原木等材料，营造了一座具有民居聚落特色、与自然环境相融的单元式驿站。

    在驿站休憩间，能感受到阳光、清风和山林，也能感受到建筑是作为自然的一部分而存在的。

    Nanyue Ancient Post Pavilion is an important channel for transportation, economic exchange and cultural dissemination in the history of Guangdong, and it is the main body of linear cultural heritage. The Meiling Post Pavilion is located in the ancient village of Meiling next to the Meiguan scenic spot in Nanxiong, Shaoguan, Guangdong.

    The design starts from three dimensions: it is beneficial to the daily life of the villagers, beneficial to the value exploration of the ancient post road in Nanyue, and the combination of locality and tradition in the local environment. The site selection integrates ancient village resources to enrich the tourism value of Meiguan Ancient Pavilion. By translating the context of vernacular architecture and traditional architecture, using modern unit design techniques, using the differential modulus system as the combination logic, and combining with local traditional materials such as blue bricks, gray tiles, logs, etc., a residential settlement is created. Features, a unit-style inn that blends with the natural environment.

    In the rest room of the inn, you can feel the sunshine, breeze and mountains, and also feel that the building exists as a part of nature.

白云山柯子岭门岗及周边景观整治工程
Baiyunshan Keziling and Surrounding Landscape Project

项目设计及完成年份：2019年5月3日－2021年5月20日
主创：陈雄
设计团队：黄俊华、郭其轶、许尧强、陈康桃、陈俊明、赖锐敏、李牧川、单超一、王立君、劳智源、肖幸怀、李东海、吴远亮
项目地点：广东 广州
建筑面积：1 565.5 m²
合作方：广州园林建筑规划设计研究总院有限公司
客户：广州市林业和园林绿化工程建设中心
摄影版权：凯剑视觉 / 李开建、李开庆

Design & Completion Time: May 3, 2019 to May 20, 2021
Leader Designer: Chen Xiong
Design Team: Junhua Huang, Qiyi Guo, Yaoqiang Xu, Kangtao Chen, Junming Chen, Ruimin Lai, Muchuan Li, Chaoyi Shan, Lijun Wang, Zhiyuan Lao, Xinghuai Xiao, Donghai Li, Yuanliang Wu
Location: Guangzhou, Guangdong
Building Area: 1 565.5 m²
Partners: Guangzhou Garden Architecture Planning and Design Research Institute Co., Ltd.
Clients: Guangzhou Forestry and Landscaping Engineering Construction Center
Photo Credits: KJ_VISION / Kaijian Li, Kaiqing Li

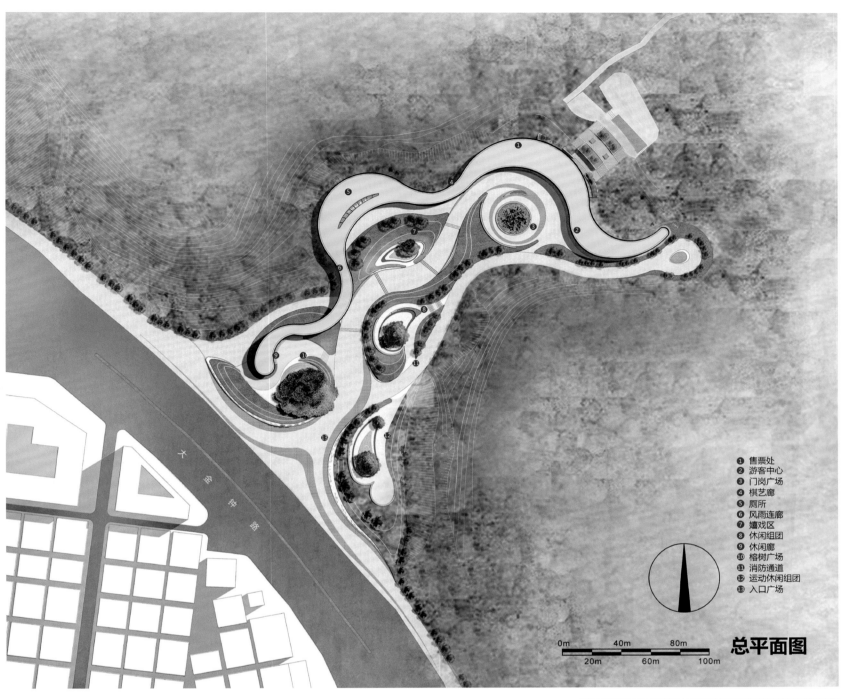

| ① 售票处
② 游客中心
③ 门岗广场
④ 棋艺廊
⑤ 厕所
⑥ 风雨连廊
⑦ 嬉戏区
⑧ 休闲组团
⑨ 休闲廊
⑩ 榕树广场
⑪ 消防通道
⑫ 运动休闲组团
⑬ 入口广场

总平面图

白云山柯子岭门岗是广州市"还绿于民"的重点工程之一。该工程设计概念取自"云山珠水"。以一条轻盈飘逸的云廊空间塑造出区别于以往具象门岗形象的一体化建筑新形象。同时多元化的平台极大地提升了市民活动的空间品质,从而营造出一个亲切自然的公园入口前导空间,兼做社区公园而受到市民喜爱,同时又展现出独特的岭南建筑气质和时代特色。

As an important green restoration initiative of Guangzhou, the project reflects the design concept of Baiyun Mountain and Pearl River. The light and elegant cloud corridor shapes a holistic new architectural image that is totally different from the valleys and multilevel platforms form a friendly and natural space leading to the entrance. This space also serves as a community park, and showcases unique Lingnan characteristics and the spirit of the times.

广州空港中央商务区项目会展中心
Guangzhou Airport Central Business District Project Convention and Exhibition Center

设计时间：2020年10月至今
建设时间：2021年5月至今
主创：陈雄
设计团队
GDAD：罗若铭、高原、龚锦鸿、卢宇、陈冠东、钟仕斌、陈俊明、梁红缘、张熠、黄晋奕、
　　　邓丽威、吴玄、郭林森
AXS：鉾岩崇、谢少明、飞永直树、牛入具之、糸濑贤司、赵雄、刘敏、小盐刚生、林映岚、
　　　郭羽扬、徐骏、陆畅、周冠龙
项目地点：广东 广州
建筑面积：35万m²（投标阶段）
合作方：株式会社佐藤综合计画（AXS SATOW）、
　　　　广州市公用事业规划设计院有限责任公司
客户：广州融创空港城投房地产开发有限公司

Design Time: October 2020 to Now; Construction Time: May 2021 to Now
Leader Designer: Xiong Chen
Design Team
GDAD: Ruoming Luo, Yuan Gao, Jinhong Gong, Yu Lu, Guandong Chen,
　　　Shibin Zhong, Junming Chen, Hongyuan Liang, Yi Zhang, Jinyi Huang,
　　　Liwei Deng, Xuan Wu, Linsen Guo
AXS: Takashi Hokoiwa, Shaoming Xie, Naoki Tobinaga, Tomoyuki Ushigome,
　　　Kenji Itose, Xiong Zhao, Min Liu, Takeo Koshio, Yinglan Lin, Yuyang Guo,
　　　Jun Xu, Chang Lu, Guanlong Zhou
Location: Guangzhou, Guangdong
Building Area: 350 000 m² (Bidding Stage)
Partners: AXS SATOW Co., Ltd.,
　　　　　Guangzhou Public Utilities Planning and Design Institute Co., Ltd.
Clients: Guangzhou Sunac Airport Urban Investment Real Estate Open Co., Ltd.

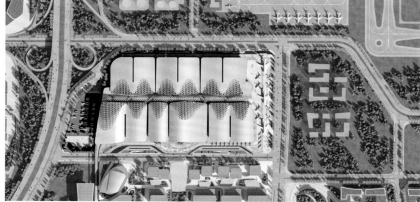

广州空港中央商务区项目会展中心,以"云山珠水、空港腾飞"为设计理念,整体造型轻盈灵动,体现空港特色,营造一种开放的姿态,打造融入城市生产生活的门户新地标。

建筑采用单层10 000 m²的单元式展馆,呈鱼骨状布局,平面集约高效,经济实用,满足会展的功能需求。屋盖结构采用梭形复合张弦桁架方案,结构美观简洁高效,经济指标在国内同类型展馆结构中属于极低水平。建筑融合屋面太阳能板等多种先进技术一体化设计,形式与功能高度统一,体现低碳可持续的目标。内部结合岭南地域特色,打造绿色生态的花园会展,提供舒适宜人的休憩洽谈环境。

Guangzhou Airport Central Business District Project Convention and Exhibition Center is designed with the concept of "clouds, mountains and waters, and the airport takes off". The overall shape is light and agile, reflecting the characteristics of the airport, creating an open attitude and creating a new landmark that integrates into the city's production and life.

The building adopts a single-storey 10 000 m² unit exhibition hall with a fishbone-like layout. The plane is intensive and efficient, economical and practical, and meets the functional requirements of the exhibition. The roof structure adopts the shuttle-shaped composite string truss scheme. The structure is beautiful, simple and efficient, and the economic index is extremely low among the structures of the same type of exhibition halls in China. The building integrates the integrated design of various advanced technologies such as roof solar panels, and the form and function are highly unified, reflecting the goal of low-carbon sustainability. The interior combines the regional characteristics of Lingnan to create a green and ecological garden exhibition, providing a comfortable and pleasant environment for rest and negotiation.

南海艺术中心
Nanhai Art Centre

项目设计及完成年份: 2021年
主创及设计团队: GDAD城市建筑工作室——陈雄、黄俊华、陈坤婷、陈康桃、徐楷莹、黄旭峰、张栋、陈俊明
项目地点: 广东 佛山
建筑面积: 123 000 m²
合作方: 方未建筑事务所——何威、严然、徐琳、徐振、李禹希

Design & Completion Time: 2021
Leader Designer & Team: GDAD Urban Architecture Design Studio - Xiong Chen, Junhua Huang, Kunting Chen, Kangtao Chen, Kaiying Xu, Xufeng Huang, Dong Zhang, Junming Chen
Location: Foshan, Guangdong
Building Area: 123 000 m²
Partners: Fangwei Architects - Wei He, Ran Yan, Lin Xu, Zhen Xu, Yuxi Li

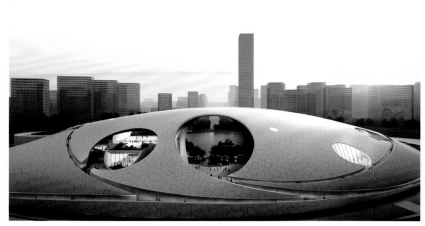

南海艺术中心位于佛山市南海区，基地地处佛山水道与千灯湖中轴线交汇处。作为沥桂新城"南海之眼"的核心，该项目将承接并引领千灯湖片区这一区域的发展建设。

方案的设计灵感来源于唐代岭南著名诗人张九龄的诗句："海上生明月，天涯共此时"。建筑面向佛山水道，以一个简洁有力的半月形造型环抱南海湖，犹如海上升起的一弯新月，朦胧轻盈，与水体相应，融合共鸣，以期唤起人们心灵深处古今与共的情愫，与环境共景，与人文共情。

南海艺术中心集文化、艺术、体育、休闲于一体，以其独特的东方浪漫情怀，为广佛同城发展提供精神驱动力，向世界讲述中国故事。

Nanhai Art Centre is located at the intersection of Foshan Waterway and the central axis of Qiandeng Lake. As the core of the "Eye of Nanhai", the project will lead the development of this area.

It is inspired by an ancient Chinese poem of Zhang Jiuling, a famous poet at Lingnan area during Tang Dynasty: "The bright moon shines on the sea, people around the world would share this moment". The building faces Foshan waterway, embracing the Nanhai Lake with a simple but powerful half-moon shape, like a crescent moon rising in the sea, hazy and light, corresponding to the water and echoing people's inner feeling. Trying to arouse the feelings from people's inner space with the environment and the humanities.

Nanhai Art Centre integrates culture, art, sports, and leisure functions together. With its unique oriental romantic feelings, it provides a spiritual driving force for the development of Guangzhou and Foshan, and tells Chinese stories to the world.

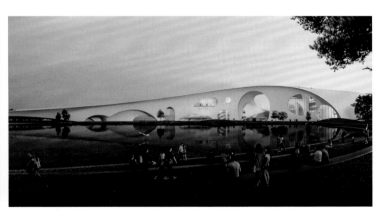

广州市设计院集团有限公司
GUANGZHOU DESIGN INSTITUTE GROUP CO., LTD.

广州市设计院集团有限公司成立于1952年，是国内首批甲级勘察设计单位，也是全球低碳城市和建筑发展（中国）倡议单位、全国优秀勘察设计企业、国家高新技术企业、全国创新型优秀企业、当代中国建筑设计百家名院、工程勘察设计行业质量管理体系升级版AAA级认证企业、广州市"百年·百品"品牌企业。秉持绿色发展理念，凭借深厚的技术储备和核心竞争优势，致力于提供绿色、智慧、高品质建筑综合解决方案。主要业务范围包括城市规划编制、工程勘察设计、工程总承包和全过程工程咨询服务，作为核心业务的建筑设计涵盖大型城市综合体、超高层、TOD、轨道交通、商业、文旅、博览、办公、酒店、教育、医疗、康养、体育、观演、居住等多种类型。汇聚1 500名行业优秀人才，打造8 000多项高品质建筑精品，遍布国内80多个大中型城市和海外10多个国家，荣获4项国家科技进步奖以及其他1 400多项各类奖项，拥有专利290多项，在绿色低碳建筑、超高层建筑、岩土和地下空间领域具有核心竞争优势，影响力位居行业同类单位前列。

Guangzhou Design Institute Group Co., Ltd., founded in 1952, is one of the first Class A survey and design institutions in China, a global low-carbon city and building development (China) initiative, a national excellent survey and design enterprise, a national high-tech enterprise, a national innovative excellent enterprise, one of the top 100 architectural design institutes in contemporary China, an AAA certified enterprise with upgraded quality management system in the engineering survey and design industry Guangzhou "100 years · 100 products" brand enterprise. Adhering to the concept of green development, and relying on deep technical reserves and core competitive advantages, we are committed to providing green, smart and high-quality architectural comprehensive solutions. The main business scope includes urban planning preparation, engineering survey and design, general contracting and whole process engineering consulting services. As the core business, architectural design covers large urban complexes, super high-rise buildings, TOD, rail transit, commerce, culture and tourism, exhibitions, offices, hotels, education, medical care, health care, sports, performances, residential and other types. It has gathered 1 500 outstanding talents in the industry, created more than 8 000 high-quality architectural products, spread over more than 80 large and medium-sized cities in China and more than 10 countries overseas, won 4 national science and technology progress awards and more than 1 400 other awards, and has more than 290 patents. It has core competitive advantages in the field of green low-carbon buildings, super high-rise buildings, rock and soil and underground space, and its influence ranks in the forefront of similar units in the industry.

地址：广州市天河区体育东路体育东横街3-5号设计大厦 Add: No.3-5 Design Building, Tiyu Dongheng Street, Tiyu East Road, Tianhe, Guangzhou
电话：020-87513031 Tel: 020-87513031
邮箱：bgs@gzdi.com Email: bgs@gzdi.com
网址：www.gzdi.com Web: www.gzdi.com

禅泉酒店
Chanquan Hotel

建成时间：2016年 Completion Time: 2016
建设地点：广东 云浮 Location: Yunfu, Guangdong
设计时间：2010年3月 Design Time: March 2010
竣工时间：2014年5月 Completion Time: May 2014
建筑面积：86 000 m² Building Area: 86 000 m²
建筑规模：323间客房 Building Size: 323 rooms

获奖情况
Awards

全国建筑行业优秀勘察设计一等奖
全国人居经典建筑规划设计方案金奖
广东省优秀工程勘察设计一等奖
National Construction Industry Excellent Survey and Design First Prize
National Habitat Classic Architecture Planning and Design Scheme Gold Award
Guangdong Province Excellent Engineering Survey and Design First Prize

禅泉酒店位于禅宗六祖惠能的故里广东省云浮市新兴县龙山风景区，由五星级度假酒店、精品酒店、天然温泉区和国际会议中心等组成。

规划设计充分尊重当地既有的风水与文脉，基地南面的主酒店大堂入口及广场，是六祖明镜广场、坛经楼、龙山国恩寺中轴线的延续。轴线两侧的一泓清澈淡雅的荷塘成为广场的天然背景。顺着主轴北望，6组园林庭院式建筑组群宁静而大气地分布其间，富有禅意和现代东方韵味。轴线继续向北延伸至露天温泉区和精品酒店区的自然园林景观，并将视线一直引向远处的群山。这样的空间布局将六祖故里龙山的气场自然地引入基地和建筑，天人合一。精品酒店设计独立的出入口，其会所与主酒店大堂隔湖相望，形成另一条重要的景观轴线和通廊。23套精品客房舒适有致地散落在溪流边、竹林中、明湖畔、草坡上，彰显顶级度假区的档次。

Chanquan Hotel is located in Longshan Scenic Area, Xinxing County, Yunfu, Guangdong, the hometown of Huineng, the sixth ancestor of Zen Buddhism. It consists of a five-star resort hotel, a boutique hotel, a natural hot spring area and an international conference center.

The planning and design fully respect the local feng shui and cultural context. The main hotel lobby entrance and plaza on the south side of the base are the continuation of the central axis of the Sixth Patriarch's Mirror Square, the Tanjing Building and the Longshan Guoen Temple. The clear and elegant lotus ponds on both sides of the axis become the natural background of the square. Looking north along the main axis, six groups of garden courtyard-style buildings are quietly and grandly distributed among them, full of Zen and modern oriental charm. The axis continues to the north to the natural garden landscape of the open-air hot spring area and the boutique hotel area, and leads the sight all the way to the distant mountains. Such a spatial layout naturally introduces the aura of Longshan, the hometown of the Sixth Patriarch, into the base and the building, so that man and nature are one. The boutique hotel is designed with independent and distinguished entrances and exits. Its clubhouse and the main hotel lobby face across the lake, forming another important landscape axis and corridor. The 23 sets of boutique guest rooms are comfortably scattered along the stream, among the bamboo forests, by the Ming Lake, and on the grassy slopes, demonstrating the grades of the top resorts.

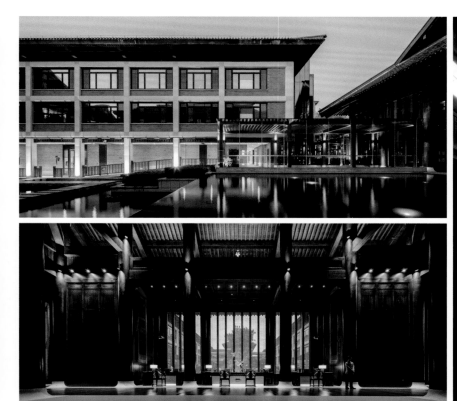

保利汕尾金町湾A006地块项目（希尔顿逸林酒店）
Poly Shanwei Jinding Bay Plot A006 Project (Hilton Yilin Hotel)

项目地点：广东 汕尾	Location: Shanwei, Guangdong
建筑功能：度假酒店、酒店式公寓	Architectural Function: Resort Hotel, Serviced Apartment
设计时间：2014年	Design Time: 2014
竣工时间：2019年	Completion Time: 2019
建筑面积：89 502 m²	Building Area: 89 502 m²
用地面积：56 956 m²	Site Area: 56 956 m²
建筑层数：地上9层、地下2层	Building Floors: Nine floors above ground, Two floors below ground
建筑高度：47.997 m	Building Height: 47.997 m
合作设计：广州市设计院集团有限公司（设计总包）	Cooperative Design: Guangzhou Design Institute Group Co., Ltd. (general design contractor)
美国道林建筑与规划设计公司（方案设计）	American Dowling Architecture and Planning and Design Company (schematic design)

保利汕尾金町湾A006地块项目（希尔顿逸林酒店）为首家入驻汕尾市的国际品牌酒店，在推进该区域商务和休闲旅游国际化发展进程中发挥不可或缺的作用，对汕尾旅游业的发展具有重要战略意义。

本项目地上西侧为8层度假酒店塔楼、东侧为9层酒店式公寓塔楼，通过中部单层酒店大堂连接。为最大限度利用海景资源，两侧高起的Y型建筑像是臂弯，将南边临海海面围合成大型室外海滨公园，独享海岸金滩的美景。

针对岭南滨海多风雨、日照强烈等环境特征，酒店大堂采用了创新的重檐屋顶（超大型混凝土结构坡屋面，对应研发了超高超大坡度斜屋面结构施工工法，已取得湖南省工程建设工法证书），起翘的屋脊，层层叠加的屋檐，形成动感的天际线，打造出独特的地标文旅形象。在兼顾造型美观的同时，有效解决滨海遮阳、隔热、通风、抗风、抗盐雾等设计难题。

Poly Shanwei Jinding Bay A006 Plot Project (Hilton Yilin Hotel) is the first international brand hotel to settle in Shanwei City. development is of strategic importance.

The west side of the project is an eight-story resort hotel tower, and the east side is a nine-story hotel-style apartment tower, connected by a single-story hotel lobby in the middle. In order to maximize the use of seascape resources, the Y-shaped buildings raised on both sides are like arms, and the southern side facing the coast is enclosed into a large outdoor seaside park, which enjoys the beauty of the golden beach on the coast.

In view of the environmental characteristics of Lingnan coastal areas such as windy, rainy and strong sunshine, the hotel lobby adopts an innovative double-eave roof (super-large folded concrete structure sloping roof, correspondingly developed a super-super-super-slope sloping roof structure construction method, which has obtained the Hunan Provincial Engineering Construction Construction method certificate), the upturned ridge, and the stacked eaves form a dynamic skyline and create a unique landmark cultural tourism image. While taking into account the beautiful appearance, it can effectively solve the design problems of coastal shading, heat insulation, ventilation, wind resistance and salt spray resistance.

珠江城
Pearl River Tower

设计团队：马震聪、黄惠菁、赵松林、周定、朱祖敬、李继路、
刘谨、周名嘉、叶充、赵力军、丰汉军、郭进军
项目地点：广东 广州
竣工时间：2013年3月
建筑面积：210 000 m²
用地面积：10 000 m²
建筑高度：309 m
合作设计：Skidmore, Owings and Merrill

Design Team: Zhencong Ma, Huijing Huang, Songlin Zhao,
Ding Zhou, Zujing Zhu, Jilu Li, Jin Liu,
Mingjia Zhou, Chong Ye, Lijun Zhao,
Hanjun Feng, Jinjun Guo
Location: Guangzhou, Guangdong
Completion Time: March 2013
Building Area: 210 000 m²
Site Area: 10 000 m²
Building Height: 309 m
Co-Designers: Skidmore, Owings and Merrill

遵循"天人合一"和"可持续发展"的理念，设计进行再创作。遴选、优化并系统性采用符合地域性气候特征的11项可持续节能技术，实现"超低能耗、超高品质"，较好地解决了超高层建筑与超低能耗无法兼顾的矛盾。

Following the concept of "harmony between man and nature" and "sustainable development", the design is recreated. Select, optimize, and systematically adopt 11 sustainable energy-saving technologies that conform to regional climate characteristics to achieve "ultra-low energy consumption and ultra-high quality", which better solves the contradiction between super high-rise buildings and ultra-low energy consumption.

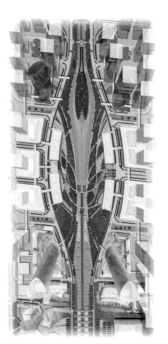

图例	LEGEND
1.前台	RECEPTION
2.前厅	PREFUNCTION
3.茶室	TEA ROOM
4.电气室	ELECTRICAL ROOM
5.储藏间	STORAGE
6.舞台	STAGE
7.空调机房	AIR CONDITIONING
8.洗手间	RESTROOM
9.贵宾休息室	VIP LOUNGE

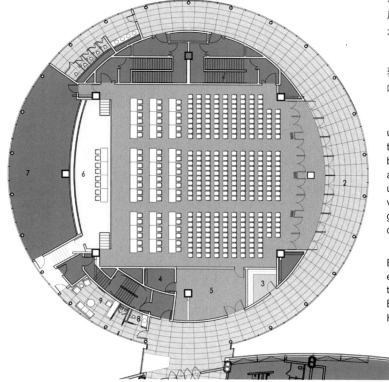

项目科研中高舒适性超低能耗冷辐射空调技术、基于光环境控制和空间高效利用的设备天花一体化体系、基于可再生能源利用的建筑、结构创新体系，反映生态友好型社会价值观和前沿技术水平，对建筑双碳技术进步起到了良好的推动作用。

2015年项目荣获全国优秀工程勘察设计奖建筑工程一等奖；获世界高层建筑与都市人居学会(CTBUH)"过去50年世界最具影响力的50栋高层建筑"，是2013年代表建筑。

In the scientific research of the project, the high-comfort ultra-low energy consumption cold radiant air-conditioning technology, the integrated system of equipment and ceiling based on light environment control and efficient use of space, and the building and structural innovation system based on the utilization of renewable energy, reflect eco-friendly social values and cutting-edge technology level, which has played a good role in promoting the progress of construction dual-carbon technology.

In 2015, it won the first prize of the National Excellent Engineering Survey and Design Award for construction engineering; won the "50 Most Influential High-rise Buildings in the World in the Past 50 Years" 2013 Your Representative Building by the World Council on Tall Buildings and Urban Habitat (CTBUH).

CONFERENCE CENTER - LEVEL 4
会议中心-4层

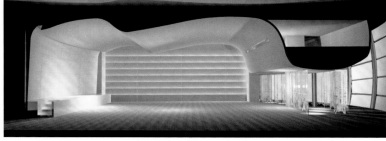

广州黄埔区委党校
Guangzhou Huangpu District Party School

设计团队：赵松林、姚迪、曾国贤、曾恒锐、贺剑龙、龙倩雯、胡婧、邝晓媚、贺楠、吴博冠、李怡婷、谢晨妤、殷俊、唐嘉希、赵梦、秦凡凡、金霄、傅东东、陈宗香、陈志超	Design Team: Songlin Zhao, Di Yao, Guoxian Zeng, Hengrui Zeng, Jianlong He, Qianwen Long, Jing Hu, Xiaomei Kwong, Nan He, Boguan Wu, Yiting Li, Chenyu Xie, Jun Yin, Jiaxi Tang, Meng Zhao, Fanfan Qin, Xiao Jin, Dongdong Fu, Zongxiang Chen, Zhichao Chen
项目地点：广东 广州	Location: Guangzhou, Guangdong
设计时间：2019年	Design Time: 2019
竣工时间：2021年	Completion Time: 2021
建筑面积：64 800 m²	Building Area: 64 800 m²
用地面积：19 540 m²	Site Area: 19 540 m²
结构形式：框架结构	Structural Form: Frame Structure

设计理念：本项目以岭南传统建筑为灵感来源，将岭南园林的布局、造景手法融入现代党校规划设计中。重塑一种清新淡雅的书香氛围，使新校区体现文化积淀和地域特色。方案规划布局对自然山体敞开，U形布局空间环抱自然景观，融入"山、水、房、廊"4大要素，串联起校园与自然的对话。充分利用地形高差特点，采用"化整为零"的设计手法，使校园空间流线充满趣味，达到移步易景的空间效果。建筑造型既具有党校建筑的庄重与典雅，又兼顾校园建筑的轻松与活力，营造具有红色文化，丰富内外空间的校园氛围。

技术特色：项目整体设计充分利用了真理公园的景观资源，使得校园最大化地享受到优美的景观视野。通透的架空层及整体舒朗、有序的布局使得校园的通风效果十分良好，使学员在校园里学习、漫步获得了舒适、与自然相融的体验。

技术创新：项目通过总图立体规划及多首层空间处理，解决了场地与周边道路存在4 m高差的问题，使项目主要出入口平接主要道路。项目北面拥有极佳景观视野资源，规划设计通过岭南式庭院布局，使校园能最大化享受景观资源，引入山景。项目采用主动式通风节能手段，通过岭南传统的空间布局及架空层处理，使校园达到了极佳的通风散热效果，中央水景庭院能有效带走夏季热量，使得校园环境舒适宜人。

Design concept: Inspired by the traditional architecture of Lingnan, this project integrates the layout and landscaping techniques of Lingnan gardens into the planning and design of the modern party school. Reshape a fresh and elegant scholarly atmosphere, so that the new campus reflects cultural accumulation and regional characteristics. The planning layout of the plan is open to the natural mountain, and the U-shaped layout space surrounds the natural landscape and integrates the four elements of "mountain, water, house and corridor" to connect the dialogue between the campus and nature. Make full use of the characteristics of terrain height difference, and adopt the design method of "breaking the whole into parts" to make the campus space flow line full of interest and achieve the spatial effect of moving and easy scenery. The architectural shape not only has the solemnity and elegance of the party school building, but also takes into account the ease and vitality of the campus building, creating a campus atmosphere with red culture and enriching the internal and external space.

Technical features: The overall design of the project makes full use of the landscape resources of Truth Park, so that the campus can enjoy the beautiful landscape to the maximum extent. The transparent overhead floors and the overall comfortable and orderly layout make the campus very well ventilated, enabling students to study and stroll on the campus to gain a comfortable and natural experience.

Technological innovation: The project solves the problem of a 4 m height difference between the site and the surrounding roads through the three-dimensional planning of the general plan and the treatment of multiple first floors, so that the main entrances and exits of the project are connected to the main roads. The north side of the project has excellent landscape view resources. Through the planning and design of the Lingnan-style courtyard layout, the campus can maximize the enjoyment of landscape resources and introduce mountain views. The project adopts active ventilation and energy-saving measures. Through the traditional spatial layout of Lingnan and the treatment of overhead floors, the campus achieves excellent ventilation and heat dissipation. The central waterscape courtyard can effectively take away summer heat, making the campus environment comfortable and pleasant.

广州珠江外资建筑设计院有限公司
Guangzhou Pearl River Foreign Investment Architectural Designing Institute Co., Ltd.

广州珠江外资建筑设计院有限公司(简称珠江设计)是提供专业技术和管理服务的综合设计公司,是以广州珠江外资建筑设计院有限公司及旗下8个子公司为主体的设计平台。广州珠江外资建筑设计院创立于1979年,是国内首家总承包(即EPC)链条内的建筑设计公司。珠江设计经过将近40年的发展,依托珠江实业集团,立足"成就蓝图之美"的本心,已完成上千个项目,项目涉及多个领域。我们的成就来自于所有"珠江设计"人对创新理念、卓越设计和优质服务的不懈追求。

珠江设计的建筑原创设计、全建制BIM正向设计、装配式、绿色节能等设计能力位于领先行列屡获奖项,并具有与多位国际设计大师合作的经验,是澳门最大的国内设计咨询单位,且项目管理服务得到行内广泛认可。

珠江设计的项目涵盖了民用建筑的大部分范畴。多个项目荣获国家大奖,创作了大量例如广州气象监测预警中心、粤澳新通道青茂口岸联检大楼、珠江颐德中心等优秀作品,并作为最重要的设计方参与完成了广州白云国际会议中心、广州大剧院、广州电视台新址工程等项目。

珠江设计将在数字化建设上努力深耕,保持创新研发的劲头,实现"十四五"期间规划的成为建筑信息化设计领跑者的奋斗目标;最终发展为全建制BIM正向设计、绿色装配式设计、建筑设计和管理服务为核心的全国领先智慧城市综合数字化设计服务商。

Guangzhou Pearl River Foreign Investment Architectural Designing Institute Co., Ltd. (PRD) is a multidisciplinary design firm that offers professional technical and management services. It is a design platform based on Guangzhou Pearl River Foreign Investment Architectural Designing Institute Co., Ltd. (PRFIADI) and its eight subsidiaries that are independent juristic persons. Established in 1979, PRFIADI is the first domestic architectural design institute included into the EPC chain. Backed by the Pearl River Industrial Group and adhering to the philosophy of "bringing the beauty of design to real life", PRD has completed over a thousand projects in diverse sectors in the past four decades. What we have achieved lies in our persistent pursuit of innovative concepts, design excellence and quality services.

As a leader in original architecture design, "all-designer, all-project, all-discipline, whole-process, full-support, whole-chain" BIM forward design, prefabricated building and energy conservation, the award-winning PRD has partnered with many international design masters. It represents the largest domestic design consulting firm in Macao, with its project management services widely recognized in the industry.

PRD has worked on most typologies of civil buildings and won various national awards. The representative award-winning projects include: Guangzhou Meteorological Monitoring and Early Warning Center, Joint Inspection Building in Qingmao Port in the new channel between Guangdong and Macao, and Pearl River Yeede Center. We also participated in the design of Guangzhou Baiyun International Convention Center, Guangzhou Opera House and the new Guangzhou Broadcasting Network Project as the lead designer.

PRD will stay committed to digital construction and innovative R&D as it works to be a pacesetter in architectural information design envisioned in the 14th Five-year Plan. Our goal is to evolve into a domestic leader and service provider in smart city integrated digital design focusing on "all-designer, all-project, all-discipline, whole-process, full-support, whole-chain" BIM forward design, green prefabricated building design, architectural design and management services.

地址:广州市环市东路362号好世界广场22楼
电话:020-83842921
邮箱:info@pearl-river.com
网址:www.pearl-river.com

Add: 22nd Floor, Haoshijie, No.362 Huanshi East Road, Guangzhou
Tel: 020-83842921
Email: info@pearl-river.com
Web: www.pearl-river.com

国家呼吸医学中心一期工程
National Respiratory Medical Center Phase I

项目地点：广东 广州	Location: Guangzhou, Guangdong
建筑面积：9 694 m²	Building Area: 9 694 m²
设计时间：2020年	Design Time: 2020
竣工时间：2020年	Completion Time: 2020

项目为国家呼吸医学中心一期工程，建成为集医疗门诊、临床医学研究、医学教育、科学研究于一体的具有国际先进水平的国家呼吸医学中心，项目原建筑主体大楼始建于1964年，原为办公大楼，按照新功能需求全方位改造，包括主体结构、机电系统及建筑外立面等全专业整合设计，建成后为广州医科大学附属第一医院承担重要功能的组成部分，同样为三级甲等医院等级。

建筑立面以"气韵相合、和而不同"的理念与沿江西路广州少年儿童图书馆等文物建筑群的色调和立面韵律相互融合、协调；同时结合广州的气候特点，引入骑楼空间、复合型立面遮阳等生态概念，以"与古为新、修旧如旧"的方式延续经典的珠江两岸街区形象。

As Phase I development, the project aims to create an internationally advanced national respiratory medical center for medical outpatient service, clinical medicine research, medical education and research. The main building, which was first built in 1964 as an office building, is fully renovated to accommodate the new program through cross-disciplinary integrated engineering of main structure, MEP and facade. Once completed, it will accommodate important functions of The First Affiliated Hospital of Guangzhou Medical University as a tertiary Level A hospital.

With the concept of "harmony and diversity", the project echoes to and harmonizes with Guangzhou Children's Library and other historical buildings along Yanjiang West Road in terms of facade tone and rhythm. Given the climatic characteristics in Guangzhou, ecological concepts like arcade spaces and compound facade shading are employed to continue the classic images of the waterfront neighbourhood by the Pearl River by "restoring the cultural site as it was".

澳门大学珠海新校区
University of Macau (UM) Hengqin Campus, Zhuhai

项目地点：广东 珠海	Location: Zhuhai, Guangdong
建筑面积：100万m²	Building Area: 1 000 000 m²
设计时间：2009年	Design Time: 2009
竣工时间：2013年	Completion Time: 2013

澳门大学新校区位于珠海横琴岛，毗邻港澳，学生规模为1万人，总建筑面积约1 000 000 m²。

作为粤澳合作重点项目，澳门大学将打造成为一所教学与科研并重的国际性大学。新校区由3个用水网分隔的岛和主体学校、行政楼等书院式功能组团组成，体现"中西荟萃、山海交融、岭南文脉、南欧风情"的建筑风格。我院承接其中约420 000 m²宿舍生活区项目的方案深化及施工图设计。

The new campus of UM is located at Hengqin Island, Zhuhai, adjacent to Hong Kong and Macao. With an planned area of 1 000 000 m², it will accommodate around 10 000 students.

As a key cooperation project between Guangdong and Macao, the new campus aims to create an international university for both taught programs and research. It comprises three islands separated by a water network, as well as academy-like functional clusters that include teaching buildings, administrative building, etc. Overall, its architectural design reflects "a mixture of Chinese and western styles, a blend of mountain and sea, the cultural context of Lingnan, and the charm of southern Europe". We provided SD detailing and CD for the 420 000 m² dormitory area.

翁源县气象防灾减灾业务技术用房
Technical Facilities for Meteorological Disaster Prevention and Mitigation, Wengyuan County

项目地点：广东 韶关	Location: Shaoguan, Guangdong
建筑面积：2 978 m²	Building Area: 2 978 m²
设计时间：2015年	Design Time: 2015
竣工时间：2017年	Completion Time: 2017

项目借鉴客家围楼等岭南建筑应对地域环境的生态设计手法，模仿当地传统建筑的营造形式和使用方式，利用现有的优美自然环境，建筑群围绕水塘及山坡展开布置，通过不同标高的庭院组合，营造出节能、生态、环保的低碳型建筑。建筑群以低矮、平和的方式嵌入山坡里面，谦卑地融入环境当中，建成后层叠的绿化屋顶与层层递进的庭院、保留的水塘一起，将原有的栖息之地还给了鸟类。

The design of the Project draws on the ecological design approaches of traditional Lingnan architecture such as Hakka enclosed housing in response to the local environment, and mimics the construction forms and uses of local traditional buildings. In the design, the buildings of the project are planned to extend around the picturesque pond and hillside, while the courtyards at different elevations are combined to realize energy-efficient, environmental-friendly and low-carbon buildings. The entire building cluster is embedded into the hillside and the local environment at large in a low-key, peaceful way. The terraced green roofs, together with courtyards at varied levels and reserved ponds, restore the original habitat of birds.

粤澳新通道（青茂口岸）
Guangdong-Macao New Passage (Qingmao Port)

项目地点：中华人民共和国澳门特别行政区、广东 珠海
珠海市部分建筑面积：65 166 m²
澳门部分建筑面积：101 244 m²
设计时间：2017年9月
竣工时间：2021年9月

Location: Macao SAR & Zhuhai, Guangdong
Building Area of Zhuhai Part: 65 166 m²
Building Area of Macao Part: 101 244 m²
Design Time: September 2017
Completion Time: September 2021

青茂口岸位于广东珠海拱北口岸西侧鸭涌河地段和澳门青州，是连接澳门和珠海的独立开放的24小时服务信息化电子口岸，采取"合作查验、一次放行"的通关模式，口岸设计通关能力为旅客200 000人/日，是粤澳重大的跨境民生工程。

项目由澳方联检大楼及名优产品展览中心、粤方联检大楼及连接通道3部分主体建筑组成。其中，珠江设计负责澳方联检大楼及名优产品展览中心、粤方联检大楼两大部分的方案设计、施工图设计、施工配合。

Qingmao Port is located in the Yachong River section on the west side of Zhuhai Gongbei Port and in Qingzhou (Ilha Verde), Macao. It is an independent 24-hour port with digitalized services connecting Macao and Zhuhai. Under the mode of "collaborative inspection and one-time release", it boasts a customs clearance capacity of 200 000 passengers per day, representing a major cross-border livelihood project of both Guangdong and Macao.

The project consists of three parts: joint inspection building on Macao side & famous high-quality product exhibition center, joint inspection building on Guangdong side, and connecting channel. We provided SD, CD and construction administration for the former two parts.

中葡商贸中心
Sinoport Plaza

项目地点：广东 珠海	Location: Zhuhai, Guangdong
建筑面积：118 854 m²	Building Area: 118 854 m²
建筑高度：170 m	Building Height: 170 m
设计时间：2017年2月—2020年12月	Design Time: February 2017 to December 2020
竣工时间：2022年6月	Completion Time: June 2022

珠海横琴中葡商贸中心将打造包含高端商务酒店、办公、休闲购物街业态的横琴口岸商业中心，包含1栋高170 m、35层的超高层塔楼、4层商业裙楼。整体造型突破传统"方盒子"，以锐角75度平行四边形为塔楼基础平面。在面向东南景观河道方向，以流畅的曲线裁切整体形态，塔楼呈现层层后退的形态，就像船上的风帆，裙楼上波浪形百叶，如船身溅起的浪花，整体宛如"乘风破浪远航的船只"，包含了助力珠海、澳门繁荣发展的深远意义。

Comprising a 170 m 35-floor super high-rise tower and a 4-floor commercial podium, Sinoport Plaza aims to create a business center at Hengqin Port that integrates high-end business hotel, office and leisure shopping street. The project features a 75-degree parallelogram foundation plan instead of the conventional "square box" one. On the southeast where the complex faces the landscaped river, smooth curves cut the tower into terraces that step backward by floor, making the whole building appear like the sail of a ship. Moreover, the wavy lamellae of the podium facade are reminiscent of the breaking waves generated by the ship. In this way, the overall imagery of the project as "a ship riding the wind and the waves" is presented to convey the far-reaching significance of boosting the prosperity of Zhuhai and Macao.

广东

广州冯智新建筑设计有限公司
Feng Architect

广州冯智新建筑设计有限公司（Feng Architect）是一家可提供项目建设全过程设计服务的公司。

公司秉着"同情理、遵道义、自在生"的设计理念，并把这一理念注入每一个设计项目中。

能留下经得起时间考验的作品是事务所的追求，也是作为建筑师的责任，在实现这一目标的过程中，愿意与其他优秀设计团队合作，共享智慧。

以积极的态度面对日新月异的计算机辅助设计技术，于2018年完成了BIM团队的转化，以求对设计质量进行更精准细致的控制，提升设计的品质。

Feng Architect is a company that can provide design services for the whole process of project construction.

Adhering to the design concept of "sympathy, compliance, and freedom", we inject this concept into every design project.

It is the pursuit of the firm and the responsibility of the architect to leave works that can stand the test of time. In the process of realizing this goal, we are willing to cooperate with other excellent design teams and share wisdom.

Facing the ever-changing computer-aided design technology with a positive attitude, the transformation of the BIM team was completed in 2018, in order to control the design quality more accurately and meticulously, in order to improve the quality of the design.

地址：广州市海珠区沥滘路100号海尚明珠智慧园10号楼		Add: Building 10, Haishangmingzhu Park, No.100 Lijiaolu, Haizhu, Guangzhou	
邮编：510650	电话：18127847800	P.C: 510650	Tel: 18127847800
邮箱：fengzhixin@fengarch.com	网址：www.fengarch.com	Email: fengzhixin@fengarch.com	Web: www.fengarch.com

广东外语外贸大学
北校区校门建筑和门前广场项目
Guangdong University of Foreign Studies North Campus Gate Building and Gate Square Project

主创设计师：冯智新
地点：广东 广州
建筑面积：8 000 m²
用地面积：1.38 hm²

Chief Designer: Zhixin Feng
Location: Guangzhou, Guangdong
Building Area: 8 000 m²
Site Area: 1.38 hm²

大学校门既是边界，也是视觉第一印象，基于场地的功能与意义，历史与现实的多维考量，塑造了上下两层各造其境、各塑其形的"双广场"解决方案。

下层广场以车流动线划分空间结构，辅以快速通行的人流动线，留有可供行人休憩的街角公园。

上层广场以学生与市民行为及所需空间的差异作为造景塑形的依据，使不同的节假日、纪念活动都有与之对应的场景。

门，不仅是一个视觉装置，更是一个可供游玩、参观的空间场所。用全新的方式来演绎学校大门，让门的形式更能体现21世纪大学的形象与内涵。

The lower ground divides the space by traffic flow and expressed Pedestrians. There is street park as a place for rest.

The upper ground designs for multiple scenes for festivals and gatherings based on citizen and students' behavioral traits. The spaces features fits for various seasonal scenes.

The Gate is not only a visual device, but also a place for sight-seeing and playing.

Architecture use a fully new approach to presents the University Gate, to deliver a 21st century image and inner space.

雁塔文化艺术中心概念规划方案
Yanta Culture and Art Center

主创设计师：冯智新	Chief Designer: Zhixin Feng
项目地点：广东 广州	Location: Guangzhou, Guangdong
建筑面积：12 340 m²	Building Area: 12 340 m²
用地面积：2.89 hm²	Site Area: 2.89 hm²
容积率：0.2	Plot Ratio: 0.2

项目位于广州增城的雁塔寺公园内，场地闹中取静，交通便利，且与增江毗邻。

寺内传统建筑以雁塔寺为中心中轴对称排列，塔是核心景区。新建区域文化艺术中心，既作为独立的市民活动场所，也与雁塔寺的佛教文化紧密相连。

虽看似不同性质的活动，但其本质是可相融和的。

此方案意在塑造"融通"的空间，使游人能感受到生活、文化、艺术、信仰之间的交融与连通，使整体空间合和为一。

新轴线与传统中轴以雁塔寺为交点，45°夹角而生成，场地内建筑，如市民广场、音乐厅、书院、博物馆等则采取"中轴自由"的形态镶嵌于内。

The project plans to locate in Wild Geese Pagoda ("Yanta" in Chinese) Temple Park in Zengcheng, Guangzhou. The place remains quiets in a bustling environment. The park has good transportation facilities and nearby the Zeng River.

The traditional buildings in the Yanta Pagoda Temple line up in symmetrical arrangement around the central axis of the temple. The pagoda is core scene. The new built regional culture and art center is a dedicated place for citizens gathering place. It also inherit the Buddhist Culture that carried out by Yanta Temple

The place facilitates different activities and harmonizes the civil lives with religious activities in a seamlessness approach.

This plan aimed at providing a "Harmony & Connection" Space for people to immerse to the harmonization of civic living, culture, art and religion and reach out to be part of the overall universe.

The new central axis meets the traditional central axis in the Yanta Temple with 45 degree angle. The buildings insides, such as Citizen Plaza, Theatre, College, Museum are liberally spread around the new central axis.

广东塑料交易所仓储中心/沟通之弧
Guangdong Plastics Exchange Storage Center/Arc of Communication

主创设计师：冯智新
项目地点：广东 广州
建筑面积：200 000 m²
用地面积：8.21 hm²
容积率：2.5

Chief Designer: Zhixin Feng
Location: Guangzhou, Guangdong
Building Area: 200 000 m²
Site Area: 8.21 hm²
Plot Ratio: 2.5

此项目的主题是沟通，包括与塑料的沟通，交易本身的沟通，与环境、时间的沟通，与城市设计的沟通。各功能区块之间的良好沟通，是项目健康运行的关键，本次运用的主要元素是"弧"，以及"弧"的不同演变。

建筑布局北高南低，在最佳用地位置设计高耸的标志性建筑，通过与用地形态有逻辑关系的"弧"塑形，构成单体及由此组合成的群体。

在建筑语言的使用中，也是以"单纯""复杂""相关联"为表达方法，从上而下呈现出"复杂—单纯"的变化。

The theme of this project is communication, including communication with plastics, communication with the transaction itself, communication with environmental time, and communication with urban design. Good communication between functional blocks is the key to the healthy operation of the project.

The main elements of this application are "arc" and the different evolutions of "arc". The building layout adopts the north high and the south low. The towering landmark buildings are designed at the best site location. Through the "arc" shaping that has a logical relationship with the land use form, the individual units are presented as curved triangles and curved rectangles. Constitute monomers and groups formed from them.

In the application of architectural language, "simple", "complex" and "related" are also used as expressions, showing "complexity-simplicity" from top to bottom.

曼飞龙国际养生度假区
Manfeilong International Health Resort

项目的规划原则是以一种业态为核心,多种业态围绕核心的混合式布局。

建筑设计理念,是传承傣族建筑文化特征,从傣民族文化内提炼建筑语汇,在现代美学原则的基础上,通过现代的材料和技术对建筑进行创新表达,表达出一种"国际的、现代的、傣族的"建筑文化,赋予传统建筑文化一种延续的生命力。

The planning principle of the project is a mixed layout with one type of business as the core and multiple business types around the core.

The concept of architectural design is to inherit the characteristics of the Dai architectural culture, extract the architectural vocabulary from the Dai nationality culture, and on the basis of modern aesthetic principles, through modern materials and modern technology to innovatively express the architecture, expressing a kind of "international, the modern and Dai" architectural cultureendows the traditional architectural culture with a continuation of vitality.

主创设计师:冯智新	Chief Designer: Zhixin Feng
项目地点:云南 西双版纳	Location: Xishuangbanna, Yunnan
建筑面积:3 000 000 m²	Building Area: 3 000 000 m²
用地面积:918 hm²	Site Area: 918 hm²
容积率:0.85	Plot Ratio: 0.85

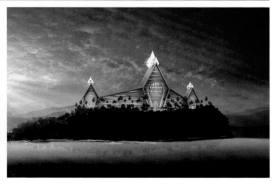

广东

广州市天作建筑规划设计有限公司
TEAMZERO ARCHITECTURE DESIGN & URBAN PLANNING CO. LTD.

广州市天作建筑规划设计有限公司（天作国际）于2002年成立，目前拥有建筑行业（建筑工程）设计甲级、城乡规划编制甲级、风景园林工程设计专项乙级和市政行业乙级4个专业设计资质，并获得国际国内各级各类奖项300多个，项目遍布全国及海外。

天作国际长期以来坚持"国际视野、本土情怀"的设计理念，依托国际国内两大技术支撑平台，联动规划、建筑、景观、市政、城市运营五大板块，为政府与商业客户提供覆盖开发咨询、规划策划、建筑设计、景观设计、市政设计等城乡开发建设各阶段各领域的一站式服务，得到业内人士的广泛认可和高度评价。

Teamzero Architecture Design & Urban Planning Co. Ltd. (TEAMZERO) Founded in 2002, TEAMZERO has achieved 4 professional qualifications during the past 18 years, in terms of Class A on both Architecture Design and Urban-Rural Planning, and Class B on both Landscape Design and Municipal Engineering. Meanwhile, we have won more than 300 rewards and a great variety of projects home and abroad.

Based on our long-term adhered philosophy of design, "Global View, Local Identity", relied on our international-domestic technical platform, and the integration of "planning-architecture-landscape-engineering", TEAMZERO provides one-stop professional services for government and business agencies, which contains development consultant, statutory and commercial planning, architecture design, landscape design, municipal engineering, etc., and covers a wide range of Mainland China and abroad, which keep TEAMZERO widely recognized and highly regarded in professional fields.

地址：广州市珠江新城华夏路28号富力盈信大厦605至612室
邮编：510623
电话：020-83840927
传真：020-83839072
邮箱：teamzero@21cn.net
微信公众号：天作设计观

Add: Room 605-612, Fuli Yingxin Building, No. 28 Huaxia Road, Zhujiang New Town, Guangzhou
P.C: 510623
Tel: 020-83840927
Fax: 020-83839072
Email: teamzero@21cn.net
WeChat Official Account: Tianzuo Design Concept

江西省文化中心
Cultural Centre, Jiangxi

项目地点：江西 南昌
项目规模：280 000 m²
设计时间：2015
委托方：江西省文化厅

Location: Nanchang, Jiangxi
Site Area: 280 000 m²
Design Time: 2015
Client: Jiangxi Provincial Department of Culture

设计理念及特色："赣水源远引客来，滕阁临渚绮席开。"设计充分研究南昌市的城市现状，把文化中心作为"一江两岸"文化生态轴线的核心节点，使赣江西岸的景观通廊更为丰富完满，并与旧城区的文化建筑群体隔江呼应，形成以赣江为中心，构建于城市水脉之上，具有文化张力的社会文化平台。设计引入城市客厅的概念，把城市的历史、文化、发展和未来综合展现在项目设计的建筑形态和人文景观之中，承载着城市的文化底蕴和人文风貌，营造具有现代都市气氛的公共空间。

Design concept and features: "The Gan River that has a faraway origin invites a great number of guests. The grand pavilion on a small island in the river offers a great banquet for all of us". Nanchang city design fully the present research situation and the cultural center as a "one river two" cultural ecology axis core node, enriched in ganjiang west bank landscape corridor is more complete, and with the echo of the culture of the old building groups across the river, formed by the gan as the center, building on the city water vein, social culture platform of cultural tension. The design introduces the concept of urban living room, integrates the history, culture, development and future of the city into the architectural form and humanistic landscape designed by the project, carries the cultural deposits and humanistic features of the city, and creates a public space with modern urban atmosphere.

澄海音乐厅
The Chenghai Concert Hall, Shantou

项目地点：广东 汕头　　Location: Shantou, Guangdong
项目规模：15 000 m²　　Site Area: 15 000 m²
设计时间：2015　　　　Design Time: 2015
委托方：奥飞娱乐　　　Client: Aofei Entertainment

设计理念及特色：音乐厅整体为椭圆形态，3个倾斜度不同的椭圆形锥台相互咬合，如同沙滩上的贝壳，体现了澄海海洋文化的独特精神，寓意着澄海人民多姿多彩的生活和富饶丰收的生产。锥台相互咬合之后，从顶视图呈现出圆与月牙状，象征着日月交辉。

Design concept and characteristics: The concert hall is in an elliptical shape as a whole, with three elliptical cone platforms with different gradients biting each other, like shells on the beach, reflecting the unique spirit of Chenghai's marine culture and implying the colorful life and productive production of Chenghai people. After the cones and abutments occlude each other, they appear round and crescent shaped from the top view, symbolizing the sun and the moon.

广州荔胜广场
Lisheng Square, Guangzhou

项目地点：广东 广州　　　　Location: Guangzhou, Guangdong
项目规模：19 000 m²　　　　Site Area: 19 000 m²
设计时间：2017　　　　　　Design Time: 2017
委托方：广州市地下铁道总公司　Client: Guangzhou Metro Corporation

项目采用双办公塔楼布局，充分考虑地铁人流、社会人流的办公流线，将B-1塔楼的大堂设于首层，B-2塔楼的大堂设于二层，且与二层地铁接驳通道联系紧密。整体功能区划分合理，走道、连廊、交通梯设置便捷。

The project adopts the layout of double office towers to fully consider the office flow line of subway and social flow. The lobby of Tower B-1 is set on the first floor, and that of Tower B-2 is set on the second floor, which is closely connected with the subway connection channel on the second floor. The overall functional area is reasonably divided, walkway, corridor, traffic ladder setup is convenient.

湛江玥珑湖养生体验中心
Yuelonghu Health Centre, Zhanjiang

项目地点：广东 湛江	Location: Zhanjiang, Guangdong
项目规模：10 200 m²	Site Area: 10 200 m²
设计时间：2013	Design Time: 2013
委托方：华邦控股集团有限公司	Client: Huabang Holding Group Co. LTD.

设计理念及特色：体验中心设计灵感来自中国山川中"山"的形态和中国汉字"围"的布局。

"山"：绵延起伏的"山"形屋顶与碧波荡漾的湖面相映成趣，营造出一幅"山之形，水之境"的诗意画面。

中国人讲究"寻好水而居，择近处而住"的理念，而围合式的布局理念则表达了东方人的谦诚与归宿，追求自然的和谐共融。为了追求更好的观景及体验效果，采用了大面积的落地玻璃，使建筑与自然相互交融联系，让老年人在临水环境中享受晚年的舒适安逸生活。

Design concept and features: The design of the experience center is inspired by the form of "mountain" in Chinese mountains and rivers and the layout of the Chinese character "wai"; "Mountain": THE rolling "mountain" shaped roof and the rippling lake are set against each other, creating a poetic picture of "the shape of the mountain, the land of the water".

Chinese people pay attention to the concept of "seeking good water and living near", while the enclosed layout concept expresses the modesty and end-result of Oriental people, and the pursuit of natural harmony. In order to pursue a better view and experience effect, a large area of floor-to-ceiling glass is adopted, so that the building and nature blend and contact each other, so that the elderly enjoy a comfortable life in their later years in the water.

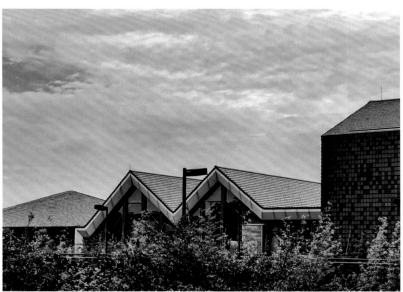

合肥天玥广场
Tianyue Square Commercial Complex, Hefei

项目地点：安徽 合肥
项目规模：230 000 m²
设计时间：2018
委托方：广州市骏誉投资有限公司

Location: Hefei, Anhui
Site Area: 230 000 m²
Design Time: 2018
Client: Guangzhou Junyu Investment Co., Ltd.

设计理念及特色：根据总体空间布局及功能需求，本方案规划总体形成"一轴三廊四区"的布局结构。配合功能定位形成时尚购物街区、商务办公区、都市乐活社区与公共活动广场4个主要功能分区，于规划区中部打造南北向多彩活力轴串联主要功能区，规划独具个性的3条主题廊道，总体形成"井"字形的活动空间系统。

Design concept and features: According to the overall spatial layout and functional requirements, the overall layout structure of "one axis, three corridors and four areas" is planned. In line with the functional positioning, four main functional zones are formed: fashion shopping block, business office district, urban Lolive community and public activity square. In the middle of the planning area, the main functional areas are connected with the north-south colorful vitality axis, and the three theme corridors with unique characteristics are planned to form a "well"-shaped activity space system.

合肥政务新区天珑广场
Tianlong Square, Hefei

项目地点：安徽 合肥
项目规模：456 800 m²
设计时间：2018
委托方：安徽骏誉投资置业有限公司

Location: Hefei, Anhui
Site Area: 456 800 m²
Design Time: 2018
Client: Anhui Junyu Investment Property Co., Ltd.

设计理念及特色：方案形成"两轴五区"的总体规划结构。考虑总体空间及功能布局，在地块内规划倒"T"字形的两大空间发展轴线。东西向魅力乐活轴结合现状道路进行景观化处理，横穿项目中部，位于5大功能分区的中部，作为项目主要的社区活动与交流场所，有利于形成办公、商业、居住3大功能使用者的空间聚集，提高街区内部的凝聚力，提升街区整体活力。南北向商业共享轴结合商业步行街与住宅区商业裙楼设置，通过商业内街的形式打造街区内部主要的商业配套服务流线。

Design concept and features: The scheme forms the overall planning structure of "two axes and five districts". Considering the overall spatial and functional layout, two spatial development axes with inverted "T" shape are planned in the plot. East-west charm lohas axis combined with the present situation of road landscape, across the central project, located in the central part of the five functional partition, as the main project community activities and communication, to form the office, commercial and residential users three functions of space, enhance the internal cohesion, block improve the whole dynamic blocks. The north-south commercial sharing axis combines the commercial pedestrian street with the commercial podium building in the residential area to create the main commercial supporting service flow line inside the block through the form of commercial inner street.

合肥复星文化金融创新城
Fosun Culture of Financial Innovation, Hefei

项目地点：安徽 合肥
项目规模：1 480 000 m²
设计时间：2015
委托方：上海星泓股权投资管理有限公司

Location: Hefei, Anhui
Site Area: 1 480 000 m²
Design Time: 2015
Client: Shanghai Xinghong Equity Investment Management Co., Ltd.

设计理念及特色：项目以"产城融合"的开发理念，打造多元创新、"产城一体"的智慧新城。方案设计对复星金融创新城的构想体现了人文精神和人性尺度。设计融合蜂巢城市的概念，创新引入"产城融合"的开发理念，在规划上以"互联网+"为出发点，引入"云谷、云链、云环"的概念，提倡生态持续、低碳出行、智慧城市。创新城不以高强度塔楼林立的城市形象为目标，方案设计通过不同单元不同产品的搭配，实现高低起伏、疏密有致的城市空间形象，并通过大尺度的绿化广场，怡人的林荫小径，多样的主题公园，营造创新的绿色生态友好城市生活及工作环境，构建智慧城市新典范。

Design concept and features: With the development concept of "integration of industry and city", the project will build a multi-innovation and intelligent new city with "integration of industry and city". The concept of Fosun Financial Innovation City reflects the spirit of humanity and the scale of humanity. The design integrates the concept of honeycomb city, innovatively introduces the development concept of "integration of industry and city", takes "Internet +" as the starting point in planning, introduces the concept of "cloud valley, cloud chain and cloud ring", and advocates ecological sustainability, low-carbon travel and smart city. Innovation city not to high strength tower of the city's image as the goal, the design by the collocation of different unit product, realize the ups and downs, the density of urban space image, and through the large-scale afforestation square, pleasant tree-lined paths, a variety of theme parks, to build innovative green eco-friendly urban living and working environment, build wisdom city new model.

贵阳中天未来方舟
Future Ark, Guiyang, Guizhou

项目地点：贵州 贵阳
项目规模：2 000 000 m²
设计时间：2014
委托方：中天城投集团贵阳房地产开发有限公司

Location: Guiyang, Guizhou
Site Area: 2 000 000 m²
Design Time: 2014
Client: Zhongtian City Investment Group Guiyang Real Estate Development Co., Ltd.

设计理念及特色：规划利用地块濒临河岸的特点，建筑沿河岸布置，自然景观视野最大化，每个商铺均能获得优秀的河道景观资源。

规划巧妙利用南北地形30 m高差的特点，采用退台式的建筑布局方式，一方面解决坡地建设中土方量大、支护费用高、平整场地少的问题；另一方面创造出商业双入口的新型商业模式，通过工程技术解决商业业态、经营等问题，增加商业沿街面，成功提升商业价值。

Design concept and features: Planning to make use of the characteristics of the land near the river bank, the buildings are arranged along the river bank to maximize the view of the natural landscape, and each shop can get excellent river landscape resources.

Planning ingenious use of the characteristics of terrain height of 30 m, desizing and desktop architectural layout, on the one hand, to solve the slope in the construction of earthwork quantity is big, the high cost of supporting, flat ground less problems, on the other hand to create the double entry of the new business model, through the engineering technology to solve business format, management and other issues, increase business street, successfully improve business value.

深圳市建筑设计研究总院有限公司
Shenzhen General Institute of Architectural Design and Research CO.,LTD.

广东

深圳市建筑设计研究总院有限公司（简称深总院/SZAD），建于1982年，伴随着深圳特区的发展，从地区性的建筑设计院发展成为立足深圳、布局全国、服务世界的城乡建设集成服务提供商。拥有建筑行业（建筑工程）甲级、城乡规划编制甲级、工程咨询资信（建筑）甲级、市政行业（给水工程、排水工程）乙级、风景园林工程设计专项乙级等多项资质，是住建部首批"全过程工程咨询试点企业"。深总院多次获评"全国建设系统先进集体""中国十大建筑设计院""中国优秀企业""中国十佳建筑设计机构"，发展规模、业务量、综合竞争实力连续多年位列全国民用建筑设计行业前列。

Shenzhen General Institute of Architectural Design and Research Co., Ltd. (SZAD) was founded in 1982. Accompanied by the fast development of Shenzhen Special Economic Zone, SZAD has been developing from a regional architectural design institute into an integrated service provider--based in Shenzhen--of both urban and rural construction, and it serves China and the world at large. SZAD has a number of qualifications such as Grade A in the construction industry (construction engineering), Grade A in urban and rural planning, Grade A in engineering consulting (architecture), Grade B in the municipal industry (water supply engineering, drainage engineering), and Grade B in landscape engineering design. It is the first "Full-process Engineering Consulting Pilot Enterprises" appointed by the Ministry of Construction. SZAD has received a series of awards, such as "National Advanced Collective of Construction System" "China's Top Ten Architectural Design Institutes" "China's Excellent Enterprises" "China's Top Architecture Design Institute". Also its development scale, business volume and comprehensive competitiveness have been in the forefront of the civil architectural design industry for many years.

地址：深圳市南山区深圳湾科技生态园10栋A座30F、B座31F
深圳市福田区振华路8号设计大厦22F
电话：0755-83785355
网址：www.szad.com.cn

Add: 30F, Block A, 31F, Block B, Building 10, Shenzhen Bay Science and Technology Ecological Park, Nanshan District, Shenzhen
22F, Design Building, No.8 Zhenhua Road, Futian District, Shenzhen
Tel: 0755-83785355
Web: www.szad.com.cn

科苑小学改扩建工程
Renovation and Expansion Project of Keyuan Primary School

设计师：吴超、王思文、洪波、姚俊伟、唐志军、闫丽君、虞子良、王丽、陈军、黄龙、苏路明、周金磊、游辉敏、徐雪峰、钟常盛

项目地点：广东 深圳

建筑面积：24 640 m²

用地面积：6 652 m²

容积率：2.97

Designers: Chao Wu, Siwen Wang, Bo Hong, Junwei Yao, Zhijun Tang, Lijun Yan, Ziliang Yu, Li Wang, Jun Chen, Long Huang, Luming Su, Jinlei Zhou, Huimin You, Xuefeng Xu, Changsheng Zhong

Location: Shenzhen, Guangdong

Building Area: 24 640 m²

Site Area: 6 652 m²

Plot Ratio: 2.97

科苑小学扩建工程构建了一个立体的学习社区，通过创造一种"疏松多孔"的空间结构，提供了灵活的室内外学习空间，并回应当地气候，探索一种新的校园空间类型以适应和引领当下不断更新的教学方式。

立面的呈现上，设计以体现材料的原真性和构造的经济性作为出发点，采用水刷石作为主体材料，灰黄调的天然质感石子肌理与周边玻璃幕墙和钢筋混凝土的现代建筑形象形成柔和的对比与互补，既承载了传统文脉，又融入了现代精神，焕发出新的生命力。

The Renovation and Expansion Project of Keyuan Primary School creates a three-dimensional learning community, provides flexible indoor and outdoor learning spaces by creating a loose and porous spatial structure, and responds to the local climate, exploring a new type of campus space to adapt to and lead the current updated teaching methods.

In the presentation of the facade, the design takes the authenticity of the material and the economy of the structure as the starting point, and uses cement as the main material. The natural texture of the grayish yellow stone texture forms a soft contrast and complement with the surrounding modern architectural image made of glass curtain wall and reinforced concrete, which not only carries the traditional context, but also integrates the modern spirit and radiates new vitality.

南京江北新区市民中心
Citizen Center in Jiangbei New District, Nanjing

设计师：孟建民、杨旭、李优、吴长华、李莉佳、徐昊、郑清、
　　　　刘文旭、李罗兵、赵仁才、章骁
项目地点：江苏 南京
设计时间：2016年
竣工时间：2020年
建筑面积：75 614 m²
用地面积：55 092 m²
结构形式：钢结构、框架结构

Designers: Jianmin Meng, Xu Yang, You Li, Changhua Wu,
　　　　　　Lijia Li, Hao Xu, Qing Zheng, Wenxu Liu,
　　　　　　Luobing Li, Rencai Zhao, Xiao Zhang
Location: Nanjing, Jiangsu
Design Time: 2016
Completion Time: 2020
Building Area: 75 614 m²
Site Area: 55 092 m²
Structural Form: Steel Structure, Frame Structure

南京江北市民中心位于江北新区核心区，定山大街与滨江大道交界处以北，通过上下错位的2个直径104 m的圆形塔楼创造出上圆遮蔽下圆的市民活动广场。其功能集城市展示、公共服务、市民活动为一体，这也是南京江北新区获批以来建设的首个荣获"鲁班奖"的建筑。

建筑设计上突出"和合而生、回归本源"。整个建筑形如中国古典宝盒，寓意幸福美满，流露出一种主题鲜明、风格独特的城市气质，让市民感受到"古朴、时尚、宽敞、明亮、温馨、绿色"的交融感和流畅感。下圆中庭布设"瞻亭赏月""珍珠泉涌""妙水叠瀑""狮岭雄姿"等极具南京园林及江北老山自然生态特征的景观节点，打造亭台廊桥、细水流石的城市庭院空间。

The project is located in the core area of Jiangbei New District, to the north of the junction of Dingshan Street and Binjiang Avenue. Two circular towers with a diameter of 104 m are misaligned up and down to create a citizen activity square with an upper circle covering the lower circle. It is the first building in Jiangbei New District, Nanjing to win the Luban Award since it was approved. It integrates urban exhibition, public service and citizen activities.

The architectural design highlights harmony and returns to the origin. The whole building is shaped like a classical Chinese treasure box with an implication of happiness, revealing an urban temperament with distinct theme and unique style, so that citizens can feel the "simple, fashionable, spacious, bright, warm, green" sense of integration and fluid. In the lower circular atrium, there are some landscape nodes with the characteristics of Nanjing Garden and the natural ecology of the old mountain in North China.

宁波市杭州湾医院
Ningbo Hangzhou Bay Hospital

设计师：孟建民、邢立华、刘瑞平、符永贤、陈一川、储琦、曾明基
项目地点：浙江 宁波
设计时间：2014年
竣工时间：2018年
建筑面积：196 585.26 m²
用地面积：69 260 m²
床位数：1 200
结构形式：钢筋混凝土框架结构

Designers: Jianmin Meng, Lihua Xing, Ruiping Liu, Yongxian Fu, Yichuan Chen, Qi Chu, Mingji Zeng
Location: Ningbo, Zhejiang
Design Time: 2014
Completion Time: 2018
Building Area: 196 585.26 m²
Site Area: 69 260 m²
Number of Beds: 1 200
Structure Form: Reinforced Concrete Frame Structure

宁波市杭州湾医院位于宁波市北部、杭州湾南岸。新区理水成网、筑湖成城，汇聚先进制造与现代服务业。在宁波市杭州湾医院的创作实践中，建筑师从医院与城市、功能、花园等层面的关系切入，围绕人的体验，探索一个面向未来、给人以关怀的疗愈环境。

方案采用集中式建筑布局，通过三列平行体量，形成简洁清晰的空间结构，便于人们定位在医院的位置。通过将手术、ICU、产房等同层布置，建立"趋热层"，提高医疗救治效率。建筑东西两侧分别形成了立体入口花园和集中式的公园绿地，并结合住院部底层架空的形式，营造"傍水筑宇、沿河而住"的江南意向。弧形墙上的方形窗洞在夜晚发出暖色的光，给人以怀抱和温暖的感受，形成独特文化意向。

Ningbo Hangzhou Bay Hospital is located in the north of Ningbo City and the south bank of Hangzhou Bay. The new area forms a network of water and a city of lakes, bringing together advanced manufacturing industries and modern service industries. In the creation and practice of Ningbo Hangzhou Bay Hospital, the architects start from the relationship between the hospital and the city, function, and garden, and explore a future-oriented and caring healing environment around human experience.

The scheme uses a centralized layout, with three rows of parallel volumes, to form a simple and clear spatial structure, which is convenient for people to locate themselves in the hospital. In addition, the design arranges operation department, ICU and delivery room on the same floor to form a "thermal layer", which can improve the efficiency of medical treatment. A three-dimensional entrance garden and a centralized park green space are formed respectively on the east side and on the west side of the building. Combined with the ground floor of the inpatient department, it creates the intention of living in the south of the Yangtze River by the water. The square window holes in the curved wall emit warm light at night, giving people a feeling of embrace and warmth, forming a unique cultural intention.

新疆大剧院
Xinjiang Grand Theater

设计师：孟建民、唐大为、韩纪升、李练英、易豫、张小丽、马净、薛岩、穆英、张景斌等
项目地点：新疆 昌吉
建筑面积：117 000 m²
用地面积：18.666 hm²
容积率：0.63

Designers: Jianmin Meng, Dawei Tang, Jisheng Han, Lianying Li, Yu Yi, Xiaoli Zhang, Jing Ma, Yan Xue, Ying Mu, Jingbin Zhang, etc.
Location: Changji, Xinjiang
Building Area: 117 000 m²
Site Area: 18.666 hm²
Plot Ratio: 0.63

新疆大剧院位于"印象西域"国际旅游城的地理中心,也是该城的核心标志与主要形象代表。总建筑面积约117 000 m²,建筑主体高度65 m,标志物部分最高79 m。新疆大剧院方案以新疆的"仙物"天山雪莲的形象为原型,建成后大剧院将专门演出大型新疆特色歌舞并举办相关文化活动,在旅游城、新疆乃至中亚地区,搭建一个具有国际水准的表演舞台,形成一个展现文化与时尚形象的窗口,产生独特的魅力与吸引力。

Xinjiang Grand Theater is located in the geographic center of "Impression of Western Regions" International Tourism City, which is also the core symbol and main image representative of the city. The total construction area is about 117 000 m², the height of the main building is 65 m, and the highest part of the marker is 79 m. The scheme of Xinjiang Grand Theater is based on the image of snow lotus in Tianshan Mountain, the "celestial object" of Xinjiang. After its completion, the Grand Theater will become a place to perform large-scale Xinjiang characteristic songs and dances and related cultural activities. It will set up an international level performance stage in the tourist city, Xinjiang and even Central Asia, showing a cultural and fashionable image window, and generating significant charm and attraction.

HHD sz 深圳华汇设计有限公司
Shenzhen Huahui Design Co.,Ltd.

深圳华汇设计有限公司成立于2003年，多年来坚持不懈地致力于为中国城市建设发展提供专业建筑设计创意与服务，尤其关注办公及产业园区、高端居住区、商业及城市综合体、文化展览及教育、城市更新及保护再生、城市设计及绿色低碳设计等多板块的设计实践。经过多年的践行，在国内与国际上完成了数百项作品，并与数十家中国优秀的城市开发与运营机构形成战略合作伙伴关系，设计作品连年荣获国内外重要奖项与荣誉。

Shenzhen Huahui Design Co.,Ltd. constantly devotes itself to provide professional service for Chinese City Construction and development, with particular attention to the design practice of large residential, Urban complex, office, space, as well as culture and educational architecture. With many years of experience it has completed thousands of works, and established partnerships with dozens of excellent developers in China. Furthermore, the works of HHD-sz have received many international and domestics awards in these years.

地址：深圳市南山区侨香路4060号香年广场C座10F-11F
电话：0755-82507103/86702519
传真：0755-88352413
网址：www.hhd-sz.com
邮箱：hhm@hhd-sz.com（市场商务）
　　　hr@hhd-sz.com（人才招聘）
　　　media@hhd-sz.com（媒体合作）

Add: 10F-11F, Block C, Future Plaza, 4060#Qiaoxiang Road, Nanshan District, Shenzhen, China
Tel: 0755-82507103/86702519
Fax: 0755-88352413
Web: www.hhd-sz.com
Email: hhm@hhd-sz.com(Market Business)
　　　 hr@hhd-sz.com(Recruitment)
　　　 media@hhd-sz.com(Media Cooperation)

西藏非物质文化遗产博物馆
Tibet Intangible Cultural Heritage Museum

项目地点：西藏自治区拉萨
设计时间：2016年
竣工时间：2018年
总建筑面积：8 000 m²
建成状态：建成

Location: Lhasa, Tibet Autonomous Region
Design Time: 2016
Completion Time: 2018
Total Building Area: 8 000 m²
Completed Status: Completed

在拉萨这座极其特殊的城市中，自然与人文、历史与当下的种种要素相互交汇叠加，形成了设计的特定条件和思考原点。

设计中强调的"天路"概念，由3个层面展开。首先是物理层面的行走路径，它提取自布达拉宫"之"字形步道的原型，经过抽象的演绎，构成了从场地入口迂回上升进入建筑，以及在博物馆内部螺旋攀升的基本空间动线。

其次是空间叠加带来的特殊体验的路径：博物馆建筑原型从大昭寺主殿演化形成，当这种内向而稳定的空间结构和由天路概念形成的参观路径相叠加后，则在人与物理空间之间形成了非常多样化的关系，或高狭、或开阔、或幽暗、或明朗，仿佛在经历一段特殊的生命旅程。

Lhasa is a very special city. Here, nature and humanity, history and modernity intertwine with each other, forming the specific conditions for our design as well as the core of our thinking.

The concept of "heavenly road" is embodied at three levels. Firstly, it refers to a walking path at physical level. Inspired by the prototype of zigzag footpath of Potala Palace, it is transformed into a spatial circulation rising from the entrance of the site to the building and spiraling up inside the museum.

The second is the path of special experience brought by spatial superposition. The archetype of the museum is evolved from the main hall of the Jokhang Temple, which is introverted and stable. Superposed with the visiting path formed by the concept of "heavenly road", a very diverse relationship between human and physical space is formed, either high or low, narrow or open, dark or bright, as if experiencing a special life journey.

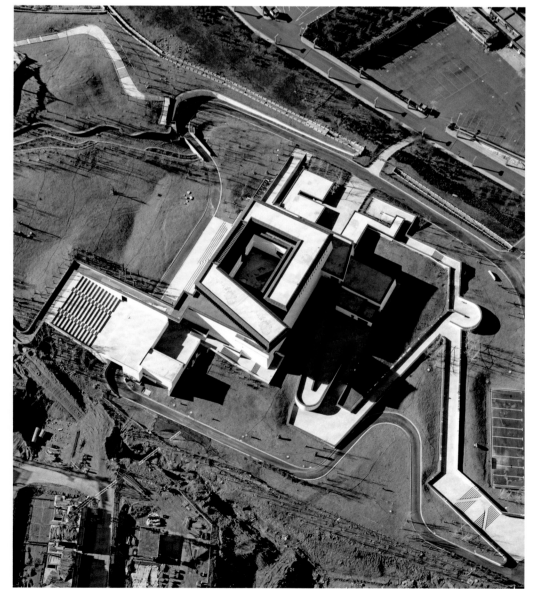

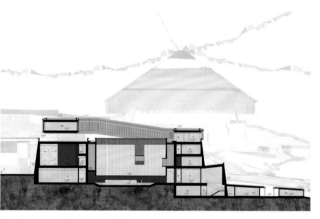

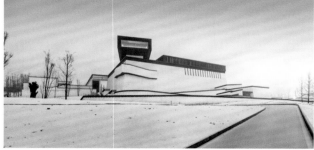

这种特殊的空间经验自然而然地引发了人们心理和情感层面的微妙变化，就成为"天路"的第三重意义——心路。

参观者经过艰苦的攀爬，一路领略藏地丰富的非物质文化遗产，最终达成和布达拉宫跨越时空的对望。这是一种对话，更是一种致敬，是向西藏伟大的自然地理和历史文化致敬，也是向每个人心中的那片圣地致敬。

The subtle psychological and emotional changes triggered by the unique spatial experience leads to a third meaning of the heavenly road, the road of heart.

After appreciating the rich intangible cultural heritage of Tibet through a hard climb, visitors will finally reach the ending point where they can overlook the Potala Palace across both time and space, establishing a dialogue as well as paying a tribute not only to Tibet's great natural landscapes, history and culture, but also to the holy land at the bottom of everyone's heart.

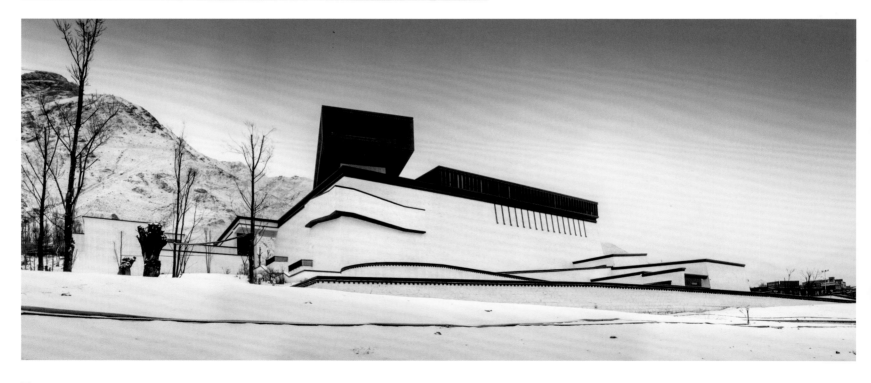

深圳改革开放主题公园项目
莲花山公园展示中心及公共卫生间
Shenzhen Reform and Opening Up Theme Park Project
Lianhuashan Park Exhibition Center & Toilet

项目地点：广东 深圳	Location: Shenzhen, Guangdong
设计时间：2018年	Design Time: 2018
竣工时间：2020年	Completion Time: 2020
总建筑面积：1 791.5 m²	Total Building Area: 1 791.5 m²
建成状态：建成	Completed Status: Completed

莲花山公园山顶公共建筑群落紧邻小平像广场，是位于中轴线上的重要公共配套建筑，包含位于山顶中轴线北端的展示中心和轴线中部的公共卫生间，南端是邓小平同志的雕像。展示中心与公共卫生间共同组成了山顶建筑群，为登山群众提供观展、交流、避雨、休憩及如厕场所。

建设用地所处环境的特殊性，构成了空间叙事依从的大语境。于山顶俯瞰，莲花山公园林木葱郁，生气勃勃，福田胜景环绕四下，而基地之中，东侧山林清幽，广场大树如盖。我们希望谨慎而克制地处理建筑介入场地的方式，使建筑以谦虚的姿态融入环境之中，成为自然的一部分。东侧主体均以吊脚楼的方式轻轻落在山坡之上，减少对山体的破坏。

The public building complex on the top of Lianhuashan Park is adjacent to Xiaoping Statue Square. It is an important public supporting building located on the central axis, including the exhibition center at the northern end of the central axis of the mountain and the public toilet in the middle of the axis, and the southern end is the statue of Comrade Deng Xiaoping. The Shenzhen Lianhua Peak public building complex including the exhibition hall and the LianHua Peak Public Restroom, provides visitors with seeing a gallery, resting place and toilet.

The particularity of the environment in which the construction land is located constitutes a large context of spatial narrative compliance. Overlooking from the top of the mountain, Lianhuashan Park is full of lush and vibrant trees, surrounded by Futian scenery, and in the base, the east side of the park is quiet, and the square is covered with trees. We hope to handle the way the building intervenes in the site with care and restraint, so that the building blends into the environment with a humble attitude and becomes a part of nature. The main body on the east side is gently landed on the hillside in the way of stilted buildings to reduce damage to the mountain.

莲花山公园展示中心
Lianhuashan Park Exhibition Center

莲花山顶公共卫生间
Lianhuashan Public Toilets

小平像广场
Xiaoping Statue Square

N

展示中心场地东侧和南侧均为山坡,缺少对场所空间的界定,我们希望设计纵横两道墙和漂浮的屋顶,对手植树庭园进行围合,塑造半流动的前场空间,在建筑和环境之间为市民提供一处静意的纪念园。展厅的屋顶漂浮于墙体之上,形成入口的挑檐,檐口轻薄,与墙体的厚重形成轻重对比。

在公共卫生间的设计中,墙同样扮演着重要的界定角色。这是一个没有"门"的卫生间,功能空间的围合界面被打开、离散,融于自然之中。在解决因此带来的私密性与公众性的矛盾上,我们根据私密性的强弱以千里江山图、格栅、石材墙面依次遮挡,通过精准计算,使私密区域避开了行人的视线。同时在长廊与卫生间之间设置高差将公共区域与相对私密的男女厕所区域显著区分出来,起到提示作用。

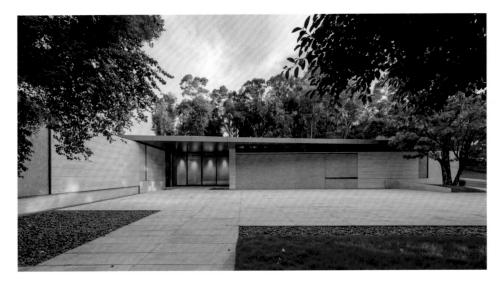

The east and south sides of the exhibition center are hillsides, which lacks the definition of the site space. We hope to design two vertical and horizontal walls and a floating roof to enclose the opponent's tree-planting garden and create a semi-fluid front field space. In between, it provides a quiet memorial garden for the citizens. The exhibition hall floats on the wall to form a cantilever for the entrance. The thin eaves form a contrast against the thick walls.

Walls also play an important defining role in the design of the public restroom. This is a bathroom without a "door", and the enclosed interface of the functional space is opened, discrete, and integrated into nature. In order to solve the contradiction between privacy and publicity caused by this, according to the strength of privacy, we block the sight by perforated metal engraved with "A Thousand Lis of Rivers and Mountain", louvers and stone walls. The calculation just happened to make the private area in the toilet avoid the pedestrian's sight. At the same time, a height difference is set between the corridor and the toilet to distinguish the public area from the relatively private male and female toilet area, which plays a prompting role.

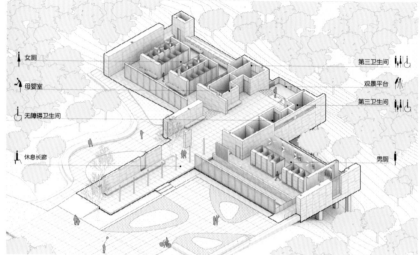

智谷AI科技中心
Artificial Intelligence Technology Valley

项目地点：湖南 长沙
设计时间：2019年
竣工时间：2020年
总建筑面积：3 433.5 m²
建成状态：建成

Location: Changsha, Hunan
Design Time: 2019
Completion Time: 2020
Total Building Area: 3 433.5 m²
Completed Status: Completed

湘江智谷AI科技中心位于长沙湘江新区人工智能科技城的杨柳公园内。作为服务片区的城市客厅，建筑将集合室内外展览、会议交流、接待及园区服务等多重意义的功能，成为未来前沿科技学习与探索的中心，同时与公园环境相结合，成为休闲活动的目的地。

建构多重尺度在这里包含了两重命题，一是建筑如何融入公园尺度，形成更多层次的空间环境；二是建筑如何通过不同尺度的空间与体量，为功能赋形，形成能容纳不同类型目的性活动的场地，因此这里所说的功能，更多的是面向未来的"空"间，而这些单元更倾向于被定义成大、中、小。

In Yangliu Park near the AI technology city of Xiangjiang New District is the Xiangjiang Artificial Intelligence Technology Valley. As the parlor serving the city, the valley offers indoor and outdoor exhibitions, meetings, exchanges, receptions and multiple services. Here we can learn and explore future frontier technology. In a picturesque park, this is also a nice place for leisure activities.

The multiple dimensions of the construction contain two different topics. One is to develop buildings into parts of the park and further evolve into multi-layer spaces. The other is to activate buildings through space and volume of all sizes to accommodate diversified activities. The function is oriented by "vacant" spaces in the future, which can be divided into small, medium and large dimensions.

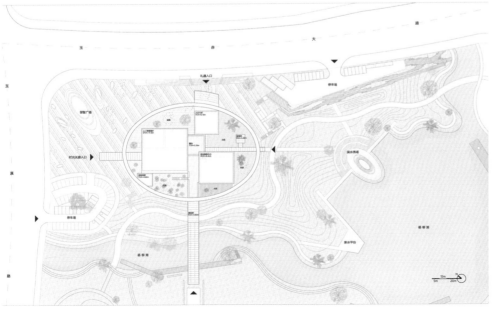

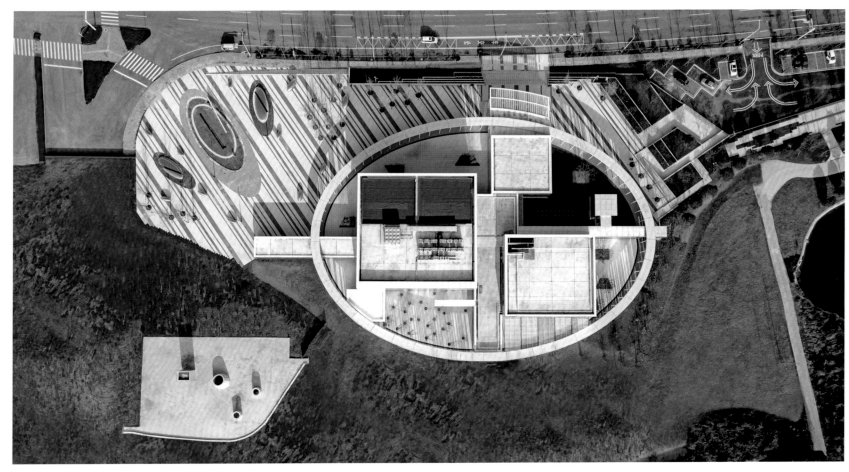

面对充满不确定性与可能性的科学研究与创新环境,我们与其说是在设计一栋建筑,不如说是提供了一套具有多重尺度意义的开放性空间构架,希望建筑在满足当下使用的同时,还能有效应对未来的需求,这不同于完全标准化工业化的建构体系,我们同时也希望通过精营巧造,为这个"科学风暴"的中心赋予更丰富的人文意涵和长效的空间品质。

Ahead of scientific research and innovation that are full of uncertainties and possibilities, our design is not a building but an unrestrained spatial framework of multiple dimensions. Except for existing demands, the architecture is projected to cope with unknown expectations. Distinct from a completely standard and industrial construction system, we thirst for an exquisite design that carries greater humanistic significance and more persistent space quality.

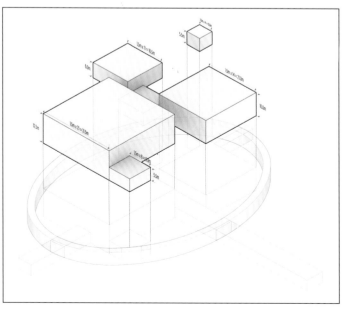

上海虹桥万创中心
Shanghai Hongqiao Wanchuang Center

设计范围：规划设计、建筑设计及室内空间设计
项目地点：上海
设计时间：2018年
竣工时间：2021年
建筑面积：117 778 m²
建成状态：建成

Design Scope: Planning and Design, Architectural Design and Interior Space Design
Location: Shanghai
Design Time: 2018
Completion Time: 2021
Building Area: 117 778 m²
Completed Status: Completed

在上海虹桥万创中心项目中，建筑师立足于探索如何在外部城市空间和内部建筑空间之间构建出丰富而有层次的关联，将城市的公共性有效地注入建筑群组内部，在为空间赋形的同时，更为场所赋能。

项目地处七宝生态商务区核心地带，建筑规模约11.8万m²，包含办公、商业、服务配套等功能。用地的西侧是横泾河，北侧是中央规划绿轴，均为自然生态的景观资源。南侧和东侧则临城市道路，是整个项目面向城市的公共界面。

In the Shanghai Hongqiao Wanchuang Center project, the architect explored building a rich and layered relationship between the external urban space and the internal architectural space and injecting the city's public nature into a building cluster. Thus, shaping the space also empowered the place.

Located in the core area of Qibao Eco Business District, this 118 000 m² project includes offices, businesses, service facilities and other functions. To the west side of the site runs the Hengjing River, while the north side connects to the central planning green axis, both of which are natural ecological landscape resources. The south and east sides are adjacent to city roads, forming the project's public interface with the city.

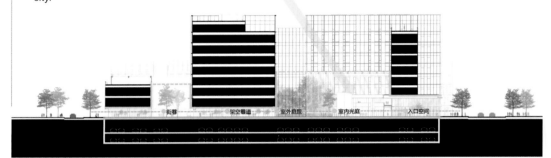

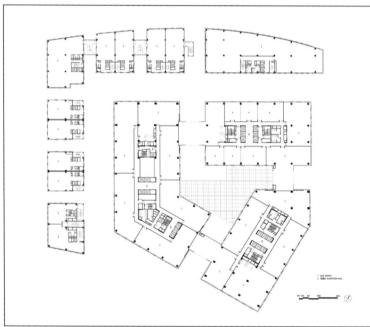

为回应城市关系，建筑师在面向城市的东侧和北侧界面设置了主楼体量。有别于常规的分散独栋式塔楼布局，3栋塔楼通过围合式布局形成1个拥有共享中庭的整体，从而形成完整、清晰、识别性强的城市形象。与主塔的内聚围和式布局形成对比，多层小体量独栋办公呈L形沿场地的南侧和西侧界面展开，面向自然景观。宜人、易于渗透的小尺度空间，将外部景观引入建筑群内部，营造出自然生态的开放式办公街区场景。

With a view to the urban relationship, the architect set up the main building volume on the east and north interfaces with the city. Different from the usual scattered high-rises with single-family units, the three towers form a whole with a shared atrium created through their enclosing layout, with a complete, clear and recognizable urban feel. In contrast with the cohesive enclosure-style layout of the main towers, the L-shaped multi-storey small offices along the south and west interfaces with a natural landscape. The pleasant and easily permeable small spaces draw the external landscape into the building cluster, creating a natural, ecological and open office block.

深圳市联合创艺建筑设计有限公司
SHENZHEN UNITED ARCHITECTURAL DESIGN CO., LTD.

深圳市联合创艺建筑设计有限公司成立于1997年,是一家拥有建筑行业(建筑工程)甲级、市政行业(道路工程)甲级、市政行业(桥梁工程)乙级、市政行业(给水工程、排水工程)乙级、风景园林工程设计专项乙级、城乡规划编制乙级、工程造价咨询企业乙级、房屋建筑工程监理甲级、市政公用工程监理甲级资质的民营综合型设计院。

公司业务范围覆盖工程建设项目全过程,涉及规划、建筑、市政、风景园林、室内等领域的咨询、工程设计、工程监理、技术支持等。服务地区覆盖中国境内27个省、市、自治区,80多个城市,以及新加坡、泰国、越南、柬埔寨、朝鲜等国外市场。

Shenzhen UNITED Architectural Design Co., Ltd. was established in 1997. It is a company with grade A in the construction industry (Construction Engineering), Grade A in the Municipal Industry (Road Engineering), grade B in the Municipal Industry (Bridge Engineering), grade B in the Municipal Industry (water supply engineering, drainage engineering), grade B in the design of landscape garden engineering, grade B in the preparation of urban and rural planning, grade B in the engineering cost consulting enterprise, Grade A in the supervision of housing construction engineering Private comprehensive design institute with grade a qualification of municipal public works supervision.

The business scope of the company covers the whole process of engineering construction projects, involving consulting, engineering design, engineering supervision, technical support, etc. in the fields of planning, architecture, municipal administration, landscape architecture, interior, etc. The service area covers 27 provinces, municipalities and autonomous regions in China, more than 80 cities, and foreign markets such as Singapore, Thailand, Vietnam, Cambodia and North Korea.

地址:深圳市福田区田面创意产业园"设计之都"10栋二层A、B、D	Add: Unit A, B, D, 2 Floor, Building 10, City of Design, Zhenhua west rd, Futian District, ShenZhen
邮编:518000	P.C: 518000
电话:0755-82795544	Tel: 0755-82795544
传真:0755-82795567	Fax: 0755-82795567
邮箱:2970278255@qq.com	Email: 2970278255@qq.com
网址:www.szlhcy.com	Web: www.szlhcy.com

横琴汽车营地(露营乐园)项目二期勘察设计施工总承包
General Contractor for Survey, Design and Construction-Hengqin Motor Campsite (Campground) Project Phase II

项目地点:广东 珠海　　Location: Zhuhai, Guangdong
建筑面积:28 000 m²　　Building Area: 28 000 m²
用地面积:2.5 hm²　　Site Area: 2.5 hm²

项目规划以"尊重自然山水格局、构建和谐形象"为原则,以"一轴、四区、多点"的功能结构布局,实现与项目一期的功能互补、风貌延续,打造和谐共生的整体营地形象。项目建筑采用现代建筑风格玻璃幕墙设计,通透大气。通过设计手法将主楼海鸥酒店和副楼星旅酒店分为两大体量,相互交融,穿插呼应,形成别具一格的风景线。

The project planning is based on the principle of "respecting the natural landscape pattern and building a harmonious image", with the functional structure layout of "one axis, four areas, and multiple points" to achieve the functional complementation and continuity of the project phase I, creating a harmonious and coexisting overall camp Image. The project building adopts modern architectural style glass curtain wall design, which is transparent and atmospheric. Through design techniques, the main building Seagull Hotel and the auxiliary building Star Travel Hotel are divided into two large volumes, which blend with each other, interspersed and echoed, forming a unique landscape.

柬埔寨七星海红树林度假区(星月海岸)一期
Chhne Dara Chan, Starcoast Tourism Town, Dara Sakor, Koh Kong Province, Cambodia

规划师与建筑师:吴立志	Planner & Architect: NG Lup Chee
项目地点:柬埔寨国公省	Location: Kok Kong Province, Cambodia
建筑面积:390 000 m²	Building Area: 390 000 m²
用地面积:51 hm²	Site Area: 51 hm²
容积率:0.6	Plot Ratio: 0.6

项目在规划上突破传统规划的规整布局,模拟红树林根系,以"红树林生命奇迹"为设计灵感,提出"生态、生命、生活"的规划理念。同时依托自然地形、景观资源特质,形成生态优先、功能复合的组团式布局结构。由点汇集,以面覆盖,辟路联通,集观光、休闲、餐饮、娱乐、文教、购物、运动等7大主要功能于一体,打造沉浸式居住体验。

In terms of project planning, the regular layout of traditional planning is broken through, and the mangrove root system is simulated. Inspired by the "Miracle of Mangrove Life", the planning concept of "ecology, life and life" is proposed. At the same time, relying on the characteristics of natural terrain and landscape resources, forming a group layout structure with ecological priority and functional composite. It integrates tourism, leisure, catering, entertainment, culture and education, shopping and sports into one, creating an immersive residential experience.

龙华区实验学校小学部设计采购施工总承包（EPC）
General Contracting of Design, Procurement and Construction of Primary (EPC)-Longhua District Experimental Primary School

项目地点：广东 深圳	Location: Shenzhen, Guangdong
建筑面积：33 400 m²	Building Area: 33 400 m²
用地面积：9 969 m²	Site Area: 9 969 m²
容积率：2.63	Plot Ratio: 2.63

项目整体设计呈双L形，加上曲线型退台的设计，使整个建筑犹如搭载莘莘学子的"智慧"方舟。同时以经典的两轴两心为设计布局，将整个校区分为教育学习区和运动休闲区，并通过开放主轴和通风主轴的设计使整个校区动静相契，为学生营造出一块寓教于乐的广阔天地。

The overall design of the project is double L-shaped, and the curved back platform design makes the whole building like a "wisdom" ark for students. At the same time, the classic two-axis and two-center design is used to divide the entire school into an education and learning area and a sports and leisure area. Through the design of the open spindle and the ventilation spindle, the movement and static of the entire campus are in harmony, creating a place for students to learn and play. It's a vast world.

上饶市立医院三江总院建设工程设计
Construction Engineering Design - Sanjiang General Hospital of Shangrao Municipal Hospital

项目地点：江西上饶	Location: Shangrao, Jiangxi
建筑面积：148 800 m²	Building Area: 148 800 m²
用地面积：4 hm²	Site Area: 4 hm²
容积率：2.29	Plot Ratio: 2.29

项目遵循"以人为本、功能完善、布局科学、便捷适用、绿色节能、适度超前"的设计理念，实现了医疗功能理性高效的空间布局，分区合理，联系便捷；各功能流线组织井然有序，实现有机分流。总体规划统一和谐，浑然一体，建筑造型结合现代建筑风格，整体简洁大气，设计达到了规划、形式与功能的完美结合。

The project follows the design concept of "people-oriented, complete functions, scientific layout, convenient and applicable, green and energy-saving, and moderately advanced", and realizes a rational and efficient spatial layout of medical functions, reasonable partitions, and convenient connections; all functional flow lines are organized in an orderly and organic manner Diversion. The overall plan is unified, harmonious, and integrated. The architectural shape combines modern architectural style, and the overall is simple and elegant. The design achieves a perfect combination of planning, form and function.

截流河(沙井段)、蚝乡路等项目拆迁安置房建设工程可行性研究和全过程设计

Demolition and Resettlement Housing Construction Project-Feasibility study and Whole Process Design -Interception River(Shajing part) & Haoxiang Rd, ect.

项目地点：广东 深圳	Location: Shenzhen, Guangdong
建筑面积：55 000 m²	Building Area: 55 000 m²
用地面积：8 112 m²	Site Area: 8 112 m²

项目在用地面积小且流线业态混杂中发掘极小地块的潜力，在满足品质生活之外实现了商业价值最大化，具有花园社区品质化、竖向布局功能化、塑造社区自豪感、社交空间精细化、社区功能全龄化、社区生活游玩化6大特点。2栋高层住宅塔楼立面处理大气。裙房首层设沿街商业；二层设社区健康服务中心、物管用房；三层设屋顶大花园。

The project explores the potential of extremely small plots in a small area of land and a mix of streamlined business formats, and maximizes commercial value in addition to satisfying the quality of life. It has the quality of the garden community, the functionalization of the vertical layout, and creates a sense of community pride and social space. There are six characteristics of refinement, full-age community functions, and recreational community life. The facades of two high-rise residential towers deal with the atmosphere. The first floor of the podium is equipped with commercial along the street; the second floor is equipped with a community health service center and property management rooms; the third floor is equipped with a large roof garden.

深圳市地平匠造规划设计有限责任公司/深圳大学景观设计研究所
Shenzhen DPJZ Planning & Design Co., Ltd. / Institute of Landscape Architecture of Shenzhen University

"地平匠造"意为以"哲匠"的方式为当前人居环境塑造最好的地平线。深圳市地平匠造规划设计有限责任公司成立于2020年，是一所以高校为背景、学术研究为主导的工作室，擅长为项目带来文化价值及社会影响。主持规划设计师梁仕然博士同时负责深圳大学景观设计研究所日常事务。

"Di Ping jiang Zao" means to create the best horizon for the current human settlement environment in the way of "philosophical artisan". Founded in 2020, it is a university-based, academic research-led studio that brings cultural value and social impact to projects. Dr. Liang Shiran, the presiding planning designer, also responsible for the daily affairs of the Institute of Landscape Architecture of Shenzhen University.

地址：深南大道7002号财富广场A座6O-T	Add: 6O-T, Block A, Fortune Plaza, No.7002 Shennan Avenue
邮编：518000	P.C: 518000
电话：15999600508	Tel: 15999600508
邮箱：doctorlouis@163.com	Email: doctorlouis@163.com

高圳车革命传统教育基地建筑设计
Architectural Design of Gaozhenche Traditional Education Base

设计师：梁仕然
项目地点：广东 茂名
建筑面积：823 m²
用地面积：2 870 m²
容积率：0.29

Designer: Shiran Liang
Location: Maoming, Guangdong
Building Area: 823 m²
Site Area: 2 870 m²
Plot Ratio: 0.29

项目坐落于广东省茂名市电白区高圳车村红色革命遗址旁。方案探索了未来红色空间的内涵，尝试以此为契机带动当地乡村振兴，缓解当前乡村空心化问题。设计以"重建村落的精神家园"为目标，延续已消逝的传统礼制建筑的纪念性特点，为村中会议、节庆活动提供具有仪式感的场所。半开放的布局使场地成为村民重要的日常社交场所。寓教育于风景的理念与手法有别于常规红色教育基地。弹性功能的室内布局有利于避免常规教育基地的闲置浪费。

项目现在不但是村民的日常核心社交场所，而且是粤西地区的"网红打卡地"，是新兴文旅路线最重要节点。

The project is located next to the historic site in Gaozhenche Village, Dianbai District, Maoming City, Guangdong Province. The plan explores the connotation of the future patriotism space, and tries to use this as an opportunity to drive the revitalization of local villages and alleviate the current problem of rural hollowing. The design aims to "rebuild the spiritual home of the village", continuing the commemorative characteristics of the traditional ceremonial buildings that have disappeared, and providing a "sense of ceremony" background for meetings and festivals in the village. The semi-open layout makes the site an important daily social spot for the villagers. The concept and method of education in landscape are different from the conventional patriotism education base. The flexible indoor layout helps to avoid idle waste in the regular education base.

The project is now not only the daily main social place of the villagers, but also the most important node of the cultural tourism route in western Guangdong.

佛山市顺德建筑设计院股份有限公司
FOSHAN SHUNDE ARCHITECTURAL DESIGN INSTITUTE CO.,LTD.

佛山市顺德建筑设计院股份有限公司创立于1958年,于1993年率先在国内改制组建成有限责任公司,于2018年以发起设立的方式,将公司整体变更为股份有限公司,是国家建设部核定的建筑行业甲级设计院,并具备城乡规划编制乙级、市政行业(道路工程、给水工程、排水工程)专业乙级、建筑行业(人防工程)乙级、风景园林工程设计专项乙级、工程咨询单位乙级资信、佛山市房屋安全鉴定单位以及建设工程总承包资质,是全国高新技术企业、中国勘察设计协会的理事单位之一。2018年获评广东省全过程工程咨询第二批试点单位及佛山市全过程工程咨询第一批试点企业。

公司配套完善,技术力量雄厚,员工450多人,于清远、江西、四川、珠海、中山、广州、海南设有公司(分院及分支机构)。拥有国家级专家、中高级职称及注册建筑师、注册结构工程师、注册设备工程师、注册城市规划师、注册岩土工程师、注册消防工程师、注册造价工程师、注册咨询工程师等人才100多人,可承接国内外各类型工业与民用建筑设计、建筑装饰工程设计、建筑幕墙设计、轻型钢结构工程设计、建筑智能化系统设计、建筑机电设备安装设计、照明工程设计、消防设施工程设计、岩土设计、绿色建筑、海绵城市、可行性研究、交通评估等咨询相应范围的甲级专项工程设计业务,可从事城市规划、市政设计、景观园林绿化设计、人防设计、造价咨询、房室安全鉴定、检测等资质证书许可范围内相应的业务,以及建筑工程总承包、项目管理和相关的技术与管理服务。具备了可持续发展的条件及潜力,在本地区同行业中处于领先地位。作为顺德城乡巨大发展的见证者和重要参与者,在顺德体制改革中积极探索走进市场的新途径,更引导专业人士将专业知识融进市场中,获得个人与企业价值和社会意义的最大化,成为转制企业的典范。

近年,顺德建筑设计院业务稳健发展,3年来业绩持续增长,更坚持发挥自身长处,聚焦城市品质提升,着力于新一代产业项目升级和集约化、城市更新及民生与文化设施配套的完善。以专业技术为城市经济服务的同时,更以质朴而雅致的风格融进现代城市多方位发展中,通过建筑去为市民提供城市现代美学与生活思考空间,顺德建筑设计院还积极投身乡村振兴、古村落活化保育与古建筑修缮与利用工作,为现代乡村提供更为便利与充满传统气息的生活空间。

多年来,佛山市顺德建筑设计院股份有限公司全面贯彻国家及地方政府部门的建设方针,严格执行国家现行的技术标准与设计规范,加强技术管理,坚持质量第一,提供优质服务,勇于开拓创新,致力于新技术、新工艺、新设备、新材料的运用。为适应国际化的管理标准,于2008年获得ISO9001国际质量认证。佛山市顺德建筑设计院股份有限公司致力于全面提升整个设计每一个环节的服务质量,历年来打造了不少优秀的设计作品,在国家、省、市各设计评优中获奖无数,得到社会的好评、国家行业机构的重视和信任。

Foshan Shunde Architectural Design Institute Co., Ltd. was found in 1958. In 1993, Foshan Shunde Architectural Design Institute took the lead in restructuring to be a limited company. Foshan Shunde Architectural Design Institute Co., Ltd. is a first class architectural design institute which is checked and ratified by national development department and is also a director unit of Chinese reconnaissance design institute.

Foshan Shunde Architectural Design Institute Co., Ltd. now has 450 designers, which has included many types of construction design engineers, such as, national specialists, high-class engineers, registered architects, registered construction designers, registered equipments designers, registered urban planners, municipal engineers, landscape architects, and interior designers. Foshan Shunde Architectural Design Institute Co., Ltd. can take full charge of all types of industrial and civil building design, urban planning, municipal engineering, building equipments installment design, landscape design and interior design.

In recent years, our institute is experiencing a stable development, with a substantial improvement in the performance 3 years in a roll. This strengthens our confidence in upgrading the urban quality and the new generation of project. It also prompts us to improve the matching facilities of urban convergence, city renewal , life and culture. We believe that we can not only provide the expertise for the city's economic improvement but also blend into the city in a humble and elegant style. We aim to provide a modern aesthetic and thinking space for the residents through the art of architecture. Meanwhile ,we also actively participate in the renewing projects in countryside or ancient villages and the restoration projects of ancient buildings, providing a convenient life space in modern villages while maintaining their traditional architectural style.

For the past many years, Foshan Shunde Architectural Design Institute Co., Ltd. insisted on quality first, enhanced technical management, got into the swing of innovation, implemented TQC quality control, and obtained ISO9001 international quality attestation in 2008. Foshan Shunde Architectural Design Institute Co., Ltd. aim at being a hundred years' company, commit itself to upgrade servings quality of all parts of the whole design and create more excellent design work for society.

地址:广东省佛山市顺德区大良立田路建筑设计大厦　邮编:528300　　Add: Archi-design Bldg., Litian Road, Daliang, Shunde, Foshan, Guangdong　P.C:528300
电话:0757-22600168　传真:0757-22600338　　　　　　　　　　　　Tel: 0757-22600168　　　Fax: 0757-22600338
邮箱:sadi@21cn.com　网址:www.sdadi.com　　　　　　　　　　　　E-mail: sadi@21cn.com　　Web: www.sdadi.com

顺德勒流龙眼历史人物纪念馆
Museum of Historical Figures in Longyan Village, Leliu Town, Shunde District

项目地点：广东 佛山	Location: Foshan, Guangdong
建筑面积：原有488 m²，加固改造后508 m²	Building Area: The Original 488 m², After the Reinforcement 508 m²
用地面积：730 m²	Site Area: 730 m²

勒流龙眼村地处顺德中部，珠三角腹地，至今一直保留传统的龙舟文化与龙眼点睛等龙眼水乡文化最长盛不衰的特色仪式，同时村内历史名人辈出，有着以梁敦彦（中国第一批赴美留学幼童，官升清末外务部尚书，对中国电讯、铁路、兴办清华大学有重大贡献）为首的各个年代历史人物。

响应乡村振兴的指导思想，本项目通过加固改造村中闲置的几间小屋与相邻空地，一方面留出展示龙眼村各历史人物相关资料的展览空间，以供后人参观瞻仰；另一方面也提供了新的场所方便开展村中各种民俗文化活动。

设计上，历史人物纪念馆以加固翻新为基础，结合"拱"元素的加入，形成丰富的空间体验与中西合璧的外立面形象，充分体现龙眼历史人物融贯中西、兼收并蓄的精神。

通过取消中间建筑物形成的内庭院，与重新利用的二层露台形成对话以及对原有建筑空间的整合重构，形成具有岭南特色的建筑布局。前广场留出大量空地以保证日后活动开展的灵活性，同时新增的连廊空间也将成为村民或游客纳凉休憩的绝佳场所。

Longyan Village, Leliu Town is located in the center of Shunde District, the hinterland of the Pearl River Delta. Up to now, traditional dragon boat culture and other ceremonies have been retained, which is the most enduring feature of water town culture of Longyan Village. At the same time, the village has a large number of historical celebrities, headed by Liang Dunyan (one of the first Chinese children to study in the United States, was promoted to the Ministry of Foreign Affairs in the late Qing Dynasty, and made significant contributions to China's telecommunications, railway, the founding of Tsinghua University).

In response to the guiding ideology of rural revitalization, the project reinforces and renovates several idle cottages and adjacent open spaces in the village. On the one hand, exhibition space is set aside to display relevant materials of various historical figures in Longyan Village for future generations to visit and admire. On the other hand, it also provides a new place to carry out various folk cultural activities in the village.

In terms of design, Museum of Historical Figures is based on the reinforcement and renovation, combined with the addition of "arch" element, forming a rich spatial experience and a facade image with combination of Chinese and Western elements, fully reflecting the eclectic spirit of historical figures in Longyan Village who integrate Chinese and Western knowledge.

By abandoning the inner courtyard formed by the middle building and creating dialogue with the reused second-floor terrace, the design integrates and reconstructs the original building space to form the architectural layout with Lingnan characteristics. A large open space is left in the front square to ensure flexibility for future activities. Meanwhile, the newly added corridor space will also become a perfect place for villagers or visitors to enjoy a cool rest.

碧江牌坊
Bijiang Memorial Archway

项目地点：广东 佛山
建筑规模：宽33.9 m，总高度16.90 m，中间主门洞净宽17.38 m，高度净高7.5 m，为机动车车行道主要通行路口；两侧侧门洞净宽2.20 m，净高3.68 m，为非机动车及行人通行路口。
结构形式：钢筋混凝土框架结构

Location: Foshan, Guangdong
Building Site: 33.9 m in width, 16.90 m in total height, 17.38 m in clear width and 7.5 m in clear height of the middle main door opening, which is the main intersection of the motorway; The side door openings on both sides have a clear width of 2.20 m and a clear height of 3.68 m. They are non motor vehicle and pedestrian crossings.
Structural Form: Reinforced Concrete Frame Structure

牌坊，中华特色文化建筑之一，是古代社会为表彰功勋、科第、德政以及忠孝节义所立的建筑物。也有一些宫观寺庙以牌坊作为山门的，还有的是用来标明地名的。牌坊又名牌楼，为门洞式纪念性建筑物，宣扬礼教，标榜功德。牌坊也有昭示家族先人的高尚美德和丰功伟绩功能，兼有祭祖的功能。

碧江牌坊以钢筋混凝土框架为骨架，外表挂花岗石，主要为"黑青麻石""泉州红"两种岩石。瓦顶采用4层7顶瓦面的设置形式，高低错落，蔚为壮观。瓦面普遍采用皇家离宫别院常用的凹黄凸绿色半边琉璃龙华脊。瓦面之间采用荔枝核色木斗拱。黄瓦绿剪边，红斗拱，使得整个牌坊古朴之中又显出祥和高贵。

碧江牌坊正面中间的拱洞上书"碧江"2字。字幅两旁有龙、凤花板，契合"顺德凤城，碧江龙头"之说。牌坊背面中间的拱洞上书"国泰家宁"4个大字，寓意着对祖国繁荣昌盛、碧江人民安居乐业的美好愿景。

Memorial archway, one of Chinese characteristic architectural culture, is a building for the commendation of feats, excellent performance in the imperial examination, moral achievements, loyalty, filial piety and integrity in ancient society. There are also some temples with archway as the gate, and some are used to mark place names. Memorial archway, also known as memorial gate, is the gate-type monumental building, which is built for preaching ethics and flaunting merits. The archway also has the function of proclaiming the noble virtue and great deeds of the ancestors of the family and offering sacrifices to the ancestors.

The archway takes reinforced concrete frame as the skeleton, decorated with two kinds of granite , namely "Heiqingma stone" and "Quanzhou red". The tile top uses scattered four-layer tile surfaces, which is magnificent. Tile surface commonly uses royal palace courtyard concave yellow convex green side glass Longhua ridge. Brackets with lychee core color are used between the tile surfaces, making the whole archway peaceful and noble in the simplicity.

"Bijiang" is written on the arch hole in the middle of the front archway. There are dragons and phoenixes on both sides of the characters, which fits the saying of "Shunde phoenix city, Bijiang dragon head". "Guotai Jiaining" is written on the arch hole in the middle of the archway back, which means a beautiful vision for the prosperity of the motherland as well as the peace and happiness of Bijiangthe people.

逢简小学牌楼修复加固项目
Restoration and Reinforcement Project for Fengjian Primary School Archway

项目地点：广东 佛山　　　　Location: Foshan, Guangdong
建筑面积：168.71 m²　　　　Building Area: 168.71 m²
用地面积：599 m²　　　　　 Site Area: 599 m²
容积率：0.28　　　　　　　　Plot Ratio: 0.28

逢简小学的校园里，有一座人称"逢简三大巴"的门楼，这里曾经做过乡公所、大队部以及小学校舍，由于安全的原因，这座有80多年历史的老楼在多年前被拆了，只是这座地标式的门楼，大家还是舍不得拆，以"校友楼"的名义留了下来，2020年，当地政府组织校舍质量安全检测，这座门楼毫不意外地被列入危楼，要么拆除，要么加固。

随着顺德逢简古雅舒闲的岭南水乡环境声名远播，越来越多的各界人士纷纷前来探访、小住甚至安居落户，新老文化碰撞、传统与现代生活模式的融合，催生了对乡村公共空间的需求。于是，借助"安全校园""乡村振兴"等机遇，逢简小学这一门楼及后方的场地即将迎来"枯木逢春"般的生机。

On the campus of Fengjian Primary School, there is a gate, which used to be a township office, brigade department and primary school. Due to security reasons, this 80-year old building was demolished many years ago. But this landmark gate was reserved in the name of "alumni building" in spite of people's reluctance to tear it down. In 2020, The local government organized a safety inspection of the school buildings, and not surprisingly, the gate was listed as dangerous building, either to be demolished or reinforced.

With the reputation of Shunde's simple, elegant and leisurely Lingnan water town, more and more people from all over the world come to visit, stay and even settle down. The collision of new and old cultures and the integration of traditional and modern life patterns have created the demand for rural public space. Therefore, with the help of "safe campus", "rural revitalization" and other opportunities, the gate of Fengjian Primary School and the site behind it will soon embrace the vitality.

佛山市顺德北滘碧江美食聚集区
Bijiang Gourmet Area in Beijiao Town, Shunde District, Foshan City

项目地点：广东 佛山	Location: Foshan, Guangdong
建筑面积：22 535 m²	Building Area: 22 535 m²
用地面积：26 664 m²	Site Area: 26 664 m²
容积率：0.85	Plot Ratio: 0.85

碧江美食聚集区的规划建筑设计，是岭南传统园林建筑的传承与创新之作。

整体规划以中国传统文化"曲水流觞"为题，以水为媒，串联各个特色庭院。庭院由特色的水庭、石庭及公共水系组成，各庭院各具特色，相互借景渗透，形成别具地方特色的水乡环境。

建筑单体以岭南特色的民居为主，点睛处结合岭南园林建筑亭、台、水榭、船舫等元素，建筑又与假山水体融为一体，别具风情。

The planning and architectural design of Bijiang Gourmet Area is the inheritance and innovation of Lingnan traditional garden architecture.

The overall planning takes the Chinese traditional culture "floating wine cup along the winding water" as the title, and connects various characteristic courtyards with water as the medium. The courtyard is composed of characteristic water court, stone court and public water system. Each courtyard has its own characteristics and penetrates with each other to form a water village environment with unique local characteristics.

The single buildings are mainly dwellings with Lingnan characteristics, together with the combination of pavilion, platform, water pavilion, boat and other elements found in Lingnan garden building, making the building integrated with the landscape, which has distinctive flavor.

顺德实验中学提质扩容设计
Quality Improvement and Expansion Design for Shunde Experimental Middle School

项目地点：广东 佛山
建筑面积：原有45 623.4 m²，新建39 961 m²
用地面积：53 107.9 m²
容积率：1.48

Location: Foshan, Guangdong
Building Area: Original 45 623.4 m², New 39 961 m²
Site Area: 53 107.9 m²
Plot Ratio: 1.48

我们所处的时代，需要更具开放性与包容性的校园空间，茂密的植被、温暖的阳光，关爱和情感的交流都成为教育"容器"中不可或缺的内容。设计以"阳光校园"为题，希望艺术的阳光能温暖校园每个角落。

设计从整体校园出发，通过学习资源中心及立体交通系统的构想，补齐校园短板，形成融合共生的校园发展策略。设计把体艺人文融合处理，架空森林、立体连廊、空中跑道的设计，既是亚热带气候建筑的地域适应，又是校园多变灵活的自由空间的载体。

建筑不仅需要提供满足教育教学与日常生活需要的功能性空间，还需要容纳可以释放天性和肆意奔跑的公共空间，塑造丰富、灵活的校园空间。

We live in an era that requires more open and inclusive campus spaces, where dense vegetation, warm sunshine, care and emotional communication have become indispensable contents in the container of education. The design takes "Sunny Campus" as the title, hoping that the sunshine of art will warm every corner of the campus.

Through the concept of learning resource center and three-dimensional transportation system, the design starts from the whole campus to complement the shortcomings of the campus, and form an integrated and symbiotic campus development strategy. The design integrates physical, artistic and cultural processing. The design of overhead forest, three-dimensional corridor and air runway is not only the regional adaptation of the building in the subtropical climate, but also the carrier of the changeable and flexible free space of the campus.

Architecture not only needs to provide functional space to meet the needs of education, teaching and daily life, but also needs to accommodate the public space that can release nature and run freely, shaping a rich and flexible campus.

悍高星际总部
HIGOLD Headquarters

项目地点：广东 佛山
总建筑面积：35 927.74 m²
用地面积：3 066.15 m²
容积率：3.0

Location: Foshan, Guangdong
Total Building Area: 35 927.74 m²
Site Area: 3 066.15 m²
Plot Ratio: 3.0

办公楼位于厂区西北端，悍高厂房二区位于东南端，西北低，东南高，形成双塔形态，平面为"L"形构成，主要以内部空间的功能设置为主要构思基准，以兼顾2个塔楼的共用空间为连接体，自然形成功能区间的"桥梁"——连廊架空体。双塔及连接体的造型，在工业区内构画了优美的天际线，在低区营造通透的视觉形象，减弱双塔连接后形成的巨型体量。整个建筑综合体外部流畅的曲线与抽象几何形体有机结合，暗喻了企业的产品特征，极其精准地表达了企业的文化背景及美学理念。外立面主要采用铝合金玻璃幕墙，通过两种不同透光率的玻璃勾画出空间体量。GRC的装饰带由高塔最顶端勾画形体的交接面，并将这一元素延伸至架空裙房，直至低塔的顶端。无论昼夜与阴晴，玻璃幕墙为主体的塔楼总是反衬着周围的天光，无论晴空万里，还是风云变幻，建筑体也时时变化出生动的形象。尤其当夜色降临，综合楼宛如由一道长练牵系琢磨剔透的水晶，其华灼灼，耀眼璀璨。

The office building is located at the northwest end of the factory, while the second factory building is located at the southeast end, which is low in the northwest and high in the southeast, forming the form of two towers. The planet takes the "L" shape. The function of the internal space is set as the main idea base, and the common space of the two towers is taken into account as the connecting body, naturally forming a bridge connecting each functional area-- corridor overhead body. The shape of the twin towers and the connecting body creates a beautiful skyline in the industrial area, and creates a transparent visual image in the lower area, weakening the giant volume formed after the connection of the two towers. The smooth curve of the exterior of the whole building complex is organically combined with the abstract geometric form, which symbolizes the product characteristics of the enterprise, and extremely accurately expresses the cultural background and aesthetic concept of the enterprise. The facade is mainly made of aluminium alloy glass curtain wall, with two types of glass with different light transmittance delineating the volume of the space. The GRC decoration ribbon outlines the junction of the form from the very top of the tower and extends this element to the elevated podium to the top of the low tower. No matter day or night, cloudy or sunny, the glass curtain wall as the main body of the tower is always against the surrounding sky light. No matter the blue sky, or the wind changes, the building body always presents a vivid image. Especially when the night comes, the complex building is like a long line holding a refined dazzling crystal.

金籁科技（惠州）磁性元器件制造项目
Jinlai Technology (Huizhou) Magnetic Components Manufacturing Project

项目地点：广东 惠州	Location: Huizhou, Guangdong
建筑面积：536 723.25 m²	Building Area: 536 723.25 m²
用地面积：58 842.44 m²	Site Area: 58 842.44 m²
容积率：3.45	Plot Ratio: 3.45
绿化率：15%	Greening Rate: 15%

项目规划为"两区四组团"。整个项目分南北2个区，4个组团分别是制造组团、生产组团、科研办公组团和生活服务组团。科研办公区以"创新科研·智慧管理"作为设计理念，主要做科技展示、技术研发等功能。生活服务区以"以人为本·智能服务"为建设理念，涵盖高管公寓、职工公寓、职工食堂、健康活动中心及其他必要生活配套功能。

制造生产组团积极响应国家"碳中和"发展战略和要求，以"环保生产·低碳运行"作为设计理念，引入先进的能源管理体系，规划能源智能管理一体化平台，植入先进的"相变储能"系统；针对制造生产组团，在屋顶布局分布式光伏发电系统，建设分布式储能系统。在项目建成运营后，对办公用电、生产用电和生活服务用电进行全面的能源使用数据收集、能源储存、能源智能监控管理，降低能耗，实现低碳运行。

The project consists of "two districts and four groups", namely the north district and the south district, manufacturing group, scientific research office group and living service group. Scientific research office area takes "innovative scientific research · intelligent management" as the design concept positioning, mainly functioning as science and technology display as well as technology research and development. Based on the construction concept of "people-orientation · intelligent service", the living service area covers executive apartment, employee apartment, employee canteen, health activity center and other necessary living supporting functions.

The manufacturing group actively responds to the national "carbon neutral" development strategy and requirements, takes "environmentally friendly production · low-carbon operation" as the design concept, introduces the advanced energy management system, plans the integrated platform of energy intelligent management, and implants the advanced "phase change energy storage" system. For the manufacturing group, distributed photovoltaic power generation system and distributed energy storage system are laid out on the roof. After the project is completed and put into use, comprehensive energy use data collection, energy storage and intelligent monitoring and management will be carried out for office, production and life service electricity consumption, so as to reduce energy consumption and realize low-carbon operation.

云米互联科技园
VIOMI IoT Tech. Park

项目地点：广东 佛山　　Location: Foshan, Guangdong
建筑面积：90 000 m²　　Building Area: 90 000 m²
用地面积：36 000 m²　　Site Area: 36 000 m²
容积率：3.1　　Plot Ratio: 3.1

本项目坐落于大湾区腹地，坐西向东，创造性地将4栋楼连成一体化，外形上酷似一艘雄伟的航空母舰，寓意航母携江水滚滚东流之势，乘风破浪，驶向未来。云米科技园设计坚持以人为中心和以创造持续发展的工业区环境为宗旨，运用先进科技，引领现代工业厂区理念，通过精心规划、精心设计为项目打造一个规划合理、设计新颖、功能齐备、设施配套、环境优美的工业厂区。

云米电器科技有限公司作为全球领先的家庭物联网的高科技旗舰企业，新总部项目采用航母外形设计，其顶部VIP会议中心造型犹如航母的指挥塔，引领着企业向更大更宽广的未来航行。会议中心造型复杂且充满科幻感、未来感，贴合了云米高新科技企业的形象，也寓意着作为"新物种"的云米即将起航。

Located in the hinterland of the Guangdong-Hong Kong-Macao Greater Bay Area, the project creatively integrates four buildings from west to east. It looks like a majestic aircraft carrier in appearance, implying that the aircraft carrier rides the wind and waves to the future with the river running east. VIOMI Tech. Park adheres to the people-oriented design and aims to create sustainable development of industrial environment. The use of advanced technology leads the concept of modern industrial plant, creating an industrial plant with reasonable planning, novel design, complete functions, facilities, and beautiful environment through careful planning and design.

VIOMI Electrical Technology Co., Ltd. is the world's leading high-tech flagship enterprise of home Internet of Things. The new headquarters project adopts the shape design of "aircraft carrier", and its top VIP conference center is shaped like the control tower of aircraft carrier, leading the enterprise to sail to a bigger and broader future. The shape of the conference center is complex full of sense of science and future, which fits the image of VIOMI high-tech enterprise, and also means that VIOMI, as a "new species", is about to set sail.

中南高科·仲恺高端电子信息产业园
Zhongnan Hightech · Zhongkai High-end Electronic Information Industrial Park

项目地点：广东 惠州
建筑面积：246 295.61 m²
用地面积：73 446 m²
容积率：3.14

Location: Huizhou, Guangdong
Building Area: 246 295.61 m²
Site Area: 73 446 m²
Plot Ratio: 3.14

根据地块条件，合理布置内部功能。在平面中设有两大主要生产区和独立的展示区。空间关系清晰，线路简洁，体现了现代化工厂高效率的特征。

建筑设计形象大气又生动，颇具时代感亦有实用性。生产厂房的结构形式为钢筋混凝土结构。地块整体的外墙造型新颖，线条流畅，使建筑物具有特有的韵律，形象突出。利用落地玻璃窗两种颜色的配套使整栋建筑错落有致，成为现代厂房标志性建筑。厂区周边草地环绕，更是好像在绿荫环抱之中。园区主入口位置设置多层厂房，并在主道路两边多层厂房位置各设置一个配套配电房，位置隐蔽且实用，在便于维修的同时又考虑到了美观。地下室位置布置在地块的高层厂房位置，使厂区形成良好的纵深感觉。展示区位于地块主入口位置，展示区的景观园林造型独特美观，醒目大气。

According to the land conditions, the design makes a reasonable layout of internal functions. In the plan, there are two main production areas and a separate display area. The clear space relationship and the simple line reflects the high efficient characteristics of modern factory.

The image of the architectural design is magnificent and vivid, with a sense of the times and practicability. The structure of the production plant is reinforced concrete structure. The whole exterior wall of the plot is novel in shape and smooth in line, which makes the building have a unique rhythm and an outstanding image. The use of French windows in two colors makes the whole building well-arranged and become a strong symbol of the modern factory building. The plant is surrounded by grass, but also seems to be surrounded by green shade. Multi-storey plant is set at the main entrance to the park, and a supporting power distribution room is set at each location of the multi-storey plant on both sides of the main road. The location is hidden and practical, which is convenient for maintenance while giving aesthetic consideration. The basement is arranged in the position of the high-rise plant in the plot, so that the factory has a good sense of depth. The exhibition area is located at the main entrance to the plot. The landscape architecture of the exhibition area is unique, beautiful and eye-catching.

广西

华蓝设计(集团)有限公司
HUALAN DESIGN & CONSULTING GROUP

华蓝设计(集团)有限公司是华蓝集团(股票代码:301027)的全资子公司,广西大型综合性甲级工程设计咨询企业之一。2012年入选中国建筑学会评选的"当代中国建筑设计百家名院",2014—2021连续8年入选美国《工程纪录》(ENR)/《建筑时报》评选的"中国工程设计企业60强",2014—2019、2021累计7年入选"中国十大民营工程设计企业"。业务分布于国内25个省(区)市,海外业务延伸至非洲和东南亚国家地区,与美国、日本、加拿大、德国、法国和澳大利亚等国的设计机构长期保持业务合作与交流。

公司通过国家"高新技术企业""博士后科研工作站"和省级"企业技术中心""工程技术研究中心"的认定,是广西高新技术企业百强,广西"城乡规划与建筑设计人才小高地"和南宁市"海绵城市建设人才小高地"建设载体单位,并获批设立省、市两级"海绵城市院士工作站"(2016—2019)。

公司在工程咨询和设计服务中,注重研究城乡发展战略与公共政策,把握地域性、民族性和时代性特征,积极开展前沿技术研究与应用。五年来,累计获得规划、设计、咨询类重点奖项350余项。

Hualan Design & Consulting Group is a wholly-owned subsidiary of Hualan Group (stock code: 301027) and one of the large-scale comprehensive Grade A engineering design consulting enterprises in Guangxi. In 2012, it was selected as one of the "Top 100 Architectural Design Institutes in Contemporary China" by the China Architectural Society; in 2014-2021, it was selected as one of the "Top 60 Engineering Design Enterprises in China" by the American Engineering Record (ENR)/Architectural Times for eight consecutive years; and in 2014-2019 and 2021, it was selected as one of the "Top 10 Private Engineering Design Enterprises in China" for 7 years. Its business is distributed in 25 provinces (autonomous regions) and cities in China, and its overseas business extends to Africa and Southeast Asian countries. It maintains long-term business cooperation and exchanges with design institutions in the United States, Japan, Canada, Germany, France and Australia.

The company has been recognized as one of the top 100 high-tech enterprises in Guangxi, the "Small Highland of Urban and Rural Planning and Architectural Design Talents" in Guangxi and the "Small Highland of Marine and Mianyang City Construction Talents" in Nanning by the national "high-tech enterprises", "postdoctoral research workstation" and the provincial "enterprise technology center" and "engineering technology research center", and has been approved to establish a provincial and municipal "Sponge City Academician Workstation" (2016-2019).

In engineering consulting and design services, the company pays attention to the research of urban and rural development strategies and public policies, grasps the regional, national and contemporary characteristics, and actively carries out research and application of cutting-edge technologies. Over the past five years, it has won more than 350 key awards in planning, design and consulting.

防城港市堤路园(园博园)项目(园博园主展馆)
Fangchenggang Dilu Park (Garden Expo Park) Project (Main Exhibition Hall of Garden Expo Park)

项目地点:广西 防城港	Location: Fangchenggang, Guangxi
建筑面积:16 000 m²	Building Area: 160 00 m²
用地面积:219.1314 hm²	Site Area: 219.1314 hm²
占地面积:17 009.55 m²	Floor Area: 17 009.55 m²
容积率:0.007	Plot Ratio: 0.007
设计时间:2016年7月	Design Time: July 2016
竣工时间:2020年6月	Complete Time: June 2020

项目以"白鹭翱翔,防城腾飞"为主题。屋面设计以白鹭张开羽翼在海面翱翔为原型,结合亚热带地区建筑大挑檐、环廊的特点打造了环绕展馆四周的景观环廊,让160 m长的延展面显得轻盈、通透。出入口处设计了78 m长、30 m宽的无柱空间,结合前广场打造出全天候的开放式"市民大舞台"。

The theme of the project is "Egret flying, Fangcheng taking off". The roof design is based on the prototype of egrets flying on the sea with their wings open. Combined with the characteristics of large cornices and ring corridors in subtropical areas, a landscape ring Corridor around the exhibition hall is created, making the 160 m long extension surface light and transparent. A 8 m long, 30 m wide columnless space is designed at the entrance and exit, and an all-weather open "citizen stage" is created in combination with the front square.

广西大学大学生创新实验中心大楼
University Student Innovation Experiment Center Building of Guangxi University

项目地点：广西 南宁
总建筑面积：37 980 m²
项目获奖：2021年度广西优秀工程勘察设计成果建筑设计一等奖

Location: Nanning, Guangxi
Total Building Area: 37 980 m²
Project Award: The project won the First Prize of the 2021 Guangxi Excellent Engineering Survey and Design Achievement Award (Architectural Design)

作为一栋多学科、多专业、多功能的高层复合式教学综合体，各种密集的功能已经将空间挤占一空。设计试图在拥挤的教室群当中，创造适合放松、交流、多层次学习的有趣场所。"空中创廊"结合了学生创新、实践、教学的功能定位，通过南北向交通的连通、墙体的开合、视线的引导、空间的缩放来塑造多层次的交往场所，而不仅局限于连廊，屋面的星空走廊、穹顶剧场、三层的空中连廊、二层观景阶梯平台、一层楼间的活动场地，均是"空中创廊"定义的延伸。

As a multi-disciplinary, multi-professional, multi-functional multi-rise complex teaching complex, a variety of intensive functions have crowded out the space. Design tries to create interesting places for relaxation, communication and multi-level learning in crowded classroom groups. "Air and corridor" combines the function of students' innovation, practice, teaching function orientation, through the north-south traffic connectivity, wall opening, the line of sight, space zoom to shape multi-level communication place, not only limited to corridor, corridor of roof stars, dome theater, three layers of air corridor, the second viewing ladder platform, one floor activity venues, is the extension of the "air and corridor" definition.

南宁市"三街两巷"项目金狮巷银狮巷保护整治改造（二期）工程设计
Nanning "Three Streets and Two Alleys" Project Golden Lion Lane Silver Lion Lane Protection and Renovation (Phase II) Project Design

项目地点：广西 南宁	Location: Nanning, Guangxi
总建筑面积：71 872.94 m²	Total Building Area: 71 872.94 m²
设计时间：2020年3月	Design Time: March 2020

项目设计过程中，坚持保护历史文化街区文化脉络，加强文化传承，注重可操作性，合理确定留、改、拆的对策，对现有建筑进行分析、分类，分级明确保护与整治方式。

空间格局延续一期明清传统民居街巷形式，并在此基础上，对整个老南宁三街两巷街巷脉络进行疏通整合，优化街巷的流线，增加巷道的转折和收放，通过院落、天井和露台的植入，提升街巷的趣味性。

整治改造建筑尽量采取原材料、原工艺、新技术相结合。为向世人讲述从前城墙的故事，项目设计利用拆除旧房留下的青砖，重新混合新砖砌筑出仓西门城门楼和城墙，内部建成一座小型的城墙历史博物馆，向市民开放。通过修旧如旧、建新如旧和整体空间的协调，拟恢复传统空间格局，体现时代特征延续的建筑风貌。

During the design of the project, we insisted on protecting the cultural context of the historical and cultural district and strengthening cultural inheritance. Pay attention to operability, reasonably determine the countermeasures for retention, modification and demolition, and analyze, classify, and classify the existing buildings to clarify the protection and rectification methods.

The spatial pattern continues the traditional residential streets and alleys of the Ming and Qing dynasties in the first phase, and on the basis of this, the entire vein of the three streets and two alleys of the old Nanning is dredged and integrated, and the flow of the streets and alleys is optimized Increase the turning and retracting of the roadway, and enhance the fun of the street through the implantation of courtyards, patios and terraces.

The renovation of buildings should be combined with raw materials, original processes and new technologies as much as possible. In order to tell the story of the former city wall, the project design uses the green bricks left by the demolition of the old houses, and remixes the new bricks to build the gate tower and the city wall of the west gate of the warehouse, the interior A small history museum of the city wall was built and opened to the public. Through the coordination of repairing the old as the old and building the new as the old, and the overall space, it is proposed to restore the traditional spatial pattern and the architectural style that reflects the continuation of the characteristics of the times.

南宁书画院装修改造工程
Nanning Calligraphy and Painting Academy Decoration and Renovation Project

项目地点：广西 南宁
建筑面积：1 487.07 m²
用地面积：1 164.18 m²
设计时间：2020年9月－2020年12月
竣工时间：2021年07月

Location: Nanning, Guangxi
Building Area: 1 487.07 m²
Site Area: 1 164.18 m²
Design Time: September 2020 to December 2020
Completion Time: July 2021

项目改造延续传统岭南院落的内核,谨慎地拓展和改造建筑空间,进一步对院落关系重构与叠加,形成完善、灵活的展览动线,营造多元化的展览空间,构建绿色生态的整体环境。

在建筑材质上借鉴书画艺术中戏剧化冲突营造的表现方式,尝试通过新与旧、粗糙与细致之间的建筑表达,呈现出具有一定张力的"冲突"与"对话",使建筑的书画艺术主题具有一定的震撼力和感染力,让这座藏于绿林深处的建筑重新焕发活力。

The project continues the core of the traditional Lingnan "courtyard", carefully expands and transforms the building space, further constructs and superposition the relationship between the courtyard, forms a perfect and flexible exhibition line, creates a diversified exhibition space, and builds a green and ecological overall environment.

In terms of architectural material, we draw lessons from the expression mode of dramatic conflict construction in the art of painting and calligraphy, and try to present the "conflict" and "dialogue" with a certain tension through the architectural expression between the new and the old, rough and meticulous, so that the theme of architectural painting and calligraphy art has a certain shock and appeal.revitalize the building hidden deep in the forest.

五象总部大厦
Wuxiang Headquarters Building

项目地点：广西 南宁	Location: Nanning, Guangxi
竣工时间：2018年11月	Completion Time: November 2018
建筑面积：222 027.99 m²	Building Area: 222 027.99 m²
用地面积：18 044.43 m²	Site Area: 18 044.43 m²

设计理念：本设计注重将广西地域特色融入设计中。建筑形态结合当地"四水归堂"特色，采用群房及塔楼屋顶放坡的建筑形式，寓意企业不仅汇聚财富，更凝聚人心。建筑外立面将壮锦图案元素融入建筑的裙楼及塔楼顶部，体现壮乡风情及壮族人民对美好生活的追求与向往。

技术难点：本项目需在有限的用地上建造容积率高达7.92的建筑群，根据五象新区的城市风貌要求，分别在场地的3个角落建造3个塔楼。通过统一规划、设计、建设和经营，将五象总部大厦建成为一个集金融、办公、会议、培训等多功能于一体的超高层综合体群体，打造新的五象新区现代商务中心区。

技术创新：建筑和结构设计的高度统一，实现了室内无结构柱的全空间；通过对地质情况的合理分析和基础选型的对比，实现主、裙楼沉降满足规范要求，从而主、裙楼不需设结构缝的效果；1号超高层底部采用型钢混凝土柱，上部采用钢筋混凝土柱，充分利用高强度材料，节省混凝土和钢材用量。

Design concept: This design focuses on integrating the regional characteristics of Guangxi into the design. The architectural form combines the local characteristics of "four water return hall", and adopts the architectural form of group room and tower roof slope, which means that the enterprise not only gathers wealth, but also condenses the people. The facade of the building combines the zhuang elements, integrating the zhuang brocade pattern elements into the podium building and the top of the tower, reflecting the zhuang township customs and the Zhuang people's pursuit and yearning for a better life.

Technical difficulties: This project needs to build buildings with a plot ratio of 7.92 on a limited land. According to the urban style requirements of Wuxiang New Area, three towers will be built in three corners of the site respectively. Through unified planning, design, construction and operation, the Wuxiang Headquarters Building will be built into a super-high-rise complex group integrating finance, office, conference and training to create a new modern business center area of the Wuxiang New Area.

Technical innovation: the height of building and structural design, realize the full space of indoor structural column, through the reasonable analysis of geological conditions and foundation selection comparison, the settlement of the main and podium to meet the specification requirements, so that the effect of structural joints; 1 # super high-rise bottom of steel concrete column, the upper reinforced concrete column, make full use of high strength materials, save the amount of concrete and steel.

贵州省建筑设计研究院有限责任公司
Guizhou Architectural Design and Research Institute Co.,LTD.

贵州省建筑设计研究院有限责任公司,创建于1952年,前身为贵州省建筑设计研究院,是我国成立最早的大型建筑设计院之一,目前是贵州省国有资产监督管理委员会监管的大型科技型国有混合所有制企业。

公司拥有国家批准的市政行业专业甲级;建筑行业甲级;风景园林工程设计专项甲级;公路行业(公路)专业乙级;市政行业(环境卫生工程)专业乙级、商务粮行业(冷冻冷藏工程)专业乙级;电力行业(变电工程、送电工程)专业丙级;房屋建筑工程监理甲级、市政公用工程监理甲级、机电安装工程监理乙级;工程勘察综合甲级;工程勘察劳务类(工程钻探);工程造价咨询甲级;城乡规划编制资质证书甲级;土地规划丙级;地质灾害危险性评估甲级;地质灾害治理工程勘查、设计、施工甲级;建筑工程施工总承包壹级;建筑装修装饰工程专业承包壹级;地基与基础工程专业承包二级、特种专业工程专业承包、测绘乙级资质,是省内专业设置较全、取得资质较高的大型勘察设计咨询企业。

Guizhou Architectural Design and Research Institute Co.,LTD., founded in 1952, formerly known as Guizhou Architectural Design and Research Institute, is one of the earliest large-scale architectural design institutes in our country, at present, it is a large state-owned mixed-ownership enterprise of science and technology type supervised by Guizhou state-owned assets supervision and Administration Commission. The company has a state-approved municipal professional class A; construction industry class A; landscape engineering design special Class A; road industry (Road-RProfessionaloClasslBss b; Municipal Industry (environmental health engineering) professional class B, business grain industry (refrigeration engineering) Professional Class B; power industry (substation engineering, Power Transmission Engineering) Professional Class C; Building Construction Supervision Class A, municipal public works supervision Class A; electrical and mechanical installation engineering supervision Class B; Engineering Survey Comprehensive Class A; engineering survey services (engineering drilling); engineering cost consultation Class A; Urban and rural planning qualification certificate level A; land planning level C; geological hazard risk assessment level A; geological Hazard Management Project Survey, design, construction level A; construction project general contract level one; Construction decoration engineering professional contract grade one, foundation and basic engineering professional contract grade two, Special Professional Engineering professional contract, surveying and Mapping Grade B qualification, it is a large-scale survey and design consulting enterprise with complete professional setup and high qualification in the province.

地址:贵州省贵阳市林城西路28号　　Add: No.28, Lincheng West Road, Guiyang, Guizhou
邮编:550081　　P.C: 550081
电话:0851-85572433　　Tel: 0851-85572433
传真:0851-85572433　　Fax: 0851-85572433
邮箱:office@gadri.cn　　Email: office@gadri.cn
网址:www.gadri.cn　　Web: www.gadri.cn

茅台学院图书馆
Library of Moutai University

设计师:程鹏、王勤书、张媛、刘刚林、俞力、田俊、周思佳、孙明恩、李曦炜、王佳炜、毛晓月、黄刚、张龙、曾曦
项目地点:贵州 仁怀
建筑面积:24 639 m²
用地面积:71.7 hm²
容积率:0.48

Designers: Peng Cheng, Qinshu Wang, Yuan Zhang, Ganglin Liu, Li Yu, Jun Tian, Sijia Zhou, Ming'en Sun, Xiwei Li, Jiawei Wang, Xiaoyue Mao, Gang Huang, Long Zhang, Xi Zeng
Location: Renhuai, Guizhou
Building Area: 24 639 m²
Site Area: 71.7 hm²
Plot Ratio: 0.48

茅台学院是中国第一家龙头白酒企业创办的高校,茅台学院图书馆为学校科研教学提供服务,同时也体现了深厚的国酒企业文化。建筑造型新颖,与校园空间层次充分呼应,以"酒樽""酒鼎"礼器造型应和茅台文化。承袭汉唐古典建筑的沉稳大气,设计利用深灰色仿石铝板模仿赤水河峡谷机理,用高粱红窗棂的嵌入映衬了茅台酒文化的地缘特征及其起于汉唐的历史渊源。

As the first university run by a leading liquor enterprise in China, library of Moutai University provides services for scientific research and teaching, as well as for the profound national liquor enterprise culture. The architectural shape is novel, which fully echoes the spatial level of the campus. The shape of "wine bottle" and "wine tripod" ritual vessels echoes moutai culture. Inheriting the calm atmosphere of the han and Tang dynasties classical architecture, using dark gray imitation stone aluminum plate to imitate chishui River canyon mechanism, the embedded sorghum red window lattice reflects the geographical characteristics of Moutai wine culture and its historical origin from the Han and Tang dynasties.

贵州医科大学附属医院第三住院综合楼
Affiliated Hospital of Guizhou Medical University

设计师：张晋、赵军龙、杨兆林、姚茂启、申晨龙、俞力、李巍、孙平
项目地点：贵州 贵阳
建筑面积：84 663 m²
用地面积：3.54 hm²

Designers: Jin Zhang, Junlong Zhao, Zhaolin Yang, Maoqi Yao, Chenlong Shen, Li Yu, Wei Li, Ping Sun
Location: Guiyang, Guizhou
Building Area: 84 663 m²
Site Area: 3.54 hm²

本项目位于市中心三甲医院已有院区内，用地十分紧张，老旧建筑多，为不影响医院正常运营且能实现循序渐进式的建设方式，设计之初就确定了运用新理念、新系统、新技术并结合医院总体规划打造高效、节能、安全、人性化关怀医院的设计理念。

本项目拥有全省最多住院床位（1 000床）、最多手术间（60间）的手术中心和最大规模的综合ICU（40床）的单体医疗建筑，为解决医院的现状问题，实现高效、节能、安全、人性化关怀医院的设计理念，采用了众多国内领先、省内第一的医疗及医疗辅助系统。例如，轨道物流传输系统、垃圾被服收集系统、静脉输液配置中心系统、医用中央纯水系统和医用中心酸化水消毒系统等，以提升医院应急能力、提高医院现代化管理水平、提高整体运行效率，节约了资源，缩减了医院消耗物资的成本，取得了良好的社会和经济效益。

The project is located in the existing hospital area of the third class a hospital in the city center. The land is very tight, and there are many old buildings. In order not to affect the normal operation of the hospital and to achieve a step-by-step construction method. At the beginning of the design, the design concept of building an efficient, energy-saving, safe and humanized hospital with new concepts, systems and technologies combined with the overall planning of the hospital was determined.

The project has an operation center with the largest number of inpatient beds (1 000 beds), the largest number of operating rooms (60 beds) and the largest single medical building of comprehensive ICU (40 beds). In order to solve the current situation of the hospital and realize the design concept of efficient, energy-saving, safe and humanized care hospital, many domestic leading and provincial first medical and medical auxiliary systems are adopted. For example: rail logistics transmission system, garbage quilt collection system, intravenous infusion configuration center system, medical central pure water system, medical central acidified water disinfection system, etc., to improve the hospital's emergency response capacity, improve the hospital's modern management level, improve the overall operation efficiency, save resources, reduce the cost of hospital consumables, and achieve good social and economic benefits.

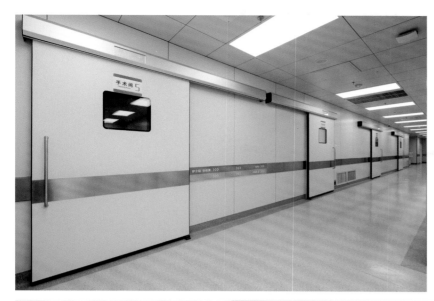

茅台集团贵定昌明玻璃瓶厂建设项目
Moutai Group Guiding Changming Glass Bottle Factory Construction Project

设计师：程鹏、任思屹、杨璞、李成波、龙桦、刘洁、谭晶、吕彦、刘刚林、胡汝君、叶梓、胡迪汉、田俊、姚耀南、严浩、黑晓亮、龙杰

项目地点：贵州 贵定

建筑面积：95 060 m²

用地面积：13.76 hm²

容积率：0.82

Designers: Peng Cheng, Siyi Ren, Pu Yang, Chengbo Li, Hua Long, Jie Liu, Jing Tan, Yan Lv, Ganglin Liu, Rujun Hu, Zi Ye, Dihan Hu, Jun Tian, Yaonan Yao, Hao Yan, Xiaoliang Hei, Jie Long

Location: Gui Ding, Gui Zhou

Building Area: 95 060 m²

Site Area: 13.76 hm²

Plot Ratio: 0.82

项目位于贵定县经开区，毗邻高速路口。总规划9座窑炉，年产3.2亿只玻璃瓶。规划以"藏风聚气、水抱山环"为主体格局，利用现状景观山体、河流水域为景观核心，结合东西向台地，集中布局生产厂房、仓储流线，利用高差，合理布置工艺生产设备，引入循环经济发展理念，构建依山傍水的整体生态工业园区。

The project is located in the Economic Development Zone of Guiding County, adjacent to the highway intersection. The project is planned to have 9 kilns with an annual output of 320 million glass bottles. The planning is based on the main pattern of wind collection, gas accumulation and water encircling mountains, using the existing landscape mountains, rivers and waters as the core of the landscape, combining the east-west terraces, centralizing the layout of production plants and warehousing streamlines, using the height difference, and rationally arranging process production equipment. Introduce the concept of circular economy development to build an overall eco-industrial park with mountains and rivers.

贵州茅台酒股份有限公司"十四五"酱香酒习水同民坝一期建设项目
Kweichow Moutai Liquor Co., Ltd. "14th Five Year Plan" Moutai Liquor Xishui Tongminba Phase I Construction Project

设计师：程鹏、任思屹、周大超、黄谦、白航宇、袁仁佶、袁中胜、周璐、周斌、张欢欢、王曦、杨亮亮、范玉林、曾承晶、张龙、倪鹏、黄世进

项目地点：贵州 习水
建筑面积：343 306 m²
用地面积：67.64 hm²
容积率：0.68

Designers: Peng Cheng, Siyi Ren, Dachao Zhou, Qian Huang, Hangyu Bai, Renji Yuan, Zhongsheng Yuan, Lu Zhou, Bin Zhou, Huanhuan Zhang, Xi Wang, Liangliang Yang, Yulin Fan, Chengjing Zeng, Long Zhang, Peng Ni, Shijin Huang

Location: Xishui, Guizhou
Building Area: 343 306 m²
Site Area: 67.64 hm²
Plot Ratio: 0.68

本项目位于贵州省遵义市习水县同民镇，同民河北侧，同民镇镇区西侧。拟建设34栋制酒厂房、1栋制曲厂房、9栋陶坛酒库、2栋酒罐库及相关配套用房。本次规划产能制酒约为1.2万吨/年，制曲约为2.9万吨/年，贮酒能力约为3.6万吨。

The project is located in Tongmin Town, Xishui County, Zunyi City, Guizhou, on the north side of Tongmin, and on the west side of Tongmin Town. It is planned to build 34 liquor making plants, 1 starter making plant, 9 pottery wine warehouses, 2 wine tank warehouses and related supporting buildings. The planned production capacity is about 12 000 tons/year for liquor making, 29 000 tons/year for koji making, and 36 000 tons for liquor storage。

中共贵州省委党校（贵州行政学院）
The Party School of the CPC Guizhou Provincial Committee (Guizhou Administrative College)

设计师：程鹏、李曦炜、朱琨、杨凡郅、曹杰	Designers: Peng Cheng, Xiwei Li, Kun Zhu, Fanzhi Yang, Jie Cao
项目地点：贵州 贵阳	Location: Guiyang, Guizhou
建筑面积：56 823 m²	Building Area: 56 823 m²

中共贵州省委党校（贵州行政学院）是贵州省培训、轮训中高级党政领导干部和国家公务员的最高学府。

设计体现干部学院治学严谨的本体特征，结合场地现有地貌及景观资源进行空间布局，打造绿色生态的公园式校园。

The Party School of the CPC Guizhou Provincial Committee (Guizhou Administrative College) is the highest institution in Guizhou Province that trains and rotates middle and senior party and government leaders and national civil servants.

The design reflects the noumenon characteristics of the Cadre College's rigorous scholarship, carries out spatial layout in combination with the existing landform and landscape resources of the site, and creates a green ecological park style campus.

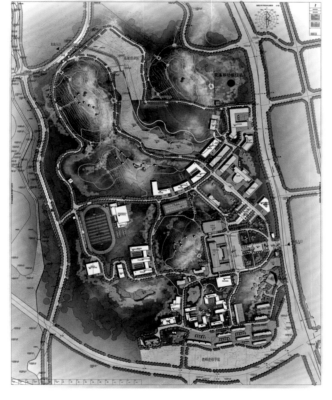

贵州

贵州省建筑科研设计院有限公司
Guizhou Architectural Research and Design Institute Co.,Ltd.

贵州省建筑科研设计院有限公司成立于1993年,前身为贵州省建筑科研设计院,2015年实现混合所有制改革后,隶属绿地控股—贵州建工集团,是建设部批准的建筑工程设计甲级资质单位,业务范围:建筑工程设计及建筑装饰工程设计、建筑幕墙工程设计、轻型钢结构设计、建筑智能化设计、照明工程设计、消防设施工程设计;市政行业道路工程设计、桥梁工程设计、给水工程设计、排水工程设计;风景园林专项设计;工程咨询;城乡规划编制;建设工程总承包、项目管理、技术与管理。

自公司成立以来,一直秉承"尊重、培育、发展、成就"人才理念,以项目管理团队为主体,立足当前,着眼长远,开展绿色建筑、BIM应用、建筑工业化等方向的课题研讨,以大格局、大经营、大合作的思维方式,找准角色定位,切实实施多元化经营战略,强调技术、管理、商务策划、资本运作为核心竞争力,以实现企业在发展上的再次突破。

历年来,通过与甲方的精诚合作,完成各类设计项目,主要有:碧桂园贵阳一号、贵安山语城、贵阳市花果园(花果园国际金融中心、小车河湿地公园)、中天未来方舟、德福中心、小石城二期、万科翡翠传奇、王振中美术馆、古城新韵-天工之城、惠水好花红千户布依寨、万象新都汇、溪山度假酒店、见龙洞路、中大国际广场、联合广场、花溪澳马荟溪城、小城故事、温泉馨苑、大正雨曦城、千户苗寨文化创意平台街、乌蒙文化风情园、荷塘月色、利尔上河城、新华路及南厂路景观整治等。此外企业还编制了"贵州省公安厅看守所""中国建设银行金库"标准图集,完成科研项目9项,其中《DZG-26改I型振动冲去灌注桩机制造及成桩工艺》获贵州省科学技术进步三等奖;《混凝土标养室调湿装置》获贵州省建设厅科技进步二等奖。

公司实施科学的现代化管理,注重新技术的推广和应用。精心设计、探索创新、争创精品,以特色的产品诚信服务于顾客和社会是公司始终不渝的宗旨。

Guizhou Architectural Research and Design Institute Co.,Ltd. was founded in 1993, Previously known as the Guizhou Province Architectural Design & Research Institute, achieved mixed ownership reformation in 2015, became the membership of Greenland Group - Guizhou Construction Engineering Group. The company owns the class-A qualification for engineering design by the Ministry of Construction. Our business including: Architectural Design, Building Decoration Engineering Design, Construction Curtain Wall Engineering Design, Light Steel Structure Design, Intelligent Building Design, Lighting design, Firefighting Device Engineering Design, Engineering Consulting (construction, municipal transportation, urban planning); Urban Planning, General Contracting, Project Management, Technology and Management of Construction Engineering.

The company has insisted its concept of "respect, cultivation, development and achievement" since founded. Taking the project management team as the main body, based on the current ground, long-term perspective, developed on the research of green building, BIM applications, architecture industrialization. According to the structure, operation and cooperation mode of thinking, identify the role on the market, implement the diversification strategy, emphasis on technology, management, business planning, take capital operation as the core competitiveness, to achieve a breakthrough in the development of the enterprise again.

In the past years, via the sincere cooperation with the party, the company has completed: Guiyang Garden 1, Gui'an hill city, Guiyang Huaguoyuan (flower orchard international financial center, car River Wetland Park), future transit ark, Zhongtian Century Park 9, 10 group, the two phase of the Little Rock Vanke Park legend the ancient city of the new day, rhyme City, Huishui haohuahong 1 000 Vientiane Xindu exchange, Buyi village, Longdong Road, new livable, see large international plaza, Huaxi garden, the story of a small town, hot spring rain, sunrise City Xinyuan, Taisho Wumeng Culture Park, lotus pond, Montreal city on the river. In addition, the company took part in the standard instruction edict of the "Guizhou Provincial Public Security Bureau detention center" and "Chinese construction bank vault" and completed 9 research projects, including "DZG-26 I vibration rushed to the pile and pile forming machine manufacturing technology" which awarded the third-prize of Guizhou provincial science and technology progress; "concrete standard curing room humidity control device" awarded the second prize of the Construction Department of Guizhou province science and technology progress.

The company carries out scientific and modern management, and emphasizes the popularization and application of new technologies. Our company's unswerving aim is to design, explore, innovate and strive for excellent products, and to serve customers and society with good products.

地址: 贵州省贵阳市云岩区春雷路67号
电话: 0851-85974608
传真: 0851-85974608
Email: gzgy_ga@163.com
网址: www.gzjky.cn

Add: No.67, Chunlei Road, Yunyan District, Guiyang, Guizhou
Tel: 0851-85974608
Fax: 0851-85974608
Email: Gzgy_ga@163.com
Web: www.gzjky.cn

开阳县冯三镇乡村振兴新华生态产业示范园
Rural Revitalization Xinhua Ecological Industry Demonstration Park in Fengsan Town, Kaiyang County

设计师：陶亮、冉芝军、冉芝航、杨小敏
项目地点：贵州 贵阳
建筑面积：12 156 m²
用地面积：33 km²

Designers: Liang Tao, Zhijun Ran, Zhihang Ran, Xiaomin Yang
Location: Guiyang, Guizhou
Building Area: 12 156 m²
Site Area: 33 km²

该项目总用地面积33 km²，依托冯三镇新华村原有1.67 km²猕猴桃产业基础，在该镇新华村和四坪村建设新华富硒生态产业示范园，计划在3年内新建5 km²猕猴桃种植基地，达到6.67 km²的猕猴桃种植基地富硒生态产业示范园。

The project covers a total land area of 33 km². Xinhua Selenium-enriched ecological industrial demonstration park will be built in Xinhua Village and Siping village of the town based on the original 1.67 km² kiwi industrial base in Xinhua Village, Fengsan Town. It is planned to build a kiwi planting base with an area of 5 km² in 3 years, and finally reach an area of 6.67 km².

雷山县大塘农业观光园
Agricultural Sightseeing Park in Datang Town, Leishan County

设计师：冉芝军、冉芝航、杨小敏	Designers: Zhijun Ran, Zhihang Ran, Xiaomin Yang
项目地点：贵州 凯里	Location: Kaili, Guizhou
用地面积：153 172 m²	Site Area: 153 172 m²

雷山县大塘农业观光园包括新桥村与咱刀村，是大塘传统村落保护和旅游开发的核心区域，新桥与咱刀村貌整治既是对村寨环境和基础设施整治改善，同时又依托旅游发展之推手，发展两村旅游设施，提升整体接待条件，为实现雷山全域旅游夯实基础。

咱刀村是长裙苗聚居地，寨内的四塘文化景观、河道景观、旱仓民族文化博物馆3个粮仓和祠堂景观，再加上新桥短裙苗是大塘旅游发展的民族文化旅游的核心资源，寨内苗族干栏式老建筑保护良好，特色鲜明，有较好的艺术价值和旅游价值。

咱刀风貌整治与区域旅游功能设计分为一期村貌整治方案设计和二期旅游设施提升概念方案，其中一期又分为房屋整治和环境整治两大部分，项目二期包括四塘文化景观及周边文化项目、多个旱仓景观、河道景观、旱仓民族文化博物馆、祠堂、生态运动公园等。

Agricultural Sightseeing Park in Datang Town, Leishan County, It includes Xinqiao Village and Zandao Village, which is the core area of traditional village protection and tourism development in Datang. The appearance renovation of Xinqiao Village and Zandao Village not only improves the village environment and infrastructure, but also relies on the promotion of tourism development to develop the tourism facilities of the two villages and improve the overall reception conditions to lay a solid foundation for the realization of all-region tourism of Leishan.

Zandao Village is an inhabitation of Changqunmiao, Sitang cultural landscape, river landscape, three granaries of Hancang National Culture Museum and ancestral temple landscape in the village, together with Duanqunmiao in Xinqiao, are the core resources of ethnic culture tourism in Datang tourism development. The old stilt houses in the village are well protected and they have distinctive features as well as good artistic value and tourism value.

Appearance renovation and regional tourism function design of Zandao Village is divided into the first phase of village appearance renovation scheme design and the second phase of tourism facilities enhancement concept scheme. The first phase is divided into housing renovation and environmental remediation, while the second phase of the project includes Sitang cultural landscape and surrounding cultural projects, a number of Hancang landscape, river landscape, Hancang national culture museum, ancestral temple, ecological sports park, etc.

石阡县楼上村特色田园乡村·乡村振兴集成示范试点
Integrated Demonstration and Pilot Projects for Rural Revitalization with Distinctive Rural Features in Loushang Village, Shiqian County

设计师：陶亮、左艳、冉芝军、冉芝航、杨小敏
项目地点：贵州 铜仁
用地面积：5.04 km²
民居改造民宿：7户，共2 700 m²

Designers: Liang Tao, Yan Zuo, Zhijun Ran, Zhihang Ran, Xiaomin Yang
Location: Tongren, Guizhou
Site Area: 5.04 km²
Home Stay Transformation: 7 Households, 2 700 m² in Total

楼上村,中国传统村落,位于贵州省铜仁市石阡县国荣乡,距县城15 km。楼上村历史悠久,文化底蕴深厚,是一个以汉族移民为主、周氏聚族而居的血缘村落。

在上位规划中,石阡县省级传统村落集聚区包括:楼上村、铺溪村、施场村等10个中国传统村落。为适应新时代的发展需要、更好地解决楼上村保护发展中的问题、有效保护楼上村丰富而珍贵的物质与非物质文化遗产、保持历史风貌、传承优秀文化,本案以已通过的上位规划为前提,对楼上村特色田园乡村规划实施情况进行评估并进行实施方案设计。

Loushang Village, a traditional Chinese village, is located in Guorong Township, Shiqian County, Tongren City, Guizhou Province. It is situated 15 km away from the county. With a long history and rich cultural heritage, Loushang Village is a lineage village predominantly inhabited by the Han ethnic immigrants and the Zhou family clan.

In the overall planning, the provincial-level cluster of traditional villages in Shiqian County includes ten traditional Chinese villages, such as Loushang Village, Puxi Village, Shichang Village, etc. In order to adapt to the development needs of the new era, address the issues in the protection and development of Loushang Village, effectively preserve its abundant and precious tangible and intangible cultural heritage, maintain its historical style, and inherit its excellent culture, this project takes the approved overall planning as a premise. It assesses the implementation of the characteristic rural planning of Loushang Village and designs an implementation plan.

贵阳永乐龙湖国家水利风景旅游城市综合体
Longhu National Water Scenic Tourism City Complex in Yongle Township, Guiyang City

设计师：冉芝军、冉芝航、杨小敏
项目地点：贵州 贵阳
用地面积：1 634.4 hm²

Designers: Zhijun Ran, Zhihang Ran, Xiaomin Yang
Location: Guiyang, Guizhou
Site Area: 1 634.4 hm²

该项目横跨南明区、双龙经济开发区、龙里县3块区域，贵阳绕城高速穿过景区境内，贵阳至龙里猫场镇二级油面公路自西往东横贯乡境中部，构成了与乡北部临近重镇东风镇、阿栗村和贵阳市区以及渔洞峡、情人谷等市级风景区的环线交通。以龙(龙洞堡)永(永乐)公路为干道，西有北京东路连接，同时有支线与各村及主要村寨和情人谷、江西坡桃园、阿栗杨梅园、永乐湖等相通。

The project extends across Nanming District, Shuanglong Economic Development Zone and Longli County. Guiyang City Highway runs through the scenic area, and the second-class oil road from Guiyang to Maochang Town, Longli County runs through the central part of the township from west to east, forming a circular traffic line with the northern part of the township, which is close to the key town Dongfeng Town, Ali Village, the urban area of Guiyang, as well as Yu Dong Gorge, Lover Valley and other municipal scenic spots. Taking Longdongba- Yongle Road as the main road, it is connected with Beijing East Road in the west, and has branch lines connecting with all villages and main stockaded villages as well as Lover Valley, Jiangxipo Peach Orchard, Ali Waxberry Orchard, Yongle Lake, etc.

高地天域
Gaodi Tianyu

设计师：杨鲲
项目地点：贵州 贵阳
建筑面积：767 488.10 m²
用地面积：185 003.25 m²
容积率：2.99

Designer: Kun Yang
Location: Guiyang, Guizhou
Building Area: 767 488.10 m²
Site Area: 185 003.25 m²
Plot Ratio: 2.99

  项目位于贵阳市云岩区，毗邻观山湖区，距离贵阳市中心大十字、喷水池约5 km。东西分别与甲秀北路、贵遵高速相邻。用地面积185 003.25 m²，其中一期用地面积78 377.54 m²，二期用地面积106 625.71 m²。用地西侧临贵遵高速公路，东侧临中坝北路，甲秀北路位于项目东侧。用地地形为山地，高低起伏，场地内最大高差达60 m。用地内地表植物茂密，该项目充分利用地形，尽可能保留原有地形地貌，将建筑布置在地势相对平缓区域，保留山顶的植被，面积达72 370 m²。其余小区绿化面积共达12 257.76 m²。整个公共绿地面积总共为84 627.76 m²。用地东部平缓地区布置住宅，东部沿街布置沿街商业。西部环山布置配套建筑（小学和幼儿园）。

  The project is located in Yunyan District, Guiyang City, adjacent to Guanshan Lake District, about 5 km away from the Grand Cross and fountain in Guiyang city center. It is adjacent to Jiaxiu North Road in the east and Guiyang-Zunyi Expressway in the west. The project covers a land area of 185 003.25 m², including 78 377.54 m² in the first phase and 106 625.71 m² in the second phase. On the west side of the land is Guiyang-Zunyi Expressway, on the east side is Zhongba North Road, and Jiaxiu North Road is located on the east side of the project. The terrain of the land is mountainous, with ups and downs, and the maximum height difference within the site is 60 m. The land is covered with thick vegetation. The project makes full use of the terrain to preserve the original landform as much as possible. The buildings are arranged in the relatively gentle terrain and the vegetation on the top of the mountain is preserved, covering an area of 72 370 m². The green area of the other residential areas is 12 257.76 m². The total public green space is 84 627.76 m². Residential buildings are arranged in the gentle area in the east of the land, and commercial buildings are arrange in the east along the street. Supporting buildings (primary schools and kindergartens) are arranged surrounding mountains.

安顺经开区领秀山水
Lingxiu Shanshui in Anshun Economic Development Zone

设计师：张洁
项目地点：贵州 安顺
建筑面积：12 330 m²
用地面积：4.99 hm²
容积率：2.5

Designer: Jie Zhang
Location: Anshun, Guizhou
Building Area: 12 330 m²
Site Area: 4.99 hm²
Plot Ratio: 2.5

项目位于安顺市西秀区以西，经济技术开发区杨湖片区，经一路与杨湖一路交汇处，周边3 km范围内分布医疗、教育、旅游、交通等功能配套设施，日常生活便利。西侧紧邻英雄水库，南面望山，览山乐水，享自然之美。

The project is located in Yanghu District, Economic and Technological Development Zone, west of Xixiu District, Anshun City, at the intersection of Jingyi Road and Yanghuyi Road. Within 3 km of the surrounding area, functional supporting facilities such as medical treatment, education, tourism and transportation are covered, making daily life convenient. The project is close to the Hero Reservoir on the west side, facing mountains on the south side, and you can enjoy the beauty of nature.

阳光城·未来悦(一期)A组团
Vision · Wonder (Phase I) Group A

设计师: 龙星翚、杨康	Designers: Xinghui Long, Kang Yang
项目地点: 贵州 遵义	Location: Zunyi, Guizhou
建筑面积: 466 025 m²	Building Area: 466 025 m²
用地面积: 175 913.31 m²	Site Area: 175 913.31 m²
容积率: 2.16	Plot Ratio: 2.16

项目位于贵州红色圣地——醉美遵义;南临贵阳、北倚重庆、西接四川。建筑依山而建,高低有致,形成一道亮丽的天际线,组团动静分区,功能配套齐全,人文娱乐应有尽有。总用地面积为175 913.31 m²,总建筑面积为466 025 m²,计容建筑面积为345 625 m²,不计容建筑面积为120 400 m²,容积率为2.16,基底面积为35 985 m²,建筑密度为22.53%,建筑高度为79.95 m,建筑层数为地上27层,地下3层。

The project is located in beautiful Zunyi, a Red Holy Land in Guizhou; It borders Guiyang in the south, Chongqing in the north and Sichuan in the west. The building is built along the mountain, forming a beautiful skyline. The group is divided into dynamic area and static area, with complete functional supporting facilities and all kinds of humanistic entertainment. It covers a total site area of 175 913.31 m², a total building area of 466 025 m², and a capacity building area of 345 625 m². The plot ratio is 2.16. The base area is 35 985 m², and the building density is 22.53%. The height of the building is 79.95 m, and there are 27 floors above ground and 3 floors under ground.

黎平天玺湾
Li Ping Cullinan Bay

设计师：李伟　　　　　　　　Designer: Wei Li
项目地点：贵州 黎平　　　　　Location: Liping, Guizhou
建筑面积：344 667 m²　　　　Building Area: 344 667 m²
用地面积：6.79 hm²　　　　　Site Area: 6.79 hm²
容积率：3.9　　　　　　　　　Plot Ratio: 3.9

黎平县依山傍水，项目依托西门河水资源顺势而为，融合现代社区绿色理念、文化韵味植入，打造亲水社区。通过建筑空间起伏和人车分流体系打造一个宜居宜游的全年龄生活社区，为当地居住品质提升创造全新保障。

Liping County is enclosed by the hills on one side and waters on the other. The project is based on the water resources of Ximen River and integrates the green concept and cultural charm of modern community to create a water-friendly community. Through the undulation of building space and the separation system of people and vehicles, an all-age living community suitable for living and traveling is created, creating a new guarantee for the improvement of local living quality.

顺海绿洲一期建设项目
Shunhai Oasis Phase I Construction Project

方案设计：杨渊、李礼	Scheme Design: Yuan Yang, Li Li
建筑：魏琰、龙星翚、杨康	Architecture: Yan Wei, Xinghui Long, Kang Yang
结构：贾文彦、韦威、张炎鑫、林菁、解东	Structure: Wenyan Jia, Wei Wei, Yanxin Zhang, Jing Lin, Dong Xie
设备：王和进、王军、侯元璐	Equipment: Hejin Wang, Jun Wang, Yuanlu Hou
项目地点：贵州 贵阳	Location: Guiyang, Guizhou
建筑面积：240 966.60 m²	Building Area: 240 966.60 m²
用地面积：5.3 hm²	Site Area: 5.3 hm²
容积率：3.5	Plot Ratio: 3.5

项目位于贵阳市乌当区土巴寨，用地性质为居住用地。以"人本、自然、文化、经济、融合、科技、安全"为中心原则，以整体社会效益、经济效益与环境效益三者统一为基准点，着意刻画优质生态环境，为居民塑造都市中自然优美、舒适便捷、卫生安全的怡然栖息之地。

建筑造型采用简洁明快的现代风格，力求创造一个富有时代感的标志性形象，为贵阳市的城市环境增添亮丽的风景。通过立面大块的虚实对比以及细腻的细部处理，充分展现建筑这一"凝固的音乐"无穷魅力的独特个性。力争创建一个富有鲜明个性特征及环境的住宅小区。

The project is located in Tuba Village, Wudang District, Guiyang City. The property of the land is residential land. With the central principle of "humanity, nature, culture, economy, integration, science and technology, safety", and the benchmark of integrating overall social benefits, economic benefits and environmental benefits, the project aims to depict the quality ecological environment, creating a natural, beautiful, comfortable, convenient, health and safety habitat for residents.

Architectural modeling adopts simple and bright modern style, striving to create an iconic image with rich sense of the times and add beautiful scenery for the urban environment in Guiyang City. Through the virtual-real contrast of the large facade and the fine detail processing, the unique charm of the architecture as "frozen music" is fully displayed. The project strives to create a residential district with distinct features and environment.

| 湖北 |

OVUD 中电光谷建筑设计院有限公司
CEC Optics Valley Architectural Design Institute Co., Ltd.

中电光谷建筑设计院有限公司(OVUD)系中电光谷(00798.HK)全资子公司,成立于2011年,拥有工程设计建筑行业(建筑工程)甲级资质、城乡规划乙级资质,景观、市政行业(给水、排水、道路、热力)工程专业乙级资质,环境工程(水污染防治工程)设计专项乙级资质。

公司集成中电光谷策划、开发、设计、运营全生命期经验,提供产业定位与咨询、概念规划、方案和施工图设计及各类专项设计、工程总承包(EPC)为一体的全流程设计咨询服务。以中电光谷产业资源共享平台为支撑,秉承"系统规划"和"综合运营"方法论体系的核心优势,通过规划设计为客户提供全生命周期、全价值链产业升级解决方案。

CEC Optics Valley Architectural Design Institute Co., Ltd. (OVUD) is a wholly-owned subsidiary of China Elect Optics (00798.HK). It was established in 2011 and has Grade-A qualification for engineering design in construction industry (construction engineering) and Grade-B qualification for urban and rural planning, Grade-B qualification for engineering design in landscape and municipal industry (water supply, drainage, road, heat), specialized Grade-B qualification for environmental engineering (water pollution prevention and control engineering), Grade-B qualification for geotechnical engineering (survey, design) in Surveying engineering.

The company integrates the full life cycle experience of CEC Optics Valley in planning, development, designing, and operation, and provides full-process design consulting services integrating industry positioning and consulting, conceptual planning, schematic and construction drawing design, various special designs, and Engineering, Procurement, and Construction (EPC) contracting. Supported by China Elect Optics industrial resource sharing platform, adhering to the core advantages of the "system planning" and "comprehensive operation" methodological system, it provides customers with full-life-cycle and full-value-chain industrial upgrading solutions through planning and design.

地址:武汉市洪山区野芷湖西路16号创意天地中电光谷建筑设计院大楼
邮编:430073
电话:027-86697739(人力资源)　027-87577715(经营)
邮箱:ovudwh@ovud.com
网址:www.ovud.com

Add: CEC Optics Valley Architectural Design Institute Co., Ltd., Wuhan Creative Capital, 16 Yezhihu West Road, Hongshan, Wuhan
P.C: 430073
Tel: 027-86697739(Human resources)　027-87577715(Operations)
Email: ovudwh@ovud.com
Web: www.ovud.com

合肥金融港中心
Hefei Financial Port Center

项目总负责:尹碧涛
项目策划:张琦
技术负责人:许可
施工图设计:高章喜、刘彪、周洁、雷明文、田继平、李青、许晓敏、刘升升、高会玲
BIM设计:白红、章梦馨、路博文、辛晓媛
项目地点:安徽 合肥
建筑面积:332 114.73 m²
用地面积:56 253 m²
容积率:4.18

Project Leader: Bitao Yin
Project Planning: Qi Zhang
Technical Leader: Ke Xu
Construction Drawing Design: Zhangxi Gao, Biao Liu, Jie Zhou, Mingwen Lei, Jiping Tian, Qing Li, Xiaomin Xu, Shengsheng Liu, Huiling Gao
BIM Design: Hong Bai, Mengxin Zhang, Bowen Lu, Xiaoyuan Xin
Location: Hefei, Anhui
Building Area: 332 114.73 m²
Site Area: 56 253 m²
Plot Ratio: 4.18

合肥金融港中心位于合肥滨湖新区徽州大道与南京路交口东北角。项目二期由5栋多层独栋办公、6栋高层办公楼、2栋高层酒店及1栋独立商业组成。建筑设计达到二星级绿色建筑设计标识。合肥金融港中心是合肥国际金融后台服务基地的重要组成部分,园区通过灵活的建筑布局,以及多个商务庭院叠合、错落、凹凸等营造一个绿色、健康的新式办公模式。

项目荣获2022年度湖北省(中小企业)勘察设计成果二等奖(公共建筑设计)。
项目荣获2022年度武汉市(中小企业)勘察设计成果一等奖(公共建筑设计)。
项目荣获2022第四届智建"SMART BIM"大赛(综合组)一等奖。
项目荣获2022"金标杯"BIM/CIM成熟度应用大赛三等成果(BIM运维成果组)。
项目荣获2022第五届"优路杯"全国BIM技术大赛优秀奖(综合组)。
项目荣获2022第三届"智建杯"智慧建造创新大奖赛铜奖(综合组)。
项目荣获2018年第四届国际BIM大奖赛"最佳运维BIM应用大奖"。
项目荣获2018年"'汉阳市政杯'武汉建筑业BIM技术应用视频大赛"铜奖。

Hefei Financial Port Center is located at the northeast corner of the intersection of Huizhou Avenue and Nanjing Road in Binhu New District, Hefei. The second phase of the project consists of 5 multi-storey office buildings, 6 high-rise office buildings, 2 high-rise hotels and 1 commercial building. The architectural design meets the two-star green building design mark. Hefei Financial Port Center is an important part of Hefei International Financial Back-office Service Base. The park creates a green and healthy new office model through flexible architectural layout and the overlapping, staggered and bumpy business courtyards. The project won the second prize (public building design) of the 2022 Hubei Province (Small and Medium-Sized Enterprises) Survey and Design Achievement Award.

The project won the second prize of the 2022 Hubei Province (Small and Medium-Sized Enterprise) Survey and Design Achievement Award (Public Building Design).
The project won the first prize of the 2022 Wuhan City (Small and Medium-Sized Enterprises) Survey and Design Achievement Award (Public Building Design).
The project won the First Prize of Fourth Smart Construction "SMART BIM" Competition 2022 (Comprehensive Group).
The project won the third prize in the 2022 "Gold Label Cup" BIM/CIM Maturity Application Competition (BIM Operation and Maintenance Achievement Group).
The project won the Excellence Award (Comprehensive Group) of the 5th "Youlu Cup" National BIM Technology Competition in 2022.
The project won the Bronze Award of the 3rd "Smart Construction Cup" Smart Construction Innovation Grand Prix in 2022 (Comprehensive Group).
The project won the "Best Operation and Maintenance BIM Application Award" in the 4th International BIM Awards in 2018.
The project won the Bronze Award of "Hanyang Municipal Cup Wuhan Construction Industry BIM Technology Application Video Competition".

黄石市科技创新中心
Huangshi Science and Technology Innovation Center

项目总负责：尹碧涛	Project Leader: Bitao Yin
项目策划：张琦	Project Planning: Qi Zhang
技术负责人：许可	Technical Leader: Ke Xu
方案设计：赵鹏、周洁、罗思、吴曾辉、王松、曾卓、陈巧情、施琛琛	Project Design: Peng Zhao, Jie Zhou, Si Luo, Zenghui Wu, Song Wang, Zhuo Zeng, Qiaoqing Chen, Chenchen Shi
施工图设计：高章喜、倪勇、刘彪、陈芳、李灵通、龙波曦、詹俊、廖华刚、田继平、但志鹏、张友良、胡亚斯、姜诗瑞、许晓敏、李青、李灿旺、张艳芳、刘升升、赵诗林、高会玲、邓仁洁	Construction Drawing Design: Zhangxi Gao, Yong Ni, Biao Liu, Fang Chen, Lingtong Li, Boxi Long, Jun Zhan, Huagang Liao, Jiping Tian, Zhipeng Dan, Youliang Zhang, Yasi Hu, Shirui Jiang, Fei Mo, Kai Zou, Xiaomin Xu, Qing Li, Canwang Li, Yanfang Zhang, Shengsheng Liu, Shilin Zhao, Huiling Gao, Renjie Deng
项目地点：湖北 黄石	Location: Huangshi, Hubei
建筑面积：191 438.31 m²	Building Area: 191 438.31 m²
用地面积：78 432 m²	Site Area: 78 432 m²
容积率：1.88	Plot Ratio: 1.88

黄石市科技创新中心位于黄石市黄金山开发新区景观主轴上，与奥体中心隔路相望。项目以"生长蜕变，链接未来"为主题，将生态连廊设计成植物经络自由生长的形态，并从展示中心延伸至各组团和各单体之间，演绎出丰富且尺度宜人的室外交流空间。建筑立面设计上也呼应了"生长"的理念，扬起的曲线勾勒出奋力向上生长的态势，塑造出极具未来感、科技感的建筑形象，象征着产业快速发展的美好寓意。基于可持续发展的原则，立面构造上局部使用了"可呼吸"式外通风双层幕墙：镀膜幕墙与光伏发电幕墙相结合的形式，幕墙上下分开设置进风口和出风口，实现空气腔之间的空气自由流通，带走多余热量，充分利用自然资源，节约能源。

项目荣获2021－2022第八届"CREDAWARD 地产设计大奖·中国"综合商办项目优秀奖。

项目一期荣获2022年IAI全球设计奖－建筑概念设计最佳设计大奖。

项目一期荣获2019年GBE建筑办公奖－最佳可持续办公建筑奖。

Huangshi Science and Technology Innovation Center is located on the landscape axis of Huang Jinshan New Industrial Region of the Huangshi Economic Development Zone, across the road from the Olympic Sports Center. With the theme of "growth and transformation, linking to the future", the project designs ecological corridors in the form of plants meridians growing freely, extending from the exhibition center to each group and each individual unit, deducing a rich and pleasant scale of outdoor communication space. The facade design of the building also echoes the concept of "growth". The rising curve outlines the trend of striving to grow upward, creating a futuristic and technologically-sounding architectural image, which symbolizes the beautiful meaning of the rapid development of the industry. Based on the principle of sustainable development, a "breathable" double-layer ventilation curtain wall is partially used in the facade structure, which is a combination of the coated curtain wall and the photovoltaic curtain wall. The air circulates freely between the rooms, takes away excess heat, makes full use of natural resources, and saves energy.

The project won the 8th CREDAWARD Real Estate Design Award China 2021-2022 Commercial & Office Project Excellence Award.

The first phase of the project won Best Design Award for Architectural Conceptual Design of IAI Global Design Award 2022.

The first phase of the project won Best Sustainable Office Building Award of GBE Building Office Award 2019.

中电光谷智造中心·武汉阳逻
CEC Optics Valley Intelligent Manufacturing Center (Yangluo, Wuhan)

项目总负责：尹碧涛	Project Leader: Bitao Yin
项目策划：张琦	Project Planning: Qi Zhang
技术负责人：许可	Technical Leader: Ke Xu
方案设计：赵鹏、周洁、王松、陈巧情、曾卓、吴曾辉、罗思、施琛琛	Project Design: Peng Zhao, Jie Zhou, Song Wang, Qiaoqing Chen, Zhuo Zeng, Zenghui Wu, Si Luo, Chenchen Shi
施工图设计：刘彪、高章喜、何涛、詹俊、田继平、许晓敏、李青、李灿旺、高会玲、高革、莫菲、邹凯	Construction Drawing Design: Biao Liu, Zhangxi Gao, Tao He, Jun Zhan, Jiping Tian, Xiaomin Xu, Qing Li, Canwang Li, Huiling Gao, Ge Gao, Fei Mo, Kai Zou
项目地点：湖北 武汉	Location: Wuhan, Hubei
建筑面积：122 189.43 m²	Building Area: 122 189.43 m²
用地面积：105 738.38 m²	Site Area: 105 738.38 m²
容积率：1.39	Plot Ratio: 1.39

"中电光谷智造中心·武汉阳逻"位于武汉市新洲区阳逻经济开发区五一大道以北，京东大道以西。项目的设计灵感来源于船帆、集装箱、灯塔等港口元素，并提取了船帆作为规划设计概念的主要元素。充分合理地利用场地进行功能与流线布局，在确保建筑使用功能合理、高效的前提下，创造一个良好的富有特色和标志性的建筑空间环境。以"和而不同、多元共生"为规划设计理念，提取历史文化元素，寓意产业示范区"扬帆起航"，产业转型升级可持续发展。功能类型囊括高层、多层研发型厂房，多层智能制造型生产厂房，单层智慧生产型厂房以及展示中心、公寓及食堂等服务配套。

项目荣获2022年IAI全球设计奖－建筑概念设计优秀奖。

项目中电光谷智造中心1.1期荣获2022－2023年度第一批湖北省建筑结构优质工程项目。

项目荣获2021－2022年度第三批湖北省建筑工程安全文明施工现场。

项目中电光谷智造中心1.1期（A1号楼）获评2022年"武汉市建筑结构优质工程"称号。

项目获评2022年武汉市建设工程安全文明施工示范项目（黄鹤杯）称号。

项目荣获2021年GBE办公建筑奖－年度最佳产业园奖。

CEC Optics Valley Intelligent Manufacturing Center (Yangluo, Wuhan) is located in the north of Wuyi Avenue and the west of Jingdong Avenue in Yangluo Economic Development Zone, Xinzhou District, Wuhan City. The design inspiration of the project comes from port elements such as sails, containers, lighthouses, etc. Sail is the main element of the planning and design concept. The design makes full and reasonable use of the site for function and traffic flow, and creates a good characteristic and iconic architectural space on the premise of ensuring the rational and efficient use of the building. Based on the planning and design concept of "harmony but difference, diversity and symbiosis", the historical and cultural elements are extracted, which means that the industrial demonstration zone "sets sail", and the industrial transformation, upgrading and sustainable development. The functional types include high-rise and multi-storey R&D plants, multi-storey intelligent manufacturing plants, single-storey intelligent manufacturing plants, as well as service facilities such as exhibition centers, apartments and canteens.

The project won the Architectural Conceptual Design Excellence Award of IAI Global Design Award 2022.

Phase 1.1 of CEC Optics Valley Intelligent Manufacturing Center won the first batch of High-Quality Construction Projects in Hubei Province from 2022 to 2023

The project won the third batch of Safe and Civilized Construction Sites of Construction Projects in Hubei Province from 2021 to 2022

Phase 1.1 (Building A1#) of CEC Optics Valley Intelligent Manufacturing Center in the project was awarded the title of "High-quality Building Structure Project in Wuhan" in 2022.

The project won the "Yellow Crane Cup Award" which is the highest honorary award for safe and civilized construction in Wuhan's construction industry in 2022.

The project won the Best Industrial Park Award of the Year of GBE Office Building Award 2021.

成都芯谷IC研发及产业基地项目二期
Chengdu Core Valley IC R&D and Industrial Base Project Phase II

项目总负责：尹碧涛
项目策划：张琦
技术负责人：许可
方案设计：赵鹏、周洁、王松、陈巧情、曾卓、吴曾辉、罗思、施琛琛、刘杰
施工图设计：陈芳、倪勇、杜森垚、杨艺、郝进、田继平、谭维、李青、刘升升、李灿旺、高会玲、刘阳
项目地点：四川 成都
建筑面积：431 494.73 m²
用地面积：192 848.25 m²
容积率：1.49

Project Leader: Bitao Yin
Project Planning: Qi Zhang
Technical Leader: Ke Xu
Project Design: Peng Zhao, Jie Zhou, Song Wang, Qiaoqing Chen, Zhuo Zeng, Zenghui Wu, Si Luo, Chenchen Shi, Jie Liu
Construction Drawing Design: Fang Chen, Yong Ni, Senyao Du, Yi Yang, Jin Hao, Jiping Tian, Wei Tan, Qing Li, Shengsheng Liu, Canwang Li, Huiling Gao, Yang Liu
Location: Chengdu, Sichuan
Building Area: 431 494.73 m²
Site Area: 192 848.25 m²
Plot Ratio: 1.49

项目西临双楠大道,东望双流国际机场,绝佳的地理位置及生态条件为建设新发展理念的公园城市示范区提供了先天条件。

设计从成都的山川地貌中汲取灵感,提出"生态都芯、海纳百川"的规划设计理念,利用"川"结构将4块用地无缝相连打造多样化社区。连续的绿地景观和地景建筑贯穿南北,在中央构建出犹如川蜀江水般灵动蜿蜒的公共活力轴,形成地块间的连接。

外部分区以现有城市道路的格局为基本骨架,在内部形成主要以车行为主的交通体系与活力景观轴相连接。高挑的办公塔楼置于地块西侧,多层办公组团围绕活力轴分布,商业空间置于首层,在确保景观资源最大化的同时,营造出立体的多维山谷布局。充分利用地块的自然资源与区位优势,打造了一个具有川蜀特色、兼具开放、合作、包容且有温度的人才聚集地,将绿色生态理念完美融入产业园区,构建科技、开放的高品质园区。

景观设计在项目规划布局基础之上,延续了"海纳百川"的整体概念。设计注重生态绿化及公共空间的打造,在贯穿地块的中央轴线上,营造多个景观节点,促进片区内的社交互动。景观设计与办公空间的屋顶及公共退台一同形成了从地面向上延伸的立体多维绿化景观。

项目荣获GBE办公建筑大奖2022年度最佳企业总部奖(建筑设计)。

The project is adjacent to Shuangnan Avenue in the west and Shuangliu International Airport in the east. The excellent geographical location and ecological conditions provide innate conditions for the construction of a park city demonstration area with a new development concept.

The design draws inspiration from the landscape of mountains and rivers in Chengdu, and proposes the planning and design concept of "ecological core, the sea is inclusive of all rivers", and the structure like Chinese character "川(chuan)" seamlessly connects the four sites to create a diverse community. The continuous green landscape and landscape buildings run through the north and south, and a public vitality axis is constructed in the center like the water of the Sichuan-Shu River, forming the connection between the plots.

The external partition is based on the existing urban road pattern as the basic skeleton, and the internal formation is mainly connected with the vehicle traffic system and the dynamic landscape axis. The tall office tower is placed on the west side of the plot, the multi-storey office groups are distributed around the vitality axis, and the commercial space is placed on the ground floor. While ensuring the maximization of landscape resources, a multi-dimensional valley layout is created to make full use of the nature of the site. Taking the advantages of resource and location, the design creates a talent gathering place with Sichuan characteristic which is open, cooperative, inclusive and warm, and perfectly integrates the concept of green ecology into the industrial park, and build an open, high-tech and high-quality park.

The landscape design is based on the project planning and layout, and follows the overall concept of "the sea is inclusive of hundreds of rivers". The design focuses on the creation of ecological greening and public space. On the central axis running through the site, multiple landscape nodes are created to promote social interaction in the area. Together with the roof of the office space and the public terrace, it forms a multi-dimensional green landscape extending upward from the ground.

The project won the 2022 Best Corporate Headquarters Award (Architectural Design) of the GBE Office Building Awards.

天津欧微优科创园(中电科创园)
Tianjin OVU Technology Innovation Park

项目总负责：尹碧涛
项目策划：张琦
技术负责人：许可
施工图设计：高章喜、倪勇、刘彪、陈芳、李灵通、杜森垚、
于群群、詹俊、但志鹏、张友良、田继平、杨哲、
樊浒、曹秀娟、许晓敏、李青、李灿旺、刘升升、
张艳芳、赵诗林、高会玲、高革、邓仁洁、罗曼、
莫菲、邹凯、刘阳
BIM设计：白红、章梦馨、路博文、辛晓媛
项目地点：天津
建筑面积：160 372.07 m²
用地面积：44 781.9 m²
容积率：2.5

Project Leader: Bitao Yin
Project Planning: Qi Zhang
Technical Leader: Ke Xu
Construction Drawing Design: Zhangxi Gao, Yong Ni, Biao Liu, Fang Chen, Lingtong Li, Senyao Du, Qunqun Yu, Jun Zhan, Zhipeng Dan, Youliang Zhang, Jiping Tian, Zhe Yang, Hu Fan, Xiujuan Cao, Xiaomin Xu, Qing Li, Canwang Li, Shengsheng Liu, Yanfang Zhang, Shilin Zhao, Huiling Gao, Ge Gao, Renjie Deng, Man Luo, Fei Mo, Kai Zou, Yang Liu
BIM Design: Hong Bai, Mengxin Zhang, Bowen Lu, Xiaoyuan Xin
Location: Tianjin
Building Area: 160 372.07 m²
Site Area: 44 781.9 m²
Plot Ratio: 2.5

项目位于天津市中心城区,处于南开区与西青区交界处。隶属于天津高新区华苑科技园。项目东临简阳快速路,西临陈台子河,南临迎水道,用地面积为44 718.9 m²,地上建筑面积112 000 m²,涵盖研发办公、展示中心及其配套酒店式长租公寓。本项目为智能产业集群的重要一环,结合四度设计原则,提出"光谷绿洲"设计理念, 以极具未来感的空间语言整合场地,打造具有智能感、生态感、体验感的全新美学作品。

项目荣获2022第二届"新基建杯"中国智能建造及BIM应用大赛－智能建造优秀BIM设计案例赛组三等奖。

项目荣获2022武汉国际创意设计大赛－年度工程设计。

项目荣获2022第五届"优路杯"全国BIM技术大赛－设计组铜奖。

项目荣获2022第三届"智建杯"智慧建造创新大奖赛－设计组金奖。

项目荣获2022湖北省第十届建筑信息模型(BIM)设计竞赛－民用建筑二等奖。

项目荣获2022年"雷霆三实杯"武汉建筑业BIM+数字化应用成果大赛三类成果。

项目荣获2022"金标杯"BIM/CIM成熟度应用大赛－BIM设计成果组二等成果。

项目荣获2022第四届智建"SMART BIM"大赛－设计组一等奖。

项目荣获2022第十一届"龙图杯"全国BIM大赛－设计组二等奖。

项目荣获2022湖北省建设工程BIM大赛－单项应用二类成果。

项目荣获2022第三届工程建设行业BIM大赛－建筑工程综合应用类二等成果。

The project is located in the central urban area of Tianjin, at the junction of Nankai District and Xiqing District. It is affiliated to Tianjin High-tech Zone Huayuan Technology Park. The project is adjacent to Jianyang Expressway in the east, Chentaizi River in the west, and Yingshui Road in the south, with a site area of about 44 718.9 m² and a construction area of about 112 000 m². This project is an important part of the intelligent industry cluster. Combining the four-degree design principle, the design concept of "Optical Valley Oasis" is proposed to integrate the site with a very futuristic spatial language to create a new aesthetic work with a sense of intelligence, ecology and experience.

The project won the third prize of the 2022 Second "New Infrastructure Cup" China Intelligent Construction and BIM Application Competition - Intelligent Construction Excellent BIM Design Case Competition Group.

The project won the 2022 Wuhan International Creative Design Competition - Annual Engineering Design.

The project won the 2022 5th "Youlu Cup" National BIM Technology Competition - Design Group - Bronze Award.

The project won the 2022 Third "Smart Construction Cup" Smart Construction Innovation Grand Prix-Design Group-Gold Award.

The project won the second prize of the 10th Hubei Provincial Building Information Modeling (BIM) Design Competition - Civil Architecture - in 2022.

The project won the three categories of achievements in the 2022 "Thunder Three Real Cup" Wuhan Construction Industry BIM+ Digital Application Achievement Competition.

The project won the 2022 "Gold Label Cup" BIM/CIM Maturity Application Competition - BIM Design Achievement Group - Second Prize.

The project won the first prize of the 4th Smart Construction "SMART BIM" Competition (Design Group) in 2022.

The project won the Second Prize of Design Group in the 11th "Longtu Cup" National BIM Competition, the Single Application Category-2 Results in Hubei Construction Engineering BIM Competition.

The project won the Second-Class Results of Comprehensive Application of Construction Engineering in the 3rd Engineering Construction Industry BIM Competition.

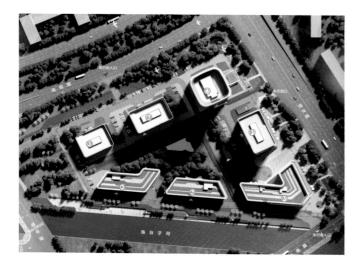

长沙中电软件园二期扩大项目
Changsha CEC Software Park Phase II Expansion Project

项目总负责：尹碧涛
项目策划：张琦
技术负责人：许可
施工图设计：严杰、高章喜、汪莉莉、于群群、何涛、陈瑶、詹俊、
张友良、田继平、但志鹏、胡亚斯、姜诗瑞、曹秀娟、
张鹏、莫菲、邹凯、李青、李灿旺、张艳芳、刘升升、
许晓敏、高会玲、高革、邓仁洁
项目地点：湖南 长沙
建筑面积：145 074.09 m²
用地面积：39 999.51 m²
容积率：2.49

Project Leader: Bitao Yin
Project Planning: Qi Zhang
Technical Leader: Ke Xu
Construction Drawing Design: Jie Yan, Zhangxi Gao, Lili Wang, Qunqun Yu, Tao He, Yao Chen, Jun Zhan, Youliang Zhang, Jiping Tian, Zhipeng Dan, Yasi Hu, Shirui Jiang, Xiujuan Cao, Peng Zhang, Fei Mo, Kai Zou, Qing Li, Canwang Li, Yanfang Zhang, Shengsheng Liu, Xiaomin Xu, Huiling Gao, Ge Gao, Renjie Deng
Location: Changsha, Hunan
Building Area: 145 074.09 m²
Site Area: 39 999.51 m²
Plot Ratio: 2.49

长沙中电软件园是湖南省政府与中国电子信息产业集团有限公司战略合作共建的国家级软件产业基地项目，总规划1 km²，分三期开发建设。其中，长沙中电软件园二期是在一期园区良好产业基础上打造的产业地标4.0升级版。二期项目位于岳麓大道以北，尖山路以东，青山路以南，园区整体规划0.27 km²，规划建筑面积850 000 m²，总投资35亿元。先导区占地面积0.13 km²，规划建筑面积约460 000 m²。项目将按照"产城融合、军民两用、科技艺术融合"的发展理念，以信息安全、军民两用产业为核心，重点布局移动互联网、智能制造、北斗应用、大数据等前沿科技领域，围绕产业价值链布局产业协同创新链。建成后将大量引入软件及电子信息，智能制造、军民两用等方向的成长性企业。

Changsha CEC Software Park is a national-level software industry base project jointly built by the Hunan Provincial Government and China Electronics Corporation through strategic cooperation. Among them, Changsha CEC Software Park Phase II is an upgraded version of Industrial Landmark 4.0 built on the basis of phase I of the park. The phase II of the project is located to the north of Yuelu Avenue, east of Jianshan Road, and south of Qingshan Road. The site area is 0.13 km², with a planned construction area of about 460 000 m². The project will follow the development concept of "industry-city integration, military-civilian dual use, technology and art integration", with information security, military-civilian dual-use industries as the core, and focus on mobile Internet, intelligent manufacturing, Beidou applications, big data and other cutting-edge technology fields, deploy the industrial collaborative innovation chain around the industrial value chain. After completion, a large number of software and electronic information, intelligent manufacturing, military-civilian dual-use and other growing enterprises will be introduced.

河北

北方工程设计研究院有限公司隶属于中国兵器工业集团有限公司,是创建于1952年的国家级综合勘察设计机构,当代中国建筑设计百家名院,总部员工1 300余人。具有国家授予的军工、城乡规划、建筑、市政等多个行业甲级咨询、设计资质;风景园林设计、建筑智能化系统设计等多项专业甲级资质;工程勘察综合甲级资质;建筑智能化工程、电子工程等专业承包壹级资质;现有北京、上海、深圳、重庆等22家分支机构。公司在国防科技工业、产业园区和校园规划设计以及科教、医疗、公共设施、居住、城市综合体、物流、园林景观等方面取得了骄人业绩;在工程勘察与岩土工程、电磁屏蔽、环境工程、生物质发电等专业技术领域树立了品牌。

Norendar International Ltd. is affiliated to China North Industries Group Corporation Limited. It is a national comprehensive survey and design institution founded in 1952. It is one of the top 100 famous architectural design institutes in contemporary China and has more than 1 300 employees in its headquarters. It has a number of professional class A qualifications in military industry, urban and rural planning, architecture, municipal and other industries awarded by the state for consulting, design, landscape architecture design, building intelligent system design and other professional class A qualifications; comprehensive class A qualifications for engineering surveys; building intelligence First-class qualification for professional contracting such as engineering and electronic engineering; there are 22 branches in Beijing, Shanghai, Shenzhen, Chongqing, etc. The company has made remarkable achievements in national defense science and technology industry, industrial park and campus planning and design, science and education, medical care, public facilities, residence, urban complex, logistics, garden landscape, etc.; in engineering survey and geotechnical engineering, electromagnetic shielding, environmental engineering , biomass power generation and other professional and technical fields have established a brand.

地址:河北省石家庄市裕华东路55号	邮编:050011	Add: No.55 Yuhua East road, Shijiazhuang, Hebei	P.C: 050011
电话:0311-86690762　传真:0311-86033237	邮箱:gmlbfy@126.com	Tel: 0311-86690762　Fax: 0311-86033237	Email: gmlbfy@126.com

中国环境管理干部学院新校区
New Campus of Environmental Management College of China

主创设计师:高明磊、白小龙、薛雪
项目地点:河北 秦皇岛
规划总建筑面积:250 000 m²;一期建成150 000 m²
总用地面积:48.35 hm²
建设用地:34.5 hm²

Chief Designer: Minglei Gao, Xiaolong Bai, Xue Xue
Location: Qinhuangdao, Hebei
The Planned Total Building Area: 250 000 m²;
The First Phase Will Be Completed With 150 000 m²
Total Area: 48.35 hm²
Site Area: 34.5 hm²

中国环境管理干部学院是全国唯一一所以环保命名的高等院校。项目用地中间蜿蜒穿过的河流将用地分为南北两区,设计利用这一特点,引入了"清洁地球"这一充满环保使命感的设计理念,将河两岸的核心建筑通过"多层桥"连成一体,呈圆形,隐喻为地球,静静的河水从中淌过,提醒人们地球需要呵护,表达出环境保护的内涵。

设计融入"环保三色"。蓝色:污废水处理和雨水收集系统形成的水源颜色;绿色:校园人工湿地、绿色建筑;黄色:太阳能利用、地源热泵、节水、废物回收的能量色。将河两侧防护绿地设计为人工湿地,污水经过湿地净化,达到排放标准,流入河内。中水处理站设在地下,湿地和中水处理站兼做相关专业学生的实习场地。校区污水达到零排放。

Environmental Management College of China is the only higher education institution in China named after environmental protection. The meandering river in the middle of the project divides the land into the north area and the south area. The design takes advantage of this feature and introduces the design concept of "clean Earth" full of environmental protection mission. The core buildings on both sides of the river are connected in a circular shape through a "multi-storey bridge", which is a metaphor for the earth. The quiet river flows through it, reminding people of the need to care for the earth and expressing the connotation of environmental protection.

"Three colors of environmental protection" are integrated into the design. Blue: the color of water source formed by sewage treatment and rainwater collection system; Green: constructed wetland and green buildings on the campus; Yellow: energy color for solar energy utilization, ground source heat pump, water saving and waste recovery. The protective green space on both sides of the river is designed as constructed wetland, through which the sewage is purified to reach the discharge standard and flows into the river. The intermediate water treatment station is located underground, and the wetland and intermediate water treatment station are also practice sites for students of related majors. The campus has reached zero sewage discharge standard.

核心区设计强调"环境融合",将多个面积不大的零散公用建筑通过桥连接成自由舒展的曲线,形成校园的核心建筑群,与湿地、河流、水生态实习场地充分融合。入口台阶与透空的大平台拉远了校区的景深,面向广场兼做文艺演出,为师生提供了交流场所,高高耸立的文峰塔是校园的制高点;阅览室、办公室、教室等公共用房面向水面打开,视野开阔;师生可以在河面上的连廊抱书阅览、沐浴清风,达到人与环境的充分融合。

The design of the core area emphasizes the "integration of the environment". Multiple scattered public buildings with small areas are connected with each other by bridges to form a freely-stretching curve, forming the core building group of the campus, which are fully integrated with the practice sites of wetland, river and water ecology. The steps at the entrance and the large permeable platform extend the landscape depth of the campus. Facing the square, they also serve as a cultural performance, providing a place for teachers and students to communicate. The towering Wenfeng Tower is the commanding height of the campus. Public rooms, such as reading rooms, offices and classrooms, are open to the water with a wide view; Teachers and students can read books and bathe in the breeze in the corridor on the river, achieving the full integration of human and environment.

官田兵工小镇
Guantian Armoury Town

设计师：高明磊、王晶玥、王茜	Designers: Minglei Gao, Jingyue Wang, Qian Wang
项目地点：江西 赣州	Location: Ganzhou, Jiangxi
新建及改造建筑面积：20 000 m²	Total Area of New and Renovated Buildings: 20 000 m²
用地面积：26.1 hm²	Site Area: 26.1 hm²

官田村位于我国著名的"将军县"江西省兴国县东北部，2013年被列入"第二批中国传统村落名录"。1931年中央军委第一个综合性兵工厂在此建立，官田被誉为"兵工始祖、军工摇篮"。村庄建设遵循上位规划，以风貌保护规划为蓝本，以红色教育培训和旅游为抓手，将官田列为井冈山红色教育线路上的重要一站。村庄以展览展示、教育培训功能为主：设置"兵工课堂"；"兵工七子"实践、实习基地；民宿文化及兵工文化研修基地。

村庄建设采用"针灸式"的设计手法，以修复9大旧址展馆为核心，整理环境，新建游客中心、兵工博物馆、民俗街、木轮发电车、兵工桥等将历史印记、军工文化和客家元素渗入到村庄各个角落，形成"馆在村中、村在馆中"的布局。兵工博物馆以江西红土颜色为基调，建筑形体顺应场地高差，匍匐于场地向传统建筑致敬，错落的斜屋顶分散了建筑体量，消解了大体量建筑对村庄肌理的破坏，并与远山取得呼应。为适应内部展陈，内设两层及三层通高的展示区，与"中央综合性兵工厂"坐落在小小山村中，"小中见大"，有异曲同工之意。

Guantian Village, located in the northeast of Xingguo County, Jiangxi Province, a famous "general county", was listed in the "Second Batch of Chinese Traditional Villages" in 2013. In 1931, the first comprehensive armoury of the Central Military Commission was established here, and Guantian was known as "the ancestor of armoury and the cradle of military industry". The construction of the village follows the upper planning, takes the landscape protection planning as the blueprint, takes red education training and tourism as the starting point, and includes Guantian as an important stop on the Jinggangshan red education route. The main functions of the village are exhibition, education and training: "armoury class", practice base of "Seven undergraduate colleges under the former Ministry of Weapons Industry" as well as training base of homestay culture and armoury culture are set up.

The construction of the village adopts the "acupuncture-style" design technique. With the restoration of 9 old museums as the core, the environment is organized. The new tourist center, armoury museum, folk street, wooden wheel power generator, and armoury bridge integrate historical marks, military culture and Hakka elements into every corner of the village, forming a "museum in the village, village in the museum". The Armoury Museum takes the red earth color of Jiangxi as the keynote. The architectural form conforms to the height difference of the site, crawling on the site to pay tribute to traditional architecture. The sloping roof disperses the building volume, dissipates the destruction of the village texture caused by the massive building and echoes the distant mountain. In order to adapt to the internal exhibition, two-storey and three-storey exhibition areas are arranged, which has the same meaning as the "central comprehensive armoury" located in a small mountain village.

村落选址与格局分析图

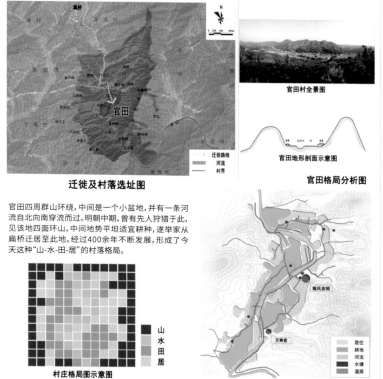

迁徙及村落选址图

官田四周群山环绕，中间是一个小盆地，并有一条河流自北向南穿流而过。明朝中期，曾有先人狩猎于此，见该地四面环山，中间地势平坦适宜耕种，遂举家从扁桥迁居至此地。经过400余年不断发展，形成了今天这种"山-水-田-居"的村落格局。

村庄格局图示意图

官田格局分析图

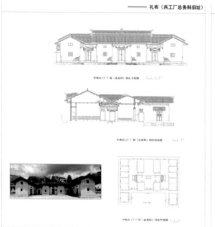

礼布（兵工厂总务科旧址）

万寿宫（工人俱乐部旧址）

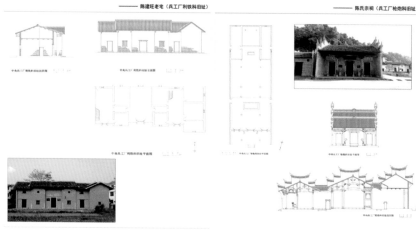

陈建旺老宅（兵工厂利铁科旧址） ・ 陈氏宗祠（兵工厂枪炮科旧址）

267

哈尔滨工业大学建筑设计研究院
The Architectural Design and Research Institute of HIT

哈尔滨工业大学建筑设计研究院创立于1958年，是全国知名大型国有工程设计机构。设计院依托百年学府哈尔滨工业大学深厚的科研资源与文化底蕴，历经半个多世纪的发展壮大，现已跻身全国行业前列，领军中国北方地区，荣获中国十大建筑设计公司、中国勘察设计协会优秀设计院、当代中国建筑设计百家名院等殊荣。

设计院业务领域涵盖工程项目建设的全过程，包括前期咨询、城市规划、建筑设计、风景园林设计、室内装饰设计、市政交通设计、工程勘察、工程监理与项目代建等。工程遍及全国省、自治区及直辖市，其中500余项工程项目获得国家金银奖等优秀设计奖励，取得了突出的行业成就。

设计院始终秉承"苛求完美、精益求精"的设计宗旨和"诚信服务、持续发展"的经营理念，充分发挥高校企业的科研、技术和人才优势，与社会各界携手合作、拼搏创新、不懈努力，为国内外客户提供高效优质服务，为社会和经济发展奉献建筑精品。

The Architectural Design and Research Institute of Harbin Institute of Technology, established in 1958, is a large domestically well-known state-owned engineering design and consulting agency. Depending on the profound scientific research resources and cultural heritage of Harbin Institute of Technology, a hundred-year old university, more than 50 years' development and growth has ranked the institute among the forefront in the industry in China and allowed it to take the lead in Northern China. The institute has been granted such honorable titles as one of China's Top Ten Architectural Design Companies, Outstanding Design Institute by CEDA and one of Contemporary China's Top 100 Famous Design Institutes.

Our institute's business fields cover the whole process of engineering project construction, including preliminary consultation, urban planning, architectural design, landscape architecture design, interior decoration design, municipal transportation design, engineering survey, engineering supervision and agency construction for projects, with projects covering all provinces, autonomous regions and municipalities in China. More than 500 projects have been granted outstanding design awards such as national gold and silver awards and prominent industry achievements have been obtained.

Our institute adheres to the design principle "Pursuing Perfection and Keeping Improving" and the business philosophy "Integrity-based Services and Constant Development" and gives full play to the scientific research, technology and talent advantages of a university enterprise, striving to provide efficient high-quality services for customers at home and abroad and magnificent buildings for social and economic development through cooperation with all sectors of society, constant innovation and unremitting efforts.

地址：黑龙江省哈尔滨市南岗区海河路202号2545信箱　　Add: Mailbox 2545, No.202 Haihe Road, Nangang District, Harbin, Heilongjiang
邮编：150090　　P.C: 150090
电话：0451-86283317（总院办公室）　　Tel: 0451-86283317
传真：0451-86283318（总院办公室）　　Fax: 0451-86283318
邮箱：harbin@hitadri.cn　　Email: harbin@hitadri.cn
网址：www.hitadri.cn　　Web: www.hitadri.cn
微信公众号：哈尔滨工业大学建筑设计研究院　　WeChat: 哈尔滨工业大学建筑设计研究院

哈尔滨工业大学建筑设计研究院科研楼
Office Building of the Architectural Design and Research Institute Building of HIT

项目地点：黑龙江 哈尔滨　　Location: Harbin, Heilongjiang
建筑面积：30 000 m²　　Building Area: 30 000 m²
设计时间：2008年　　Design Time: 2008
竣工时间：2009年　　Completion Time: 2009

本项目位于哈尔滨工业大学第二校区，主要功能包括办公、会议、展示以及餐厅、健身房、地下车库等辅助功能。设计充分体现科研办公建筑的特点，强调空间的通透性、多样性、流动性，以人为中心，提供舒适、优美的环境，从而激发其创作灵感，在满足日常高效率工作节奏的基础上，体现人性化的深刻内涵。办公楼外观朴素典雅，细节力求精致、谐调，延续校园建筑的文脉。

The project is located in the second campus of Harbin University of technology. Its main functions include office, conference, exhibition, restaurant, gym, underground garage and other auxiliary functions. The design fully embodies the characteristics of scientific research office building, emphasizes the permeability, diversity and mobility of space, provides a comfortable and beautiful environment with people as the center, so as to inspire its creative inspiration, and reflects the profound connotation of Humanization on the basis of satisfying the daily high-efficiency work rhythm. The appearance of the office building is simple and elegant, and the details strive to be exquisite and harmonious, continuing the context of the campus building.

大连体育中心体育场、体育馆
Dalian Sports Center Stadium and Gymnasium

项目地点：辽宁 大连
建筑面积：119 622 m²（体育场）；72 823 m²（体育馆）
用地面积：80 650 m²（体育场）；39 848 m²（体育馆）
设计时间：2008年
竣工时间：2013年
容积率：0.75（体育场）；0.86（体育馆）

Location: Dalian, Liaoning
Building Area: 119 622 m² (Stadium) ; 72 823 m² (Gymnasium)
Site Area: 80 650 m² (Stadium) ; 39 848 m² (Gymnasium)
Design Time: 2008
Completion Time: 2013
Plot Ratio: 0.75 (Stadium) ; 0.86 (Gymnasium)

大连体育中心体育场,总座席数6万座,属国内特级体育场,内设田径及足球标准场地,曾作为第十二届全运会足球比赛场地。主体建筑结构采用钢筋混凝土框架体系,屋顶采用钢结构桁架体系,罩棚采用ETFE充气枕结构。大连体育中心体育馆,是目前国内可以举办NBA篮球比赛的仅有几个场馆之一,规模庞大,属国内特级特大型体育馆,功能多样,充分体现了集观赛与文艺表演于一身的综合性体育演艺中心的全新设计理念。其造型围绕一条空间的螺旋线展开,形体流畅完整,体现了滨海城市大连的独特地域文化特征。

Dalian Sports Center Stadium, with a total seating capacity of 60 000, is a special level stadium in China. It has track and field and football standard venues. It was once used as the football match venue of the 12th National Games. The main building structure adopts reinforced concrete frame system, the roof adopts steel truss system, and the canopy adopts ETFE inflatable pillow structure. Dalian Sports Center Gymnasium is one of the few venues that can hold NBA basketball games in China at present. It is a large-scale super large gymnasium with various functions. It fully embodies the new design concept of a comprehensive sports performance center that integrates watching and performing arts. Its shape revolves around a spiral line of space, and its shape is smooth and complete, reflecting the unique regional and cultural characteristics of Dalian Binhai.

第十三届全国冬季运动会冰上运动中心
Ice-Sports Center of the 13th National Winter Games

项目地点：新疆维吾尔自治区 乌鲁木齐
建筑面积：78 334 m²
用地面积：43 023 m²
设计时间：2012年
竣工时间：2014年
容积率：0.24

Location: Urumchi, the Xinjiang Uygur Autonomous Region
Building Area: 78 334 m²
Site Area: 43 023 m²
Design Time: 2012
Completion Time: 2014
Plot Ratio: 0.24

　　本工程是我国目前规模最大的冰上运动建筑综合体。设计基于新疆的自然环境与历史文脉，以天山雪莲与丝绸之路为创作主题，将体育建筑与自然景观有机融合，体现了鲜明的当代体育建筑地域特色，塑造了丝路花谷的建筑群体意向，通过优化建筑形体实现节能降耗，采用围合式的建筑布局抵御冷风侵袭，利用高效导风的流线型屋盖形态减少屋面雪荷。

　　The project is the largest ice sports building complex in China. Based on the natural environment and historical context of Xinjiang, taking Tianshan snow lotus and the Silk Road as the creative theme, the design organically integrates the sports building and natural landscape, embodies the distinctive regional characteristics of contemporary sports buildings, shapes the architectural group intention of the Silk Road Flower Valley, realizes energy conservation and consumption reduction by optimizing the building shape, adopts the enclosed architectural layout to resist the cold wind, and makes use of high efficiency The streamline roof shape of wind guide reduces the roof snow load.

寒地建筑科学研究中心
Harbin Institute of Technology, Cold Ground Building Science Laboratory Building

项目地点：黑龙江 哈尔滨
建筑面积：9 768.97 m²
用地面积：2 597.36 m²
设计时间：2012年－2013年
竣工时间：2014年

Location: Harbin, Heilongjiang
Building Area: 9 768.97 m²
Site Area: 2 597.36 m²
Design Time: 2012-2013
Completion Time: 2014

本项目是我国东北地区首个寒地建筑科学重点实验室，提供建筑声、光、热及材料等实验场所。实验室通过模块化布局，建构了完备的建筑工程设计与科学研究体系。本工程结合建筑功能要求与场地限制，合理组织各建筑要素，形成有机整体，在绿色建筑技术措施等方面多有创新。建筑形式体现寒地建筑特征，采用先进碳化木装饰，营造舒适宜人的环境氛围；形体塑造简洁完整，特色鲜明。

This project is the first key laboratory of cold Building Science in Northeast China, providing experimental sites of building sound, light, heat and materials. Through modular layout, the laboratory has built a complete system of architectural engineering design and scientific research. The project reasonably organizes all building elements to form an organic whole in combination with building functional requirements and site restrictions, and has many innovations in green building technical measures, etc. The architectural form reflects the architectural features of cold area, adopts advanced carbonized wood decoration to create a comfortable and pleasant environment atmosphere; the shape is simple and complete with outstanding features.

黑龙江

方舟国际设计有限公司
Fangzhou International Design Group Co.,Ltd.

方舟国际设计有限公司成立于2001年，是全国知名的民营工程设计机构。公司荣获全国勘察设计行业建国70年优秀勘察设计单位、全国建筑设计行业诚信单位、全国创优型企业等殊荣。具有建筑行业（建筑工程）设计甲级、施工图审查甲级、风景园林专项设计乙级、城乡规划设计乙级、市政专项设计乙级、土地规划乙级、工程咨询等多项资质。我公司现有员工200余人，其中专业技术人员占总人数90%以上，各类国家一级注册人员70余名，高级技术职称人员90余名，中级技术职称人员80余名。各专业配套齐全，拥有一批学术造诣深、实践经验丰富的专业带头人和中青年技术骨干。公司下设建筑设计分院、景观规划院、创研中心、审定中心等部门。业务领域涵盖前期咨询、城市规划、建筑设计、风景园林设计、市政设计、室内设计、全过程工程咨询、EPC工程总承包等。

公司一直践行"敬天爱人、精益设计"的核心价值观，秉承专业化和精细化设计的方针，设计产品类型包含教育、医疗、文体、商业、科研、居住、绿色建筑、城市更新等，作为核心产品类型的教育建筑与医疗建筑设计，技术水平牢牢占据着国内领先地位。公司立足黑龙江省，面向全国，业务范围覆盖国内十几个省份的百余个城市；历年完成的工程项目中，获得国家及省部级单位奖项30余项，建筑工程设计奖项200余项，在行业内具有较好口碑及声誉。

Established in 2001, Fangzhou International Design Group Co. Ltd. is a well-known private engineering design institution. It has won many awards such as outstanding survey and design unit in the national survey and design industry during the 70 years after the foundation of China, the national architectural design industry integrity unit, the national outstanding enterprise. With Class A qualification for construction industry (construction engineering) design and construction drawing review, Class B qualification for landscape specialized design, urban and rural planning and design, municipal specialized design, and land planning, as well as qualification for engineering consulting. Our company has more than 200 employees and professional and technical personnel account for more than 90%. Among them, there are more than 70 national first-class registered personnel, more than 90 personnel with senior technical title, more than 80 personnel with intermediate technical title. We have a complete set of specialized persons, such as professional leaders with deep academic achievements and rich practical experience and young and middle - aged technical backbones. The company has 5 branches of architectural design, landscape planning, innovation and research center, approval center and other departments. The business field covers preliminary consulting, urban planning, architectural design, landscape design, municipal design, interior design, whole process engineering consulting, EPC project general contracting, etc.

The company has been practicing the core values of "lean design keep deference to nature, show concern for human", adhering to the principle of specialized and refined design. The design products include education, medical treatment, recreation and sports, commerce, scientific research, residence, green building, urban renewal, etc. The design for the core product types - educational building and medical building - takes the leading position in China. The company based in Heilongjiang Province, facing the whole country, with the business scope covering more than a dozen domestic provinces, more than 100 cities; Over the years, the project has won more than 30 national and provincial awards, more than 200 architectural engineering design awards, and has a high public praise and reputation in the industry.

地址：黑龙江省哈尔滨市道外区红旗大街991号	Add: No.991, Hongqi Street, Daowai District, Harbin, Heilongjiang
邮编：150030	P.C: 150030
电话：0451-87858518	Tel: 0451-87858518
传真：0451-87858504	Fax: 0451-87858504
邮箱：fz2001@fzjz.net	Email: fz2001@fzjz.net
网址：www.fzjz.net	Web: www.fzjz.net

哈尔滨市儿童医院松北院区、市妇幼保健院松北院区、市红十字中心医院松北院区

Songbei Hospital of Harbin Children's Hospital, Songbei Hospital of Harbin Municipal Maternal and Child Health Care Hospital, Songbei Hospital of City Red Cross Central Hospital

项目地点：黑龙江 哈尔滨
建筑面积：74 800 m²
总用地面积：100 000 m²

Location: Harbin, Heilongjiang
Building Area: 74 800 m²
Total Site Area: 100 000 m²

设计理念
Design Concept

秉承可持续发展、资源共享、人性化和绿色生态的设计原则，重视人群特点，提倡因地制宜，营造出舒适、便捷、优美的诊疗环境。

汲取中国传统群落空间文化精华，理性布局，秩序分明，高低错落，营造富有层次的个性动态空间。展示出栩栩如生的灵动姿态，犹如一只即将展翅飞舞的蝴蝶，寓意了妇幼保健医院迎接生命、珍爱生命的内涵。

Adhering to the design principles of sustainable development, resource sharing, humanized design and green ecology, the hospital attaches importance to the characteristics of the population and advocates adapting to local conditions to create a comfortable, convenient and beautiful diagnosis and treatment environment.

Absorbing the essence of Chinese traditional community space - rational layout, clear order - the design aims to create a dynamic space rich in hierarchical personality. It shows the vivid and flexible posture, like a butterfly about to spread its wings and fly, implying the connotation of maternal and child health hospital to give life and cherish life.

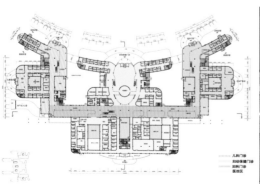

伸马商业综合体
Shenma Commercial Complex

项目地点：黑龙江 哈尔滨　　Location: Harbin, Heilongjiang
总建筑面积：30 860 m²　　Total Building Area: 30 860 m²
总用地面积：7 770 m²　　Total Site Area: 7 770 m²

考虑到用地局促、周边日照遮挡和城市形象等因素，建筑沿街呈一字形布置，体块高低错落，相互穿插咬合。

整体建筑通过"飘带"将各体块融合串联起来，强调线条流动感，弱化转角处体量关系，减少对城市的压迫感，形成较为舒展、简洁明快、活泼的立面形象。

Considering the confined land space, surrounding sunshine shading, urban image and other factors, the buildings are well arranged in a linear shape along the street, and the blocks are interlaced with each other.

The overall building integrates and connects each block in series through streamers, emphasizing the flow sense of lines, weakening the relationship between volumes at corners, reducing the sense of pressure on the city, and forming a relatively open, concise and lively facade image.

黑河学院综合体育馆
Comprehensive Gymnasium of Heihe College

项目地点：黑龙江 黑河
总建筑面积：12 460 m²
总用地面积：60 422 m²

Location: Heihe, Heilongjiang
Total Building Area: 12 460 m²
Total Site Area: 60 422 m²

规划尊重校园的现有环境，功能布局合理，形态活泼自由。建筑造型设计遵循简洁、庄重的原则，将古典气息赋予现代设计中，充分展现了体育建筑的力与美。外观设计美观、端庄、自由，体现校园文化内涵，形态与周边的建筑相协调，活泼轻松。注重空间的塑造，创造出宜人的空间环境，体现开放的校园文化。

The planning respects the existing environment of the campus, the functional layout is reasonable, and the form is lively and free. The architectural modeling design follows the principle of simplicity and solemnity, endows classical atmosphere to modern design, and fully demonstrates the power and beauty of sports architecture. The appearance design is beautiful, dignified and free, reflecting the connotation of campus culture. And the form is in harmony with the surrounding buildings, lively and relaxed. The design also pays attention to the shaping of space to create a pleasant space environment, reflecting the open campus culture.

华润·置地公馆
China Resources Land Residence

设计师: 刘清君、曹洋、李研
项目地点: 黑龙江 哈尔滨
建筑面积: 78 400.51 m²
用地面积: 2.27 hm²

Designers: Qingjun Liu, Yan Cao, Yan Li
Location: Harbin, Heilongjiang
Building Area: 78 400.51 m²
Site Area: 2.27 hm²

本项目地块处于哈尔滨市南岗区发展大道与东方大街交会处。项目北临哈西新区的高铁站，选址于城市老城区与哈西新区的交界位置，交通十分便利。用地面积22 689.30 m²，总建筑面积78 400.51 m²，为商业服务业设施用地。主要功能为商业及LOFT办公，其中高层办公4栋，多层商业2栋。

规划理念：以城市设计的手法，整合城市总体与布局、人文及自然环境的要求，协调规划区域内外部的环境关系，建立新的城市中心空间秩序，营造自然、活跃、生机盎然的特色空间。

The project plot is located at the intersection of Development Avenue and Dongfang Avenue, Nangang District, Harbin City. The project is in the north of the high-speed railway station of Haxi New District, located at the junction of the old city and Haxi New District, with convenient transportation. The project covers a land area of 22 689.30 m² and a total construction area of 78 400.51 m², which is used for building commercial service facilities. The main functions are commerce and LOFT office, including 4 high-rise office buildings and 2 multi-storey commercial buildings.

By means of urban design, the planning concept integrates the requirements of the overall urban layout, humanity and natural environment, coordinates the relationship between the internal and external environment of the planning area, establishes a new spatial order of the city center, and creates a natural, active and vibrant characteristic space.

佳木斯大学实训楼AB栋
Training Building AB of Jiamusi University

设计师：李韬、闫春雷、朱涵宇、张岩、王琨
项目地点：黑龙江 佳木斯
建筑面积：33 300 m²
用地面积：2.98 hm²

Designers: Tao Li, Chunlei Yan, Hanyu Zhu, Yan Zhang, Kun Wang
Location: Jiamusi, Heilongjiang
Building Area: 33 300 m²
Site Area: 2.98 hm²

融入校园现状肌理，创造新的空间体验——现代校园应是一个有机的整体，故在规划中应避免将各个校园空间割裂成片段，应注重建筑外部空间秩序，与原有校园之间实现新与旧对话。本方案即是对原有校园中心环境的一次修补与更新，通过合理布局让建筑群体形成半围合的开放空间，囊括景观广场及保留绿化，为来自校园各方向的师生提供公共活动的休闲场所。

联动周边教学区域，延续场地环境文脉——建筑的功能空间和室外环境设计相结合，场地内的原始地形、现有绿化树木，都在建筑和场地设计中作为文脉的延续被保留和尊重，同时通过空间与景观的塑造，引导人流，形成联动空间。

Integrating into the current campus texture to create new space experience - Modern campus should be an organic whole, so the planning should pay attention to the building external space order, and create new and old dialogue with the original campus instead of dividing each campus space into segments. This scheme is a repair and renewal of the original central environment of the campus. Through reasonable layout, the building group forms a semi - enclosed open space, including landscape squares and retaining greenery, providing a leisure place for public activities for teachers and students from all directions of the campus.

Connecting the surrounding teaching area to continue the environment context of the site - The functional space of the building is combined with the outdoor environment design. The original terrain and the existing green trees in the site are preserved and respected as the continuation of the context in the design of the building and the site. At the same time, through the shaping of space and landscape, the flow of people is guided to form an interconnected space.

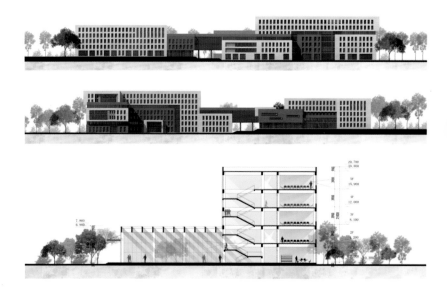

江湾学校
Jiangwan School

项目地点：黑龙江 哈尔滨　　Location: Harbin, Heilongjiang
总建筑面积：24 000 m²　　Total Building Area: 24 000 m²
总用地面积：37 300 m²　　Total Site Area: 37 300 m²

根据地域特征和孩子的活泼天性，为了塑造一个具有童趣的学习乐园，我们从儿童启蒙教育的"积木游戏"中汲取灵感，将积木这一颇具特色形象引入校园的方案设计中，从规划形态、建筑空间、立面形象上，通过积木搭接，创造不同尺度的、有趣的空间体验，为孩子创造一方寓教于乐的广阔天地。

According to regional characteristics and children's lively nature, in order to create a childlike learning paradise, we draw inspiration from the "game of building blocks" in children's elementary education, and introduce blocks with distinctive features into the campus design. From the planning form, architectural space and facade image, we create interesting space experience of different scales through building blocks in order to build a wide edutainment world.

绥化市廉政教育中心
Clean Government Education Center in Suihua City

项目地点：黑龙江 绥化
建筑面积：35 675 m²
用地面积：30 005.62 m²
设计时间：2020年1月4日
竣工时间：2020年12月31日
容积率：1.03

Location: Suihua, Heilongjiang
Building Area: 35 675 m²
Site Area: 30 005.62 m²
Design Time: 2020.1.4
Completion Time: 2020.12.31
Plot Ratio: 1.03

规划设计：通过合理的建筑布局，创造出一套"三进式"院落空间，以院落组织人们在场所中的工作与生活。创造一个形态端庄、功能独立且联系方便，既有东方文化延展又有新时代特色、开合有度的园林式廉政教育中心。

建筑设计：设计采用"新中式"风格，建筑形式典雅大方且富有书卷气息，与"廉政教育"的建筑主题紧密相扣，同时新中式的建筑风格也体现了设计对传统文化的传承与发扬，体现了新时代背景下中华民族的道路自信和文化自信。

Planning and design: Through reasonable architectural layout, a set of "ternary" courtyard space is created to organize people's work and life in the site. It aims to create a garden-style clean government education center with dignified form, independent function and convenient contact, which has both oriental culture extension and new era characteristics.

Architectural design: The design adopts the "new Chinese" style. The architectural form is elegant, magnificent and knowledgeable, closely linked with the architectural theme of "clean government education". At the same time, the new Chinese architectural style also reflects inheritance and development of traditional culture in the design as well as the road confidence and cultural confidence of the Chinese nation under the background of the new era.

绥化市图书馆
Suihua Library

设计师：刘远孝、闫春雷
项目地点：黑龙江 绥化
建筑面积：29 800 m²
用地面积：34 800 m²

Designers: Yuanxiao Liu, Chunlei Yan
Location: Suihua, Heilongjiang
Building Area: 29 800 m²
Site Area: 34 800 m²

本项目旨在建设一个具有强烈人文情怀的、现代化、智能化、个性化、人性化且反映北方寒地特色的图文信息中心。设计表达了"天圆地方"的传统宇宙观，蕴天地之精神，自强不息，厚德载物。

图书馆内部空间强调了公共性、私密性的对比以及丰富性、趣味性的表达。通过方庭、圆廊等空间变化，层层叠叠，交融转换；也运用"中国龙"坡道、游廊等来提升空间趣味，春秋书屋和冬夏小院寓意时空变换；屋顶的风车形布局，代表了生生不息的运动。

This project aims to build a modern, intelligent, personalized and humanized graphic information center with strong humanistic feelings and reflecting the characteristics of the cold region in the north. The design expresses the cosmic view of "orbicular sky and rectangular earth", containing the spirit of heaven and earth-constant self-improvement as well as self-discipline and social commitment.

The interior space of the library emphasizes the contrast of publicity and privacy as well as the expression of richness and amusement. The adoption of quadriporticus, round gallery and other space changes, as well as "Chinese dragon" ramps and galleries enhances the spatial interest. The Spring and Autumn Library and the Winter and Summer Courtyard imply the transformation of time and space. The wind model layout of the roof represents the endless movement.

哈尔滨市血液中心
Harbin City Blood Center

设计师：赵慧、刘昕竹、姜晓光
项目地点：黑龙江 哈尔滨
建筑面积：12 500.00 m²
用地面积：4 958.37 m²

Designers: Hui Zhao, Xinzhu Liu, Xiaoguang Jiang
Location: Harbin, Heilongjiang
Building Area: 12 500.00 m²
Site Area: 4 958.37 m²

本项目为高层公共建筑，建筑高度36 m，地上部分9层，地下部分1层。由于项目建设用地较为局促，规划将新楼设计成与老楼贴临建设。设计保留了场地南侧的原有出入口，作为将来院区主要的人行出入口使用，另外在东侧开设了一个新的出入口，主要供后勤及车辆出入使用，很好地实现了人车分流，也避免办公与后勤的流线交叉。

建筑单体设计采用简约现代风格，外立面采用米黄色石材，使得建筑彰显严谨理性的同时富有文化气息与亲和力。

This project is a high-rise public building with a height of 36 m, consists of 9 floors above ground and 1 floor underground. As the construction land of the project is relatively limited, the new building is designed to be attached to the old building. The design retains the original entrance on the south side of the site as the main entrance for people in the future. In addition, a new entrance is set up on the east side, which is mainly used for logistics and vehicles, so as to realize the separation of people and vehicles and avoid the crossing of streamline between office and logistics.

The single design adopts simple and modern style, and the facade is made of beige stone, which makes the building manifest rigorous rationality and rich cultural flavor and affinity.

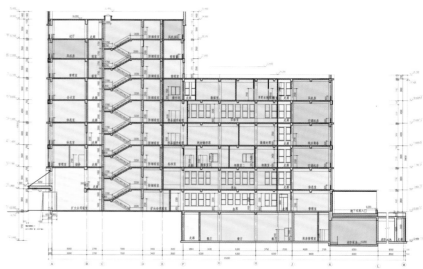

哈尔滨麦硕国际教育社区
Harbin Maestro International Education Community

设计师：闫春雷、柴方建、李志东、陈刚、刘昕竹、曹洋、李岳、董士权、赵佳启、李迪、杨志涛、闫炳池
项目地点：黑龙江 哈尔滨
建筑面积：102 021 m²
用地面积：8.77 hm²

Designers: Chunlei Yan, Fangjian Chai, Zhidong Li, Gang Chen, Xinzhu Liu, Yang Cao, Yue Li, Shiquan Dong, Jiaqi Zhao, Di Li, Zhitao Yang, Bingchi Yan
Location: Harbin, Heilongjiang
Building Area: 102 021 m²
Site Area: 8.77 hm²

项目将国际公学教育模式与现代教育理念相结合，打造了一座富于古典精神和贵族气质的具有文化和精神感召力的"未来学校"。校园集中式布局，留出尽可能多的空间遍植高大的树木，营造绿树掩映、环境宜人的低密度绿色校园。

设计注重教室外的"非正式学习空间"的设计，让教育无所不在。3层廊道体系、首层风雨廊体系及地下廊道体系3条动线，有效集成整合和优化校园内各类资源，实现了校内空间的互联、融合和共享。

The project combines the international public school education model with modern education concepts to create a future school with cultural and spiritual appeal rich in classical spirit and noble spirit. The centralized layout of the campus leaves as much space as possible for planting tall trees to create a low-density green campus with a pleasant environment surrounded by green trees.

The design focuses on the design of "informal learning space" outside the classroom, so that education is everywhere. The corridor system of the second floor, the wind and rain gallery system of the first floor and the underground corridor system are three moving lines, which effectively integrate and optimize all kinds of resources on the campus, realizing the interconnection, integration and sharing of campus space.

江苏

东南大学建筑设计研究院有限公司
ARCHITECTS & ENGINEERS CO.,LTD. OF SOUTHEAST UNIVERSITY

东南大学建筑设计研究院有限公司(以下简称东大院)始建于1965年,隶属于教育部和东南大学,综合实力位居高校设计企业前列,是国内外知名的建筑设计公司。

东大院现有院士工作站1个,站内有2名合作院士;全国工程勘察设计大师1名,江苏省设计大师8名,江苏省有突出贡献中青年专家1名,江苏省优秀勘察设计师32名,全国优秀青年建筑师16名,同时本校4名院士担任公司顾问,各专业注册师350余人。

主要业务有建筑设计、城市规划、建筑智能、绿色建筑、室内设计、风景园林、建筑遗产保护、电力、公路、市政设计、BIM应用等。

近年来,东大院创新创优能力不断提升,行业领先地位进一步巩固,设计作品遍及全国30个省、直辖市、自治区,荣获历届国家和部委、省、市的优秀设计奖1 000余项。

Founded in 1965, ARCHITECTS & ENGINEERS CO.,LTD. OF SOUTHEAST UNIVERSITY (hereinafter referred to as the East University) is a well-known architectural design company at home and abroad, which is subordinate to the leadership of the Ministry of Education and Southeast University. Its comprehensive strength ranks among the first-class college design enterprises.

The East University has 1 academician workstation, 2 cooperative academicians, 1 national engineering survey and design master, 8 design masters in Jiangsu Province, 1 young and middle-aged expert with outstanding contributions in Jiangsu Province, 32 outstanding survey designers in Jiangsu Province, 16 national outstanding young architects, 4 academicians of the University serving as company consultants, and more than 350 professional registrars.

The main businesses include architectural design, urban planning, building intelligence, green building, interior design, landscape architecture, architectural heritage protection, power, highway, municipal design, BIM application and other related majors.

In recent years, the innovation and excellence creation ability of the East Grand Courtyard has been constantly improved, and its leading position in the industry has been further consolidated. Its design works have spread to 30 provinces, municipalities and autonomous regions across the country, and it has won more than 1 000 excellent design awards from previous national, ministerial, provincial and municipal levels.

地址:江苏省南京市四牌楼2号	Add: No.2 Sipailou, Nanjing, Jiangsu
邮编:210096	P.C: 210096
电话:025-83793178	Tel: 025-83793178
传真:025-57713341	Fax: 025-57713341
邮箱:ad@adriseu.com	Email: ad@adriseu.com
网址:http://adri.seu.edu.cn/	Web: http://adri.seu.edu.cn/

第十届江苏省园艺博览会博览园主展馆
Main Exhibition Pavilion of the Expo Park for the 10th Jiangsu Horticultural Exposition

设计师:王建国、葛明、徐静、朱雷、韩重庆、陆伟东、程小武、李亮、孙小鸾、许轶、赵晋伟、王玲、章敏捷、王志东、丁惠明
项目地点:江苏 扬州
建筑面积:14 154 m²
用地面积:3.1 hm²
容积率:0.40

Designers: Jianguo Wang, Ming Ge, Jing Xu, Lei Zhu, Chongqing Han, Weidong Lu, Xiaowu Cheng, Liang Li, Xiaoluan Sun, Yi Xu, Jinwei Zhao, Ling Wang, Minjie Zhang, Zhidong Wang, Huiming Ding
Location: Yangzhou, Jiangsu
Building Area: 14 154 m²
Site Area: 3.1 hm²
Plot Ratio: 0.40

主展馆坐落于博览园入口展示区，是园区内主要的地标建筑和展览建筑，并在世园会期间作为中国馆使用。建筑设计汲取扬州当地山水建筑和园林特色的文化意象，以"别开林壑"之势表现扬州园林大开大合的格局之美。

主要展厅采用现代木结构技术，最大限度简化装修，展现结构的自然之美。主要木构件均由工厂加工生产、现场装配建造，不仅是一种绿色建造方式，符合节能环保要求，而且还有效提升了施工效率。项目获得2020年度全国绿色建筑创新奖一等奖。

As the main landmark building, the main exhibition pavilion is located in the entrance area of the Expo Garden, and is used as the China Pavilion during the World Horticultural Exposition. Drawing on the cultural images of the local landscape architecture and gardens in Yangzhou, the main exhibition pavilion expresses the beauty of the large opening and closing pattern of Yangzhou gardens with the image of "sudden view of forest ridges".

The main exhibition halls adopt Modern wooden structure technology. The main components are manufactured by factories and assembled on site. It is not only a green construction, which meets the requirements of energy saving and environmental protection, but also effectively improves the construction efficiency. The project won the first prize of the 2020 National Green Building Innovation Award.

青岛市民健身中心
Qingdao Citizen Fitness Center

设计单位：东南大学建筑设计研究院有限公司 （合作设计单位：青岛市城市规划设计研究院） 建筑师：高庆辉、万小梅、刘宾、袁玮 设计团队：石峻垚、赵效鹏、艾迪、崔慧岳、薛丰丰、李宝童、吴文竹等 结构工程师：孙逊、韩重庆、杨波、张翀、唐伟伟等 项目地点：山东 青岛 建筑面积：214 000 m² 用地面积：36.5 hm² 容积率：0.58	Design Enterprise: ARCHITECTS & ENGINEERS CO.,LTD. OF SOUTHEAST UNIVERSITY (Co-designer: Qingdao Urban Planning and Design Institute) Architects: Qinghui Gao, Xiaomei Wan, Bin Liu, Wei Yuan Design Team: Junyao Shi, Xiaopeng Zhao, Aidi, Huiyue Cui, Fengfeng Xue, Baotong Li, Wenzhu Wu, etc. Structural Engineers: Xun Sun, Chongqing Han, Bo Yang, Chong Zhang, Weiwei Tang, etc. Location: Qingdao, Shandong Building Area: 214 000 m² Site Area: 36.5 hm² Plot Ratio: 0.58

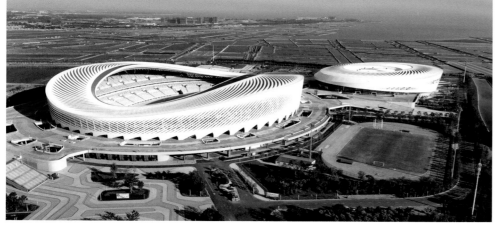

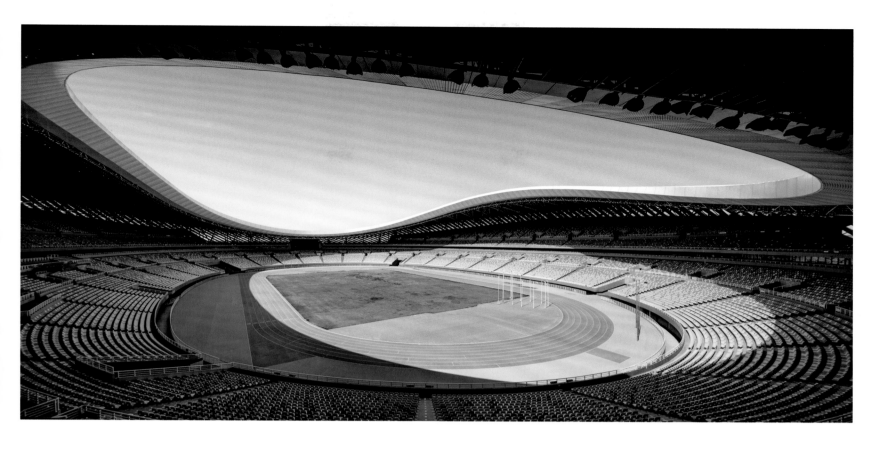

该项目包含6万座体育场及1.5万座体育馆,倡导生态优先、轻触湿地、集约建构和诗意营造,获2019－2020中国建筑学会建筑设计奖公共建筑一等奖、2019年全国优秀工程行业奖优秀勘察设计建筑工程一等奖、2020亚建协（ARCASIA）专门化建筑荣誉提名奖（金奖空缺）、国际建协（UIA）第27届世界建筑师大会参展作品。项目建成后承办过2020中国男篮职业联赛（CBA）北方主赛区、2019中加女篮与中澳男篮对抗赛等国内外赛事。

The project includes 60 000 stadiums and 15 000 stadiums, with priority to ecology, light touch to wetlands, intensive construction and poetic creation. Won the first prize of Public Building of architectural Design Award of Architectural Society of China 2019-2020, the first prize of Excellent Survey, Design and Architectural Engineering of National Excellent Engineering Industry Award in 2019, ARCASIA awards for Architecture 2020, Buildings Honorary Mention(Gold award vacancy)Specialized, the works exhibited in the 27th World Architects Conference of UIA.It has hosted the 2020 North Division of China Men's Basketball Professional League (CBA), the 2019 China - Canada Women's Basketball Match with China - Australia men's Basketball match and other domestic and international competitions.

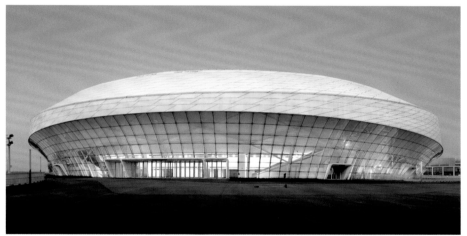

金陵大报恩寺遗址博物馆
Site Museum of Jinling Grand Bao'en Temple

设计师: 韩冬青、陈薇、王建国、马晓东、孟媛
项目地点: 江苏 南京
建筑面积: 60 849 m²
用地面积: 7.5 hm²
容积率: 0.51

Designers: Dongqing Han, WeiChen, Jianguo Wang,
　　　　　Xiaodong Ma, Yuan Meng
Location: Nanjing, Jiangsu
Building Area: 60 849 m²
Site Area: 7.5 hm²
Plot Ratio: 0.51

　　金陵大报恩寺遗址博物馆是为保护和展示中国明代皇家寺庙大报恩寺遗址而建设的重大文化设施——金陵大报恩寺遗址公园的一期工程。

　　设计理念有两个创新点:第一,在严格保护遗址本体的前提下,以立体的空间组织呈现寺庙遗址的多层次历史信息及其与城池山川格局关系;第二,以新的技术手段创造具有历史文化意韵的场所特质。

　　项目建成后因其遗产保护与公共文化生活的有机融合而受到专业领域和社会的极大关注。

Site Museum of Jinling Grand Bao'en Temple is the first phase of the project of Grand Bao'en Temple Heritage Park, which is planned as a significant culture facility and aims for protecting and exhibiting this royal Temple of Ming Dynasty (early of the 15th century) in China.

The museum is designed following two principles. First, the ruins shall be placed under strict conservation, while allowing a three-dimensional spatial presentation of the historical information embedded in the ruins from different levels and of its spatial ties with the ambient surroundings; and second, build the ruins as a venue with historical and cultural insights using new technological means.

The project has attracted great attention and won the praises of both professionals and the public due to the organic integration of heritage conservation and public cultural activities.

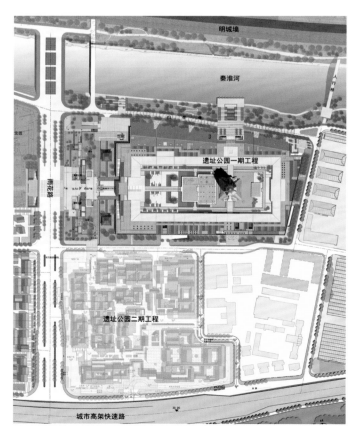

深圳清真寺
The Mosque of Shenzhen

设计师：马晓东、韩冬青、谭亮
项目地点：广东 深圳
建筑面积：10 864.92 m²
用地面积：0.66 hm²
容积率：1.6

Designers: Xiaodong Ma, Dongqing Han, Liang Tan
Location: Shenzhen, Guangdong
Building Area: 10 864.92 m²
Site Area: 0.66 hm²
Plot Ratio: 1.6

软件谷学校
Software Valley School

设计师：谭亮、高崧、孙菲、邹康	Designers: Liang Tan, Song Gao, Fei Sun, Kang Zou
项目地点：江苏 南京	Location: Nanjing, Jiangsu
建筑面积：86 852.22 m²	Building Area: 86 852.22 m²
用地面积：5.1 hm²	Site Area: 5.1 hm²
容积率：1.26	Plot Ratio: 1.26

软件谷学校在片段化的城市用地中将城市街区形态与传统院落空间引入校园，形成基本原型单元——"四方院"。"院"的聚落不仅重塑了城市街区的整体风貌，还承载了"教与学"的核心功能，将公共社区的意识融入校园日常生活。院落中"正式"与"非正式"的行为空间相汇交融，成为一个自然开放、富于归属感的校园聚落共同体。

In the fragmented urban land, the software valley school introduces the urban block form and traditional courtyard space into the campus, forming the basic prototype unit - "Square courtyard". The "courtyard" settlement not only "stitches" the originally isolated and fragmented urban land, but also carries the core function of "teaching and learning". It integrates the sense of public community into the daily life of the campus, the "formal" and "informal" behavior spaces merge together to form a natural and open community with a sense of belonging.

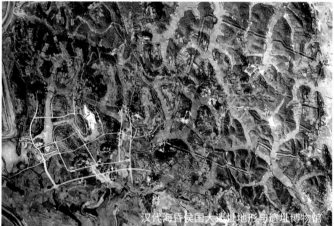

南昌汉代海昏侯国遗址博物馆
Site Musuem of Hai-Hun Kingdom Relics of Han Dynasty

设计师：张彤、袁玮、齐康、张四维、严希、李宝童、韩重庆、孙逊、李亮、黄凯、王若莹、丁惠明、杨妮、王志东、屈建球、周桂祥、张磊、崔岚、杨冬辉、李伟强、周杰

项目地点：江西 南昌
建筑面积：39 250 m²
用地面积：11.88 hm²
容积率：0.29

Designers: Tong Zhang, Wei Yuan, Kang Qi, Siwei Zhang, Xi Yan, Baotong Li, Chongqing Han, Xun Sun, Liang Li, Kai Huang, Ruoying Wang, Huiming Ding, Ni Yang, Zhidong Wang, Jianqiu Qu, Guixiang Zhou, Lei Zhang, Lan Cui, Donghui Yang, Weiqiang Li, Jie Zhou

Location: Nanchang, Jiangxi
Building Area: 39 250 m²
Site Area: 11.88 hm²
Plot Ratio: 0.29

南昌汉代海昏侯国遗址位于南昌市新建区铁河乡和大塘坪乡内，鄱阳湖西岸，是中国目前发现的面积最大、保存最好、格局最完整、内涵最丰富的典型汉代列侯国都城聚落遗址。

遗址博物馆东距墎墩墓700 m，距紫金城城址西侧边界约1 300 m，其设计以考古研究为依据，以遗产价值为导向，以真实性、完整性、永续传承、多方受益四项原则为指导，场地设计实现了风土地貌的"整体保护"和"最小干预"。遗址博物馆通过嵌藏于地形并喻显汉文化特征的"虬龙潜野"形态和分区清晰、动线流畅的空间组织，为海昏侯国遗址与文物的展示、馆藏、研究和交流提供设施先进、功能完备的一流场馆。

Hai-Hun Kingdom Relics of Han Dynasty is located in Tiehe Township and Datangping Township, Xinjian District, Nanchang, on the west bank of Poyang Lake. It is the largest, best - preserved, most complete, and most connotative typical Han Dynasty capital city settlement site found in China.

The Site Museum is located 700 m away from the Tomb of Jundun in the east and about 1 300 m away from the western boundary of the Zijin City site. Benefiting from the four principles as a guide, this design achieves "total protection" and "minimal intervention" in the aerodynamic landscape. Through the "Horrowed Dragon's Hidden Field", which is embedded in the terrain and shows the characteristics of Han culture, and reveals the spatial organization with clear divisions and smooth moving lines, the Site Museum provides advanced facilities and complete functions for the display, collection, research and exchange of Hai-hun Kingdom relics.

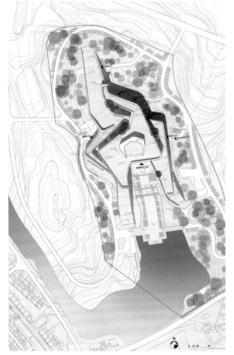

启迪设计集团股份有限公司
Tus Design Group Co., Ltd.

启迪设计集团股份有限公司（以下简称启迪设计）创建于1953年，是一家以覆盖人居环境全过程的投融资、咨询、设计、建造、运维等多元化全产业链集成服务为核心，以"全过程咨询、工程建设管理、双碳新能源、城市更新、数字科技"5大板块为支撑的城乡建设科技集团。

启迪设计具备建筑工程、城乡规划编制、人防工程、风景园林、建筑智能化系统工程设计5项甲级资质在内的多项甲级资质和施工资质，汇聚了千余名优秀人才，建成了华东、华北、华南、西南、华中等多区域的全国性服务网络，为客户提供全方位的整体解决方案及长期的专业服务。

公司牢记传承、保护和发扬中华民族建筑文化的使命，坚持科技创新引领发展，致力于将启迪设计发展成为全国勘察设计行业领先，特色领域技术优势明显，综合实力雄厚的全国一流城乡建设科技集团。

Tus Design Group Co., Ltd. was founded in 1953, It is an urban and rural construction technology group with diversified whole industry chain integration services covering the whole process of human settlements, such as investment and financing, consulting, design, construction, operation and maintenance, as the core, and supported by five major plates of "whole process consulting, engineering construction management, dual carbon new energy, urban renewal, digital technology".

The company has a number of class A qualifications and construction qualifications, including class A in the construction industry. It has gathered more than 1 000 excellent talents and built a national service network in East China, North China, South China, Southwest China, central China and other regions to provide customers with comprehensive overall solutions and long-term professional services.

Bearing in mind the mission of inheriting, protecting and carrying forward the architectural culture of the Chinese nation, the company adheres to scientific and technological innovation to guide development, and is committed to developing edification design into a national first-class urban and rural construction science and technology group with strong comprehensive strength, leading the survey and design industry in the country, obvious technical advantages in characteristic fields.

地址：江苏省苏州工业园区星海街9号	Add: No.9, Xinghai street, Suzhou Industrial Park, Jiangsu
邮编：215021	P.C: 215021
电话：0512-65150100	Tel: 0512-65150100
邮箱：service@tusdesign.com	Email: service@tusdesign.com
网址：www.siad-c.com	Web: www.siad-c.com

枫桥工业园
Fengqiao Industrial Park

设计师：蔡爽、贾韬、李新胜、张慧、陆春华、闫莲、石晓燕、杨璐、徐明、唐旭、王锐、马琦、汤若飞、孙光、张志豪
项目地点：江苏 苏州
建筑面积：83 678.49 m²
用地面积：31 605.77 m²

Designers: Shuang Cai, Tao Jia, Xinsheng Li, Hui Zhang, Chunhua Lu, Lian Yan, Xiaoyan Shi, Lu Yang, Ming Xu, Xu Tang, Rui Wang, Qi Ma, Ruofei Tang, Guang Sun, Zhihao Zhang
Location: Suzhou, Jiangsu
Building Area: 83 678.49 m²
Site Area: 31 605.77 m²

项目顺应城市发展战略，融合景观、建筑设计，围绕园区健康产业定位，致力于打造长三角生物与生命科技产业高地，成为产业升级和城市更新形象展示区。

The project conforms to the urban development strategy, integrates landscape and architectural design, focuses on the positioning of the park's health industry, and is committed to building a highland of biological and life technology industry in the Yangtze River Delta, becoming an image display area for industrial upgrading and urban renewal.

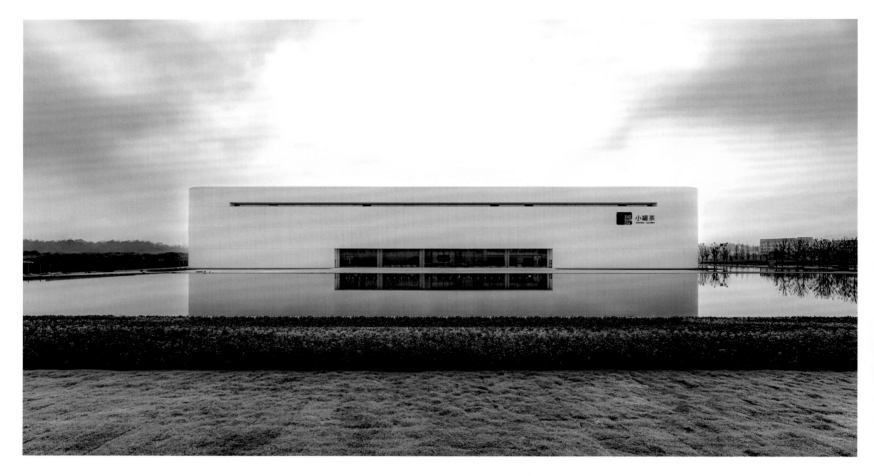

黄山小罐茶超级工厂
Huangshan Xiaoguancha Central Factory

设计师：胡哲、靳建华、王迎旭、金京洙、侯宇楠、朱晓磊、叶永毅、陈晶晶、庄岳忠、吴卫平、姚彬峰、赵德栋、凌贺君、曹雪平、蔡汉星
项目地点：安徽 黄山
建筑面积：105 919.08 m²
用地面积：116 470.24 m²

Designers: Zhe Hu, Jianhua Jin, Yingxu Wang, Jingzhu Jin, Yu'nan Hou, Xiaolei Zhu, Yongyi Ye, Jingjing Chen, Yuezhong Zhuang, Weiping Wu, Binfeng Yao, Dedong Zhao, Hejun Ling, Xueping Cao, Hanxing Cai
Location: Huangshan, Anhui
Building Area: 105 919.08 m²
Site Area: 116 470.24 m²

项目坐落在黄山脚下，是以弘扬中国茶文化为目标，以打造高端商品为目的的研究及生产总部。我们将"把简单的事情做到极致"这种沉稳精致的匠人精神通过设计展现出来。秉承"向方形致敬"的基本原则，运用最简单的方形构成园区的建筑及景观：通过对产品气质的理解，将方形的直角优化为圆角，且在项目初期就认真研究并选择了质感优良的立面材料，通过多方面的综合考虑使得建筑本身更温润细腻；规划布局着力思考与解决工业建筑与城市的关系，沿园区西侧主入口做较大水面，凸显了建筑形象，又丰富了城市主干线景观，空间对城市开放，成为黄山经济开发区的主要展示窗口。整个园区布局高低错落有致，形成良好的天际线关系，与周边山体的形态自然融合。

The project is located at the foot of Mount Huangshan. In the aim to promote Chinese tea culture and build a research and production headquarters for high-end products, we try to make simple things to the extreme through design, as the craftmanship of the branded. With the basic principle of "paying tribute to the square", the simplest square is used to form the buildings and landscape of the park. Through the understanding of the product temperament, the right angle of the square is optimized into rounded corners. At the beginning of the project, it has been carefully studied and selected with excellent texture materials of façade. comprehensive consideration makes the building itself warm and touching. The strategy of planning focus on solving the relationship between industrial buildings and the city, by making a water square along the main entrance on the west side of the park. Not only highlights the architectural image, but also enriches the city. It becomes the main display window of the Huangshan Economic Development Zone. The project forming a good relationship with the skyline, which is more naturally integrated with the shape of the surrounding mountains.

苏州丁家坞精品酒店
Suzhou Dingjiawu Boutique Hotel

设计师：查金荣、蔡爽、汪泱、张慧、范静华、李新胜、苏鹏、张志刚、殷文荣、张广仁、袁泉、陆凤庆、孙文、李杰、车伟
项目地点：江苏 苏州
建筑面积：20 935.83 m²
用地面积：47 877.7 m²

Designers: Jinrong Zha, Shuang Cai, Yang Wang, Hui Zhang, Jinghua Fan, Xinsheng Li, Peng Su, Zhigang Zhang, Wenrong Yin, Guangren Zhang, Quan Yuan, Fengqing Lu, Wen Sun, Jie Li, Wei Che
Location: Suzhou, Jiangsu
Building Area: 20 935.83 m²
Site Area: 47 877.7 m²

项目在设计时考虑含蓄而隐逸的东方文人精神与温良娴雅的生活理念，力图打造恬淡精致的度假体验。场地设计上试图做到清静、回归自然，利用山谷本身丰富的空间体验来塑造景观，创造建筑、人、自然和谐共生的亲密关系。

In the design of the project, the implicit and secluded Oriental Literati Spirit and gentle and elegant life attitude are considered, and the project tries to create a quiet and exquisite holiday experience. The site design tries to be quiet, return to nature, use the rich spatial experience of the valley itself to shape the landscape, and create a close relationship of harmonious coexistence between architecture, people and nature.

苏州北部文体中心
Suzhou North Cultural & Sports Center

设计师：蔡爽、贾韬、叶露、闫莲、陆春华、汪泱、石晓燕、钱成如、
杨阳、张黎忠、郑海霞、沈舟、孙文、徐辉、唐海兵
项目地点：江苏 苏州
建筑面积：48 000 m²
用地面积：23 830.03 m²

Designers: Shuang Cai, Tao Jia, Lu Ye, Lian Yan, Chunhua Lu, Yang Wang, Xiaoyan Shi, Chengru Qian, Yang Yang, Lizhong Zhang, Haixia Zheng, Zhou Shen, Wen Sun, Hui Xu, Haibing Tang
Location: Suzhou, Jiangsu
Building Area: 48 000 m²
Site Area: 23 830.03 m²

项目位于苏州工业园区，总建筑面积约48 000 m²，其中地上面积23 500 m²，地下面积24 500 m²。整个文体中心由剧院、综合体育馆、游泳馆、综合图书馆4个相对独立的单体组成。项目设计达到国家绿色建筑二星标准。功能配置上，室内主要有图书馆、剧场、多功能厅、文化展览空间、青少年活动中心等公共文化设施和综合体育馆、游泳馆等公共体育设施，室外有多功能运动场，并巧妙地嵌入了群众大舞台。

The project is located in Suzhou Industrial Park, with a total construction area of about 48 000 m², of which the aboveground area is about 23 500 m² and the underground area is 24 500 m². The whole cultural and sports center is composed of four relatively independent units: Theater, comprehensive gymnasium, natatorium and comprehensive library. Project design meets the national green building two-star standard. In terms of functional configuration, there are mainly public cultural facilities such as libraries, theatres, multi-function halls, cultural exhibition spaces, youth activity centers, and public sports facilities such as comprehensive gymnasiums and natatoriums. There is a multi-function playground outdoors, which is cleverly embedded in the mass stage.

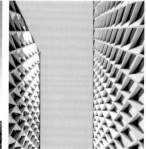

中银大厦
Bank of China Building

设计师：戴雅萍、查金荣、唐韶华、靳建华、李新胜、刘元杰、陈凯、
　　　　王莺、张志刚、陈磊、钱沛如、陈苏、沈丽芬、庄岳忠、王海港
项目地点：江苏 苏州
建筑面积：99 640.47 m²
用地面积：25 096 m²
容积率：3.18%
合作单位：贝氏建筑事务所

Designers: Yaping Dai, Jinrong Zha, Shaohua Tang, Jianhua Jin, Xinsheng Li, Yuanjie Liu, Kai Chen, Ying Wang, Zhigang Zhang, Lei Chen, Peiru Qian, Su Chen, Lifen Shen, Yuezhong Zhuang, Haigang Wang
Location: Suzhou, Jiangsu
Building Area: 99 640.47 m²
Site Area: 25 096 m²
Plot Ratio: 3.18%
Partner: PEI Architects

项目建筑设计的基本理念在于利用其基地东北两侧河流两岸美丽的景观及宽广的视野，采用简洁的建筑体型和外表来呈现这个在中国具有主导地位的金融机构的实力。贝氏建筑事务所将苏州园林的技法运用到世界各地的建筑项目中，包括这座地处姑苏的中银大厦。建筑与江南水景相互交融，传统的灰和白的主导色被充分利用，并采用装设性很强的石材铺地模式，苏式建筑的神韵被自然挥发出来。

The basic concept of the architectural design of the project is to use the beautiful landscape and broad vision on both sides of the river on the northeast side of the base, and adopt simple architectural shape and appearance to show the strength and dignity of this leading financial institution in China. Bayes applied the techniques of Suzhou gardens to construction projects all over the world, and it was no more than the Bank of China building located in Suzhou. The traditional dominant tone of gray and white is fully utilized, the stone paving mode with strong installation is adopted, and the mutual integration of architecture and Jiangnan waterscape, the charm of Soviet style architecture is naturally volatilized.

星湖大厦
Xinghu Building

设计师：查金荣、唐韶华、张稚雁、陈凯、王莺、郝怡婷、赵舒阳、王笑颜、沈茂松、钱沛如、庄岳忠、戴雅萍、赵宏康、武川川、王海港
项目地点：江苏 苏州
建筑面积：136 574.94 m²
用地面积：12 884 m²
建筑高度：222.8 m

Designers: Jinrong Zha, Shaohua Tang, Zhiyan Zhang, Kai Chen, Ying Wang, Yiting Hao, Shuyang Zhao, Xiaoyan Wang, Maosong Shen, Peiru Qian, Yuezhong Zhuang, Yaping Dai, Hongkang Zhao, Chuanchuan Wu, Haigang Wang
Location: Suzhou, Jiangsu
Building Area: 136 574.94 m²
Site Area: 12 884 m²
Building Height: 222.8 m

项目位于苏州工业园区湖东CBD，其位置决定它将成为高层建筑群的一个部分。建筑通过高度与形体的控制，使城市界面统一协调。方案在建筑体形上用裙房加点式塔楼的方式，放弃板式高楼带来的拥堵感。建筑以一种凌然而立的姿态与在附近的高层建筑作出呼应，形象大方稳重，能够和周边其他建筑协调。

The project is located in Hudong CBD of Suzhou Industrial Park. Its location determines that the building will become a part of the high-rise building group. The height and shape of the building are controlled to make the urban interface unified and coordinated. In the scheme, podium buildings and dotted towers are used to give up the sense of congestion caused by high-rise slab buildings. The building responds to the high-rise buildings nearby with a posture of being towering. The image is generous and stable, and can coordinate with other surrounding buildings.

大兆瓦风机新园区项目
Large MW Wind Turbine New Park Project

设计师：查金荣、蔡爽、李新胜、吴卫保、汪泱、叶露、胡旭明、石晓燕、
　　　　王云峰、钱如成、王开放、邓春燕、武川川、殷文荣、高展斌
项目地点：江苏 江阴
建筑面积：18 425.2 m²
用地面积：40 002.86 m²

Designers: Jinrong Zha, Shuang Cai, Xinsheng Li, Weibao Wu, Yang Wang, Lu Ye, Xuming Hu, Xiaoyan Shi, Yunfeng Wang, Rucheng Qian, Kaifang Wang, Chunyan Deng, Chuanchuan Wu, Wenrong Yin, Zhanbin Gao
Location: Jiangyin, Jiangsu
Building Area: 18 425.2 m²
Site Area: 40 002.86 m²

　　项目位于江苏省江阴市，规划为二期厂房1栋、餐厅综合楼1栋，在满足自身生产的同时预留远期生产研发的功能。该项目建成后将生产风能发电核心构件，配套辅助设施完备。多条生产线的生产用房与行政接待、研发办公、员工餐厅等组成了全新二期厂区。启迪设计在项目前期对冗杂功能进行了反复研究，通过前期策划、产线分析、布局研究明确了设计需求。

　　The project is located in Jiangyin City, Jiangsu Province. It is planned to be a phase II factory building and a restaurant complex building, which not only meet their own production, but also have the functions reserved for long-term production and R & D. After the completion of the project, it will produce the core components of wind power generation function, and the supporting auxiliary facilities are complete. The production rooms of multiple production lines, administrative reception, R & D office, staff restaurant, etc. form a new phase II plant. Edification design has repeatedly studied the miscellaneous functions in the early stage of the project, and clarified the design requirements through early planning, production line analysis and layout research.

骊住建材（苏州）有限公司新工厂
The New Factory Project of Lixin Building Materials (Suzhou) Co., Ltd

设计师：潘磊、吴海波、郭欣雨、严龙、沈亚军、季泽、张传杰、徐亦华、孙文隽、李立、孙谭秋、高展斌、陆景、李杰、王萌萌

项目地点：江苏 苏州

建筑面积：46 000 m²

用地面积：33 000 m²

Designers: Lei Pan, Haibo Wu, Xinyu Guo, Long Yan, Yajun Shen, Ze Ji, Chuanjie Zhang, Yihua Xu, Wenjun Sun, Li Li, Tanqiu Sun, Zhanbin Gao, Jing Lu, Jie Li, Mengmeng Wang

Location: Suzhou, Jiangsu

Building Area: 46 000 m²

Site Area: 33 000 m²

骊住建材（苏州）有限公司新工厂项目位于苏州市高新区，规划厂房1栋、餐厅1栋以及其他零星建筑，在满足自身生产的同时预留二期生产研发功能。多条生产线的生产用房与行政接待、研发办公、员工餐厅等组成了可容纳400人的全新厂区。

The New Factory Project of Lixin Building Materials (Suzhou) Co., Ltd. is located in Suzhou High Tech Zone. It is planned to build a factory, a restaurant and other sporadic buildings. It can meet its own production and at the same time serve the function reserved for phase II production and R & D. The production rooms of multiple production lines, administrative reception, R & D office, staff canteen, etc. form a new plant area that will accommodate 400 people.

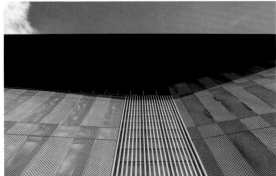

潘祖荫故居修缮整治工程
Pan Zuyin's Former Residence

设计师：蔡爽、吴树馨、孟庆涛、顾思港、殷茹清、朱晓蕾、钱盼、石文韬、周婧、徐佳宇、吴卫平、杨应秋、王笑颜、张广仁
项目地点：江苏 苏州
建筑面积：861 m²
用地面积：650 m²

Designers: Shuang Cai, Shuxin Wu, Qintao Meng, Sigang Gu, Ruqing Yin, Pan Qian, Wentao Shi, Jin Zhou, Jiayu Xu, Weiping Wu, Yingqiu Yang, Xiaoyan Wang, Guangren Zhang
Location: Suzhou, Jiangsu
Building Area: 861 m²
Site Area: 650 m²

作为苏州古城区优秀历史建筑的代表，潘祖荫故居是苏式传统民居精湛技艺及设计精神的载体。潘祖荫故居进路关系明确，组团清晰，作为重要历史人物的居住建筑，其体量大于周边普通民宅。本次修缮充分尊重其原有格局及形制，守住传统，还原历史格局。

As a representative of the excellent historical buildings in the ancient city of Suzhou, pan Zuyin's former residence is the carrier of the exquisite skills and design spirit of the traditional Soviet style dwellings. Pan Zuyin's former residence has clear access relations and clear clusters. As an important historical figure, the residential building has a larger volume than the surrounding ordinary houses. This renovation fully respects its original pattern and system, preserves the tradition and restores the historical pattern.

树山村改造提升工程
Shushan Village Reconstruction and Upgrading Project

设计师：查金荣、蔡爽、吴树馨、孟庆涛、顾思港、朱晓蕾、石文韬、
　　　　殷茹清、钱盼、柴庆霖、周婧、汪奕潇
项目地点：江苏 苏州
用地面积：15 000 km²

Designers: Jinrong Zha, Shuang Cai, Shuxin Wu, Qingtao Meng, Xiaolei Zhu, Wentao Shi, Ruqing Yin, Pan Qian, Qinglin Chai, Jing Zhou, Yixiao Wang
Location: Suzhou, Jiangsu
Site Area: 15 000 km²

设计过程中充分挖掘当地地域特色，尊重村内固有的历史文化特色及景观特色，保留村内体现农耕文化的景观，并使用与之相符的石、竹、木等原生态材料作为景观及建筑要素。重视村内的山水资源，注意打造及保护景观视通廊，美化村内池塘，美化花溪等水域资源。

During the design process, we should fully tap its regional characteristics, respect the inherent historical and cultural characteristics and landscape characteristics of the village, retain the farming culture of the village, and use the corresponding raw ecological materials such as stone, bamboo, wood as landscape and architectural elements. Pay attention to the landscape resources in the village, build and protect the landscape corridor, beautify the pond in the village, beautify the water resources such as Huaxi.

常州市三江口公园
Changzhou Sanjiangkou Park

设计师：毛永青、叶凡、王加伟、钱怡婷、吴尚、廖嘉、陈池、许彩芬、万旭平、付光、刘淼珺、孙昊、李冰晖、张越、张佳敏

项目地点：江苏 常州

用地面积：246 000 m²

Designers: Yongqing Mao, Fan Ye, Jiawei Wang, Yiting Qian, Shang Wu, Jia Liao, Chi Chen, Caifen Xu, Xuping Wan, Guang Fu, Miaojun Liu, Hao Sun, Binghui Li, Yue Zhang, Jiamin Zhang

Location: Changzhou, Jiangsu

Site Area: 246 000 km²

将"太阳辐射的光波"作为设计概念，以三江口公园作为"太阳的能量之核"向四周区域扩散，绿色、文化、休闲3条"光波"相互交融，贯穿始末，交织成不同的功能空间。

The light wave radiated by the sun is taken as the design concept, and Sanjiangkou park is taken as the energy core of the sun to spread to the surrounding areas. The three light waves of green, culture and leisure blend with each other, running through the whole and interweaving into different functional spaces.

中国科学技术大学苏州研究院仁爱路校区（166、188地块）
Suzhou Research Institute of University of Science and Technology of China

项目地点：江苏 苏州
建筑面积：146 000 m²
用地面积：20 200 m²

Location: Suzhou, Jiangsu
Building Area: 146 000 m²
Site Area: 20 200 m²

中国中医科学院大学
China Academy of Chinese Medical Sciences

项目地点：江苏 苏州
建筑面积：320 000 m²
用地面积：406 666 m²
联合体成员：清华大学建筑设计研究院有限公司

Location: Suzhou, Jiangsu
Building Area: 320 000 m²
Site Area: 406 666 m²
Consortium Members: Architectural Design & Research Institute of Tsinghua University Co., Ltd.

昆山杜克大学二期
Duke Kunshan University Phase II

项目地点：江苏 昆山
建筑面积：95 600 m²
合作单位：Perkins&Will、同济大学设计院

Location: Kunshan, Jiangsu
Building Area: 95 600 m²
Partner: Perkins&Will, Design Institute of Tongji University

中国常熟世联书院
China Changshu World Union Academy

项目地点：江苏 苏州
建筑面积：56 443 m²
用地面积：41 825 m²
合作单位：莫平建筑设计顾问（北京）有限公司

Location: Suzhou, Jiangsu
Building Area: 56 443 m²
Site Area: 41 825 m²
Partner: Moping Architectural Design Consulting (Beijing) Co., Ltd.

苏州城发建筑设计院有限公司
SUZHOU URBAN-DEVELOPMENT ARCHITECTURAL DESIGN INSTITUTE CO.,LTD.

苏州城发建筑设计院有限公司成立于2006年2月,业务范围贯通全产业链,涵盖城市规划、建筑工程设计、建筑装饰工程设计、建筑幕墙工程设计、轻型钢结构工程设计、建筑智能化系统设计、装配式设计、BIM技术咨询、海绵城市设计、绿色建筑设计等领域,提供概念方案规划、方案设计、施工图设计等全项目周期服务。在历年省、市勘察设计行业各项评比中名列前茅。

SUZHOU URBAN-DEVELOPMENT ARCHITECTURAL DESIGN INSTITUTE CO.,LTD. was established in February 2006, and its business scope covers the whole industry chain, including urban planning, building engineering design, building decoration engineering design, building curtain wall engineering design, light steel structure engineering design, building intelligent system design, assembly design, BIM technology consulting, sponge city design, green building design, etc., providing concept planning, scheme design, construction drawing design and other services. We provide full project cycle services such as scheme design and construction drawing design. In the past years, the provincial and municipal survey and design industry ranked top in the various evaluation.

地址:苏州市东环南路1号和诚大厦6楼　　Add: 6F HeCheng EDIFICE NO.1,South East Outer Ring Road,Suzhou
邮编:215000　　　　　　　　　　　　　　P.C: 215000
电话:0512-62873866　　　　　　　　　　Tel: 0512-62873866
传真:0512-67773881　　　　　　　　　　Fax: 0512-67773881
邮箱:mail.szudad.com　　　　　　　　　Email: mail.szudad.com
网址:www.szudad.com　　　　　　　　　Web: www.szudad.com

苏州城发建筑设计院设计研发中心
Suzhou Urban-development Architectural Design Institute Design & Research Center

设计师:夏平、李丽、张薇薇	Designers: Ping Xia, Li Li, Weiwei Zhang
项目地点:江苏 苏州	Location: Suzhou, Jiangsu
建筑面积:24 887 m²	Building Area: 24 887 m²
用地面积:0.7 hm²	Site Area: 0.7 hm²
容积率:1.99	Plot Ratio: 1.99

利用起承转合与平面构成的设计语言，极大地丰富了建筑表皮细节，光影变换之间营造出具有力量感的建筑之美。两个主体建筑形成大气沉稳的整体，大面积的立面材质采用白色铝板和纯净透亮的玻璃幕墙，现代简约。整个项目侧重于建筑功能最大化、建筑立面最优化，为企业和员工提供优质高效的综合性办公总部。

Using the design language of rise and fall and planar composition, the building skin is greatly enriched, creating an architectural aesthetic with a sense of power between light and shadow transformation. The two main bodies form an atmospheric and calm whole, with large areas of white aluminum panels and pure and translucent glass curtain walls, modern and simple. The whole project focuses on maximizing the building functions and optimizing the building facade to provide a high-quality and efficient comprehensive office headquarters for enterprises and employees.

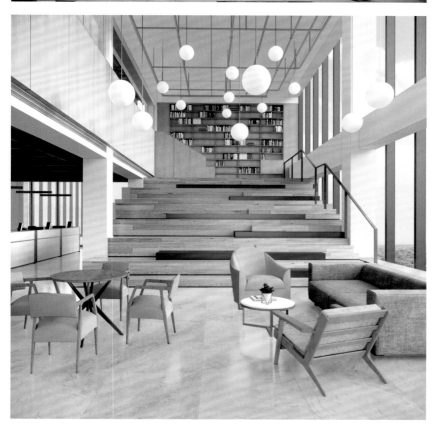

仁恒仓街商业广场
YANLORD Suzhou Cangjie Commercial Plaza

设计师：夏平、魏轶炫、达峰
项目地点：江苏 苏州
建筑面积：138 421.91 m²
用地面积：3.92 hm²
容积率：0.795

Designers: Ping Xia, Yixuan Wei, Feng Da
Location: Suzhou, Jiangsu
Building Area: 138 421.91 m²
Site Area: 3.92 hm²
Plot Ratio: 0.795

本工程地块位于苏州古城东隅，临近古城墙。本设计方案以保护为目的，能提升古城周边的区域品质，保持古城风貌的协调。古迹建筑与公共建筑之间的关系以及突出运河是此项目的重点——运河零售街依托运河设置，打造天然水滨商业；广场及商业街与古城门相对，营造古城商业氛围；博物馆周边形成花园餐饮区，并与车辆落客区相邻，方便游客参观。

The project site is located in the eastern corner of the ancient city of Suzhou, adjacent to the ancient city wall. This design scheme aims to preserve and enhance the quality of the area around the ancient city and maintain the harmony of the ancient city's appearance. The relationship between monumental buildings and public buildings and highlighting the canal is the focus of this project - the canal retail street is set up by the canal to create natural waterfront commerce; the plaza and commercial street are opposite the ancient city gate to create the commercial atmosphere of the ancient city; the garden dining area is formed around the museum and adjacent to the vehicle drop-off area to facilitate visitors' visit.

旭辉铂悦犀湖
CIFI Lake Mansion

设计师：夏平、魏轶炫、蒋华
项目地点：江苏 苏州
建筑面积：312 200 m²
用地面积：13.10 hm²
容积率：1.60

Designers: Ping Xia, Yixuan Wei, Hua Jiang
Location: Suzhou, Jiangsu
Building Area: 312 200 m²
Site Area: 13.10 hm²
Plot Ratio: 1.60

对于这块独墅湖最后的滨水之地，整个湖面都是项目基地的后花园，如何最大限度地去感知这片城市水域是本案规划的核心主题。建筑应与环境彼此依赖，相互呼应，将湖景特色打造到极致，成为最理想的人居环境和人人向往的滨水住宅。我们希望本案的建筑对于环境是一个渐渐消隐的形象，它不压迫也不侵扰城市空间，同时还能反映湖水的质感，延续滨水的居住感觉，因此高层的设计选择了用玻璃和铝材质的结合来打造这个形象。

For this last waterfront area of Dushu Lake, the entire lake is the back garden of the project base, and how to maximize the perception of this urban water is the core theme of this case planning. The architecture and the environment should depend on each other and echo with each other, so as to make the lake-view feature the most ideal living environment and the waterfront residence to which everyone aspires. We hope that the building in this case is a fading image to the environment, which does not oppress or intrude into the urban space, but also reflects the texture of the lake to continue the feeling of living on the waterfront, therefore, the combination of glass and aluminum is chosen for the design of the upper floors to express this image.

太仓天镜湖展示中心
YANLORD Tianjing Lake Exhibition Center

设计师：魏轶炫、柴继峰	Designers: Yixuan Wei, Jifeng Chai
项目地点：江苏 太仓	Location: Taicang, Jiangsu
建筑面积：30 279.67 m²	Building Area: 30 279.67 m²
用地面积：1.48 hm²	Site Area: 1.48 hm²
容积率：1.34	Plot Ratio: 1.34

本项目入选具有建筑界"奥斯卡奖"之称的美国"Architizer A+Awards"低层商业办公类别的决赛名单，成为该类别的全球前5强项目。建筑方案以"涟漪"为灵感来源，由几条相互交叠的圆弧曲线构成，双螺旋的形态形成了极具特色的内部空间。项目的真正难点在于如何从概念效果到落地，设计师通过BIM设计将建筑、结构、水、暖、电、内装、幕墙、景观进行精细化建模，形成全专业综合BIM模型，协同设计，最终实现了"大、静、空"的展示空间效果。

The project was selected as a finalist in the Architizer A+Awards COMMERCIAL Office - Low Rise (1-4 Floors) category, making it one of the top 5 projects worldwide in this category. The architectural scheme is inspired by "ripples" and is composed of several overlapping circular curves, with a double helix form forming a distinctive interior space. The real difficulty of the project lies in the implementation of the conceptual effect. Through BIM design, architectural, structure, water supplyment and drainage, HVAC, electricity, interior decoration, curtain wall and landscape are finely modeled to form a comprehensive BIM model of the whole profession, and the collaborative design finally realizes the effect of "big, quiet and pure" exhibition space.

玉成实验小学
Yucheng Experimental Primary School

设计师：夏平	Designer: Ping Xia
项目地点：江苏 苏州	Location: Suzhou, Jiangsu
建筑面积：61 200 m²	Building Area: 61 200 m²
用地面积：5.99 hm²	Site Area: 5.99 hm²
容积率：0.63	Plot Ratio: 0.63

将新时代新型小学、幼儿园全面素质教育与地方传统文化教育相结合，在这里，我们希望传统的教学空间被延伸到户外，重拾对场地的人文关怀。校园内外空间合为一体，共同探索成长的多维度，打造一个可以在阳光下奔跑的快乐校园。校园中绿树婆娑、书香飘逸，建筑风格兼具历史感与时代感，彰显学校以中国传统玉文化为载体，寻求内涵发展、特色发展、创新发展、品牌发展的理念。

This project combines comprehensive quality education of new elementary school and kindergartens in the new era with local traditional culture education, where we hope that the traditional teaching space is extended to the outdoors and the humanistic care of the site is regained. The space inside and outside the campus is combined into one, exploring the multidimensionality of growth together and giving every childhood and youth a happy campus where they can run under the sun. The campus is lined with green trees and fragrant books, and the architecture has both a sense of history and a sense of the times. With the traditional Chinese jade culture as the carrier, the school seeks connotation development, characteristic development, innovation development and brand development.

黄桥总部经济园
Huangqiao Headquarters Economic Park

设计师：夏平、魏轶炫、张洁
项目地点：江苏 苏州
建筑面积：171 817 m²
用地面积：5.97 hm²
容积率：1.96

Designers: Ping Xia, Yixuan Wei, Jie Zhang
Location: Suzhou, Jiangsu
Building Area: 171 817 m²
Site Area: 5.97 hm²
Plot Ratio: 1.96

在高楼林立的城市森林中,我们如何觅得那渐行渐远的自然体验?

充分引入水景,将建筑环绕中央水景布置,草坡使得建筑与场地融为一体,水边漫步,草坡憩息,与自然相融为一。空中大厅结合屋顶花园模糊了室内外的边界,生态景观由外而内缓缓渗透,极目远眺,远景近景尽收眼底。阳光、水、空气、绿化,在一座生态绿色建筑中,我们期待生活与美好的不期而遇。

In the urban forest of high-rise buildings, how can we find the experience of nature that is fading away?

The waterscape is fully introduced, the building is arranged around the central waterscape, and the grass slope makes the building and the site blend into one. Walking by the water, resting on the grass slope, blurring the boundary between indoor and outdoor with the sky hall combined with the roof garden, the ecological landscape slowly penetrates from outside to inside, and the distant and near scenery can be seen at a glance. Sunlight, water, air, greenery, in an ecological green building, we look forward to an unanticipated encounter between life and beauty.

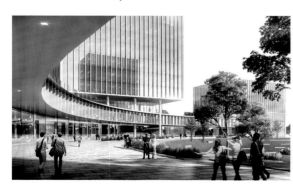

辽宁

 中国建筑东北设计研究院有限公司
CHINA NORTHEAST ARCHITECTURAL DESIGN & RESEARCH INSTITUTE CO.,LTD.

中国建筑东北设计研究院有限公司始建于1952年，隶属于世界500强企业——中建集团。公司经营业绩遍布国内及海外57个国家，业务布局覆盖城乡规划、勘察设计、顾问咨询、建筑施工、新兴业务等建筑工程全产业链，拥有城乡规划编制甲级、建筑工程甲级、市政行业甲级、工程勘察综合类甲级等10余项甲级设计资质和建筑工程施工总承包一级工程资质。现有员工3 000余人，其中全国工程勘察设计大师2人，辽宁省工程勘察设计大师20人，享受国务院特殊津贴专家22人，是东北地区历史最久、规模最大、品牌最强、一体化程度最高的大型综合建筑设计企业。

Established in 1952, China Northeast Architectural Design and Research Institute Co., Ltd. ("CNADRI") is subordinated to China State Construction Engineering Corporation ("CSCEC"), a world TOP 500 enterprise. CNADRI has undertaken the design projects across all provinces, municipalities and autonomous regions in China and 57 countries and regions in the world. CNADRI's business covers the whole industrial chain of construction engineering, including urban and rural planning, engineering survey and design, consultation, building construction, and emerging businesses. CNADRI has more than 10 Class-A design qualifications, such as Class-A for urban and rural planning compilation, Class-A for construction engineering, Class-A for municipal engineering design and Class-A for comprehensive engineering survey, as well as Class-I qualification for EPC. CNADRI and its subsidiaries have more than 3 000 employees in total. Over the years, CNADRI has cultivated numerous experts, including two National Engineering Survey and Design Masters, 20 Engineering Survey and Design Masters of Liaoning Province, and 22 experts enjoying the Special Allowance of the State Council. CNADRI is a large-scale comprehensive building design enterprise with the longest history, the largest scale, the strongest brand and the highest degree of integration in northeast China.

地址：辽宁省沈阳市和平区光荣街65号
电话：024-81978000
传真：024-81978000
邮箱：dbyzgb@cscec.com
网址：https://nein.cscec.com/

Add: No.65, Guangrong Street, Heping District, Shenyang, Liaoning
Tel: 024-81978000
Fax: 024-81978000
Email: dbyzgb@cscec.com
Web: https://nein.cscec.com/

中建东北院总部大厦
The Headquarters Building Project of China Northeast Architectural Design and Research Institute Co., Ltd.

项目地点：辽宁 沈阳
设计时间：2017年－2020年
竣工时间：2022年
建筑面积：77 329.5 m²
用地面积：5 980 m²

Location: Shenyang, Liaoning
Design Time: 2017-2020
Completion Time: 2022
Building Area: 77 329.5 m²
Site Area: 5 980 m²

中建东北院总部大厦项目是公司在新形势下，为适应市场发展需求，提高企业竞争力，进行转型升级过程中的重点工程。新建办公楼总建筑面积77 329.5 m²，以注重建筑与城市之间的关系、关注地块历史文脉与沿革、满足功能空间使用的灵活性、体现绿色建筑与可持续发展性和创造独特的建筑形象并体现项目的文化特色为设计理念。设计过程中，公司集内部设计资源优势，在规划与建筑方案、景观环境、室内装修、建筑幕墙、夜景亮化设计等方面充分发挥企业自身技术优势，尤其在施工图设计过程中采用三维协同设计，运用BIM技术提高设计与服务质量。

The headquarters building project of China Northeast Architectural Design and Research Institute Co., Ltd. is a key project in the process of transformation and upgrading in order to meet the market development needs and improve the competitiveness of the company under the new situation. The total construction area of the new office building is 77 329.5 m², the design concept of which is to focus on the relationship between the building and the city, to pay attention to the historical context and evolution of the land parcel, to satisfy the flexibility of the use of functional space, to reflect the green building and sustainable development, to create a unique architectural image and embody the cultural characteristics of the project. During the design process, the company takes advantage of internal design resources to give full play to the technical advantages of the enterprise in planning and architectural schemes, landscape environment, interior decoration, building curtain wall, night scene lighting design, etc., especially in the process of construction drawing design, using three-dimensional collaboration design and BIM technology to improve design and service quality.

北山四季越野滑雪场项目
Beishan Four Seasons Cross-country Ski Resort Project

设计师：李力红、王洪礼、马佳、刘一、张晋蓉
项目地点：吉林 吉林
建筑面积：30 000 m²

Designers: Lihong Li, Hongli Wang, Jia Ma, Yi Liu, Jinrong Zhang
Location: Jilin, Jilin
Building Area: 30 000 m²

该项目利用吉林北山风景区山腹中的既有人防隧道，改扩建成越野滑雪场。多项创新技术应用实现了绿色建造、节能运行的目标。该项目是与隧道技术结合的生态覆土类建筑，既满足2022冬奥会国家队冬季两项运动员夏季训练以及省市运动队四季训练使用需求，同时面向公众开放500 m观光雪道以推广冰雪运动。该项目为亚洲首座、世界第四座越野滑雪隧道。洞内专业雪道1.3 km，洞外1.7 km。

In 2017, it was expanded into a cross-country ski resort using the existing manned tunnel in the hinterland of the mountain in Jilin Beishan Scenic Area. A number of innovative technology applications have achieved the goal of green construction and energy-saving operation. It is an ecological overburden building combined with tunnel technology. It not only meets the summer training of the national team's winter biathletes and the four-season training of provincial and municipal sports teams, but also opens 500 m sightseeing trails to the public to promote ice and snow sports. The project is the first cross-country ski tunnel in Asia and the fourth cross-country ski tunnel in the world. The professional snow tunnel inside the cave is 1.3 km, and the cave is 1.7 km away.

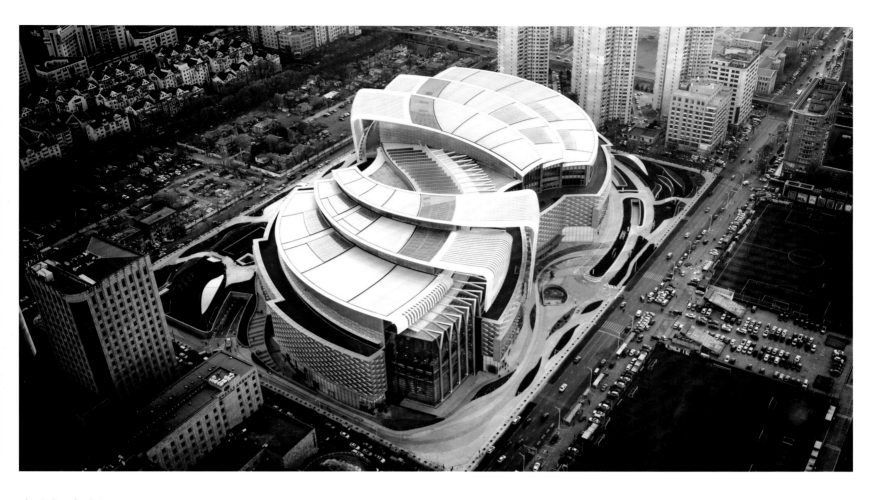

大连恒隆广场
Dalian Henglong Plaza

设计师：陈天禄、高嵩、朱江、赵成中、苏晓丹、于欣、赵磊、张学刚、
　　　　田丰、宁兆恒、刘旭、张守平、李少博、赵金文、邹冬
项目地点：辽宁 大连
建筑面积：371 900 m²
用地面积：6.34 hm²
容积率：3.5
联合设计：凯达环球建筑设计有限公司

Designers: Tianlu Chen, Song Gao, Jiang Zhu, Chengzhong Zhao, Xiaodan Su, Xin Yu, Lei Zhao, Xuegang Zhang, Feng Tian, Zhaoheng Ning, Xu Liu, Shouping Zhang, Shaobo Li, Jinwen Zhao, Dong Zou
Location: Dalian, Liaoning
Building Area: 371 900 m²
Site Area: 6.34 hm²
Plot Ratio: 3.5
Joint Designer: Aedas

　　大连恒隆广场是一座集购物、餐饮、娱乐、休闲、体验、商务洽谈等功能于一体超大商业综合体，为消费者带来一站式的生活体验。它的出现，树立了大连高端商业项目的新标杆，设计所采用的"双鱼、航船、浪花"等海洋元素，契合"时尚之都、浪漫大连"的滨海城市文化，目前仍然是东北地区已建成最大规模的单体商业建筑。本项目由我院与凯达环球建筑设计有限公司联合设计完成。

　　Dalian Olympia 66 - the largest single commercial building in northeastern China, is a large commercial combination of shopping, catering, entertainment, leisure activities, business negotiation and other functions, providing consumers a one-stop living experience. Its emergence has set a new benchmark for high-end commercial projects in Dalian, where the marine elements such as "Double-fish, Ships, and Waves" adopted in the design are in line with the coastal city culture of "Fashion Capital, Romantic Dalian". This project is jointly designed and completed by our institute and Aedas.

大连中心·裕景
Eton Place · Dalian

设计师：魏立志、苏晓丹、赵成中、苏志伟、高嵩、朱江、张学刚、
赵磊、曹立强、田丰、金鹏、谭明、陈天禄、赵海波
项目地点：辽宁 大连
建筑面积：475 869 m²
用地面积：4.1 hm²
容积率：9.95
建筑高度：383 m
联合设计：美国NBBJ建筑设计公司

Designers: Lizhi Wei, Xiaodan Su, Chengzhong Zhao, Zhiwei Su, Song Gao, Jiang Zhu, Xuegang Zhang, Lei Zhao, Liqiang Cao, Feng Tian, Peng Jin, Ming Tan, Tianlu Chen, HaiBo Zhao
Location: Dalian, Liaoning
Building Area: 475 869 m²
Site Area: 4.1 hm²
Plot Ratio: 9.95
Building Height: 383 m
Joint Designer: NBBJ

大连中心·裕景位于城市商业中心，是一座规模庞大、功能多元的城市综合体，是目前东北地区已建成的第一高楼，代表着"北方明珠、浪漫之都"的城市新高度，成为城市天际线的最高点。项目整体设计以"山峰与河谷"为题，2栋塔楼的顶部及一个角部通过切削的方式形成了尺度巨大、雕塑感强烈的造型。

本项目由我院与美国NBBJ建筑设计公司联合设计完成。

Eton Place · Dalian, the highest building in Northeast China, located in the commercial center of Dalian City, is the large-scale, multi-functional city complex super high rise building. Eton Place · Dalian becomes the most glorious architecture in the skyline of Dalian, called the the pearl of the north and the city of romance, which it stands for the new height of. "Mountain Top and River Valley", as the overall design theme, makes large-scale shape with a strong sense of sculpture, which is by cutting a form through the top and one corner of the two towers.

This project is jointly designed and completed by our institute and NBBJ.

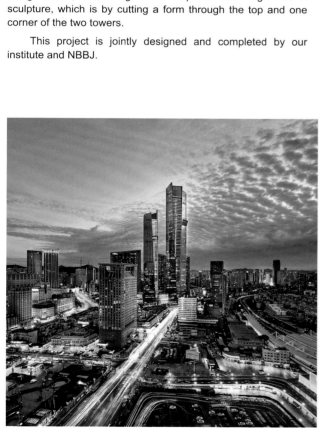

润友科技长三角(临港)总部项目
RunYou Technology (Lingang) Headquarters Project

设计师：曹辉
设计时间：2020年
竣工时间：在建
项目地点：上海
建筑面积：56 236 m²
用地面积：12 047.8 m²
容积率：3.0

Designer: Hui Cao
Design Time: 2020
Completion Time: Under Construction
Location: Shanghai
Building Area: 56 236 m²
Site Area: 12 047.8 m²
Plot Ratio: 3.0

润友科技长三角(临港)总部项目，在高度限制的场地上，强调场地上的横向力量，折叠上升的体块，摈弃多余的修饰，纯粹的形式给予人们以精神力量。悬挑的玻璃体延伸出"眺望滴水湖"的盒子，"虚实交错"的形体削弱了建筑体量感，两个竖向主体在基底及顶层相连，在半空中围合出气势恢宏的室外中庭空间，以促进办公与社会活动之间创造性的"异花授粉"。立面气泡窗悬挂在建筑之外，呈现梦幻的视觉感受，提高人们的感知度，获得更深层的思考。

The building form is generated by the intersections of the volumes. With the limited height restriction, the building is designed to emphasize its horizontality. With minimal decoration and pierced elements, the architecture is designed in a way to indicate its spiritual power. The gigantic transparent glass block slides out from the base of the building, with a view overlooking the lake. The intersection of the volumes imposes lightness and an open posture toward the environment. The two main vertical bodies are connected at the base and the top, enclosing a magnificent outdoor atrium, where multi-level terraces are created here to foster social activity and creative cross-pollination. The bubble - shape windows are adopted on the building facade, which dramatically contrasts with the traditional appearance of the public building. This new image aims to add to people's perceptions of the environment.

包钢集团设计研究院（有限公司）
Baogang Group Design & Research Institute Co., Ltd.

包钢集团设计研究院（有限公司）创建于1954年5月1日，前身为包钢设计处，是国家第一个五年计划为建设包头钢铁公司而组建的冶金设计单位。拥有与包钢生产全流程息息相关的30多个专业的专业技术团队和控股公司：北京清润国际建筑设计研究有限公司。现持有冶金行业甲级、建筑行业（建筑工程）甲级、电力行业（火力发电）专业乙级等多项设计资质。

Baotou Steel Group Design and Research Institute (Co., Ltd.) was founded on May 1, 1954, formerly known as Baotou Steel Design Office, is the country's first five-year plan to build Baotou Iron and Steel Company and the establishment of metallurgical design units. Has more than 30 professional and technical teams closely related to the whole process of Baotou steel production and holding company: Beijing Qingrun International Architectural Design and Research Co., Ltd. and now holds a grade A in the metallurgical industry; Construction industry (construction engineering) Grade A; Power industry (thermal power generation) professional grade B and other design qualifications.

地址：内蒙古包头市昆都仑区钢铁大街89号	Add: No.89, Iron and Steel Street, Kundulun District, Baotou, Inner Mongolia
邮编：014010	P.C: 014010
电话：0472-2127483	Tel: 0472-2127483
传真：0472-2125268	Fax: 0472-2125268
邮箱：bgsjy@btsteel.com	Email: bgsjy@btsteel.com
网址：www.bgsjy.cn	Web: www.bgsjy.cn

垂直森林－山水林居
Shanshuilinju, Vertical Forest

设计师：马树新、李晓京、张海鹰、龚艳红、崔佳、邵文娟
项目地点：河北 沧州
建筑面积：54 172.63 m²
用地面积：19 101.39 m²
容积率：1.994

Designers: Shuxin Ma, Xiaojing Li, Haiying Zhang, Yanhong Gong, Jia Cui, Wenjuan Shao
Location: Cangzhou, Hebei
Building Area: 54 172.63 m²
Site Area: 19 101.39 m²
Plot Ratio: 1.994

该项目地块总用地面积19 101.39 m²，包括住宅、配套公建、地下车库。"垂直森林"建筑是绿色建筑、生态建筑、可持续建筑的最佳载体，具有极高的科技含量，把空气、绿色与生命力带入建筑，把装配式、集成式、低碳式等高技术带入建筑，把私密性、庭院与居室的空间有机交融带入建筑。户型设计采用"九宫格"式布局，以客厅为中心，客厅空间与餐厨空间、庭院空间融为一体，每户的露台庭院都是向上挑空的，挑空高度为6.3 m~9.45 m，可以种大树，又不计算建筑面积。每户院子都是私密的，在院子里有十足的安全感。

The total site area of the project is 19 101.39 m², including residential, supporting public construction and underground garage. "Vertical forest" building is the best carrier of green buildings, ecological buildings and sustainable buildings, with high scientific and technological content. It is to bring air, green and vitality into the building, to bring high-tech such as prefabricated, integrated, and low-carbon into the building, and to bring the organic blend of privacy, courtyard and living room space into the building. The apartment design is based on the layout of the nine-palace grid, the living room as the center, the living room space and the kitchen space, the courtyard space is integrated, and the terrace courtyard of each household is upward and empty, and the height of the empty space is 6.3 m - 9.45 m, which can plant large trees, but does not calculate the construction area. Each yard is private and there is a complete sense of security in the yard.

包钢宾馆
Baogang Hotel

设计师：马树新、张云、王凡
项目地点：内蒙古 包头
建筑面积：12 179.49 m²
用地面积：10 151.54 m²
容积率：1.2
家具配套商：楷模家居

Designers: Shuxin Ma, Yun Zhang, Fan Wang
Location: Baotou, Inner Mongolia
Building Area: 12 179.49 m²
Site Area: 10 151.54 m²
Plot Ratio: 1.2
Furniture Supplier: Coomo Home

包钢宾馆分为1、2号楼建筑群，1号楼地上3层，为砖混结构，是20世纪50年代所建住宅楼，于20世纪90年代改建，作为宾馆客房使用。2号楼群为砖混与框架混合结构，于20世纪90年代新建。

此次改造是对1号楼、2号楼进行装修翻新、消防改造，并依据使用和功能要求展开建筑、结构、给排水、暖通空调、建筑电气、装修的专业设计。

Baogang Hotel is divided into Buildings 1 and 2, Building 1 has three floors above ground, which is a brick-concrete structure, and the residential building built in the 1950s was rebuilt in the 1990s as a hotel room. Building 2 is a mixed brick and frame structure, which was newly built in the 1990s.

This renovation is to renovate and renovate Building 1 and Building 2, fire protection transformation, and carry out professional design of buildings, structures, water supply and drainage, HVAC, building electrical and decoration according to the use and functional requirements.

超级碗
Super Bowl

设计师：马树新、李晓京
项目地点：河北 沧州
建筑面积：30 000 m²

Designers: Shuxin Ma, Xiaojing Li
Location: Cangzhou, Hebei
Building Area: 30 000 m²

该项目用地面积32 248.21 m²，主楼设计为4层，建筑外立面采用玻璃幕墙，设计简洁大方，形成极具现代感与前瞻性的建筑风貌。以绿色生态、适用美观为导则，深入研究建筑功能与周边环境之间的关系。平面以圆形为设计元素，结合功能布局与文化传承，将内院作为空间纽带，形成室内空间与外部环境的有机融合，营造具有向心力凝聚力的场所精神。整体建筑空间体现展览建筑的开放性与包容性。

设计灵感来源于中国天坛与宋瓷天目盏，将现代钢结构技术与中国木构文化相结合，用细部代替装饰，体现精致美、材料美、线条美。

The project has a land area of 32 248.21 m², the main building is designed as a four-storey, the façade of the building adopts glass curtain wall, the design is generous, forming a very modern and forward-looking architectural style. Based on the principle of green ecology and aesthetic application, we will deeply study the relationship between building functions and the surrounding environment. The plane takes the circle as the design element, combines the functional layout and cultural inheritance, takes the inner courtyard as the spatial link, forms the organic integration of the indoor space and the external environment, and creates a place spirit with centripetal cohesion. The overall architectural space reflects the openness and inclusiveness of the exhibition building.

The symbolic thread of the building is inspired by the Chinese Temple of Heaven and the Song Porcelain Tianmu Cup. It combines modern steel structure technology with Chinese wooden structure culture, and replaces decoration with fine details to reflect exquisite beauty, beautiful materials, and beautiful lines.

状元楼
Zhuangyuan Building

设计师：马树新
项目地点：河北 沧州

Designer: Shuxin Ma
Location: Cangzhou, Hebei

该项目占地1 578 m²，建筑面积为5 600 m²，位于状元湖北侧，状元湖公园内。建筑高度为46.5 m，地上7层，地下1层。地下一层为博物馆，一层为图书馆，平台之上2~7层为观光塔。

建筑风格为仿古建筑，正六边形塔身。塔顶将庑殿顶与歇山顶有机结合，从正北、东南、西南方向观看，塔顶为庑殿顶；从正南、西北、东北方向观看，塔顶为歇山顶。状元楼利用多个歇山顶造型，形成韵律美，突出"众"字，寓意状元为人上人，与状元楼名称暗合。设计通过采用现代钢结构技术和中国传统木构文化的结合，设计美学与和谐对称的关系组织，打造体现东方神韵的阁楼建筑。

The project covers an area of 1 578 m², with a construction area of 5 600 m² located on the north side of Zhuangyuan Lake, in Zhuangyuan Lake Park. The building has a height of 46.5 m with seven floors above ground and one floor underground. The first basement floor is a museum, the first floor is a library, and the 2-7 floors above the platform are sightseeing towers.

Style antique building, hexagonal tower, the top of the tower is the organic combination of the top of the temple and the top of the mountain, from the north, southeast, southwest view, the top of the tower is the top of the temple. Viewed from the south, northwest and northeast, it is the summit of the tower. The Zhuangyuan Pagoda uses a number of peaks to form a rhythmic beauty, highlighting the word "crowd", implying that the Zhuangyuan is a superior person, which is implicitly consistent with the name of the Zhuangyuan Pagoda. Using the combination of modern steel structure technology and traditional Chinese wood culture, the design aesthetic and harmonious symmetry relationship organization embodies the oriental charm of the attic building.

一号高炉展厅设计
No.1 Blast Furnace Exhibition Hall Design

设计师：王朝宇、马树新、钮建慧
项目地点：内蒙古 包头
建筑面积：270.69 m²

Designers: Zhaoyu Wang, Shuxin Ma, Jianhui Niu
Location: Baotou, Inner Mongolia
Building Area: 270.69 m²

为了应对厂区高粉尘的环境，包钢一号高炉展厅设计采用实体屋面、玻璃墙面的设计。整个方案采用通透的处理手法，让原有浮雕及新做的半透明浮雕与其身后的一号高炉融为一体，用一号高炉给整个参观做背景，增加历史印记的现场感，让历史的印记镌刻在一号高炉上，也留在每位参观者的心中。

Baotou Steel No.1 blast furnace exhibition hall design in order to cope with the high dust environment of the factory area, the design of solid roof and glass wall is adopted. The whole scheme adopts a transparent treatment method, so that the original relief and the newly made translucent relief are integrated with the No.1 blast furnace behind it, and the No.1 blast furnace is used to make the background for the entire visit, increasing the sense of the scene of the historical imprint, so that the historical imprint is engraved on the No.1 blast furnace, so that the historical imprint remains in the heart of each visitor.

山东

山东省建筑设计研究院有限公司
Shandong Provincial Architectural Design & Research Institute Co., Ltd.

山东省建筑设计研究院有限公司是山东省建筑设计研究院改制企业，是国家甲级勘察设计单位、全国重信用守合同单位、全国建筑设计行业诚信单位、当代中国建筑设计百家名院。公司主要从事建筑、规划、市政工程设计、工程勘察、工程测量、鉴定加固改造、工程咨询，以及施工图审查等业务。拥有19种设计、咨询、勘察、监理等资质，并取得了对外承包工程资格证书。公司坚持"精诚、创新、和谐、共进"的核心理念，设计成果遍布国内31个省、市、自治区，并在亚洲、非洲、大洋洲、北美洲等十几个国家承揽过项目。累计完成6 000余个工程设计项目，设计总面积达1.4亿m²。代表作品有山东剧院、山东省体育中心、京沪高铁济南西客站、山东大学齐鲁医院、山东省立医院、江苏省人民医院等。

Shandong Provincial Architectural Design & Research Institute Co., Ltd. is a restructuring enterprise of Shandong Provincial Architectural Design and Research Institute. It is a national Grade - A survey and design unit, a national credit and contract-keeping unit, a national integrity unit in architectural design industry, and one of the famous Top 100 contemporary Chinese architectural design institutes. The company is mainly engaged in construction, planning, municipal engineering design, engineering survey, engineering measure, identification and reinforcement, engineering consulting, and construction drawing review. It has got 19 kinds of qualifications for design, consulting, survey, and supervision. Moreover, it has obtained the qualification certificate for foreign contracted projects. The company adheres to the core concept of "sincerity, innovation, harmony and common progress", whose design works are distributed in 31 provinces, municipalities and autonomous regions in China. In addition, the company has contracted projects in more than ten countries including Asia, Africa, Oceania and North America. A total of more than 6 000 engineering design projects have been completed, with a total design area of 140 million m². Its representative works include Shandong Theater, Sports Center of Shandong Province, Jinan West Railway Station for Beijing-Shanghai Highspeed Railway, Qilu Hospital of Shandong University, Shandong Provincial Hospital, Jiangsu Provincial People's Hospital and so on.

地址：山东省济南市经四路小纬四路2号	Add: No.2 Xiaowei 4th Road, Jingsi Road, Jinan, Shandong
邮编：250001	P.C: 250001
电话：0531-87913010	Tel: 0531-87913010
传真：0531-87913010	Fax: 0531-87913010
邮箱：sdad1953@163.com	Email: sdad1953@163.com
网址：www.sdad.cn	Web: www.sdad.cn

日照市奎山综合客运站及配套工程设计项目
Design Projects of Kuishan Comprehensive Passenger Station in Rizhao City and Its Supporting Facility Engineering

设计师：王卓然、王晋、李培民
项目地点：山东 日照
建筑面积：111 662 m²
用地面积：45 hm²

Designers: Zhuoran Wang, Jin Wang, Peimin Li
Location: Rizhao, Shandong
Building Area: 111 662 m²
Site Area: 45 hm²

以"山之厚重、水之灵动"为设计理念，建筑厚重沉稳如山、景观灵动多变如水。站前广场整体抬高，形成了景观平台，平台以"流水"作为概念原型，条形划分的景观场地自平台倾斜而下，宛如流水。高铁站坐北朝南，西侧为长途客运站，东侧为公交综合楼，三大建筑呈"品"字形对称布局。同时，站前广场设U形高架平台，将三大体量有机整合为统一的整体。

With the design concept of "heavy mountain, agile water", the building is heavy and calm like a mountain, and the landscape is agile like water. The square in front of the station is raised as a whole, forming a landscape platform, which uses "flowing water" as its conceptual prototype. The strip-shaped landscape site slopes down from the platform, like flowing water. The high-speed railway station is facing south, with a long-distance terminal on the west side and a bus complex on the east side. The three buildings are symmetrically arranged in a shape of a Chinese character "品". At the same time, the U-shaped elevated platform is set in the square in front of the station, which integrates the three major buildings into a unified whole.

深圳大学学府（深圳大学总医院）
Shenzhen University (Shenzhen University General Hospital)

设计师：王岗、贾敬龙、徐备、田少斌
项目地点：广东 深圳
建筑面积：135 000 m²
用地面积：8.98 hm²
容积率：1.2

Designers: Gang Wang, Jinglong Jia, Bei Xu, Shaobin Tian
Location: Shenzhen, Guangdong
Building Area: 135 000 m²
Site Area: 8.98 hm²
Plot Ratio: 1.2

基地位于深圳南山区，呈不规则多边形且地势平坦，院内南侧有保留的教学楼和行政楼各1幢，项目选址符合城市规划的选址原则。规划总用地面积8.98 hm²，总建筑面积135 000 m²，住院病床一期800床、二期500床，合计1 300床，手术间数19间，ICU床位数25床，停车场停车泊位843个，其中地上75个、地下768个。基地约束因素分析是形成原创方案的关键，典型外部造型科学体现了功能、景观与造型的逻辑关系；通风采光与绿色节能是人性化关怀的重要前提和保证。医技分离设置是前瞻性医学流程思考的结果。装修与BIM应用展现精美医疗空间品质。采用一体化设计系统打造医院精品工程。设计总承包与限额设计是设计、施工、投资控制的必要保证。

The base is located in Nanshan District, Shenzhen. The terrain is irregular and flat, and there is a reserved teaching building and administrative building on the south side of the campus. The site selection of the project is in line with the principle of urban planning. It is planned to cover a total site area of 8.98 hm² and a total building area of 135 000 m², with a total of 1 300 beds—800 beds in the first phase and 500 beds in the second phase, 19 operating rooms, 25 ICU beds, 843 parking berths—75 aboveground berths and 768 underground berths. The analysis of the constraint factors in the base is the key to the formation of original planning. Typical external modeling reflects the logical relationship between function, landscape and modeling; ventilation and lighting and green energy conservation are important prerequisites and guarantees for humanized care. The separation of medical technology buildings is the result of forward-looking medical process thinking. Decoration and BIM applications reproduce the quality of the beautiful medical space. The integrated design system is adopted to build a quality hospital project. General contracting of design and quota design are necessary guarantees for design, construction and investment control.

东营市人民医院急诊急救中心暨内科病房综合楼
Emergency Center and Internal Medicine Ward Comprehensive Building in Dongying People's Hospital

设计师：李维东、索慧波、郑林进、杨海涛、史大洋、王健栎、李鹤、邵洪波、任立全
项目地点：山东 东营
建筑面积：960 000 m²
用地面积：8.97 hm²
容积率：2.65

Designers: Weidong Li, Huibo Suo, Linjin Zheng, Haitao Yang, Dayang Shi, Jianyue Wang, He Li, Hongbo Shao, Liquan Ren
Location: Dongying, Shandong
Building Area: 960 000 m²
Site Area: 8.97 hm²
Plot Ratio: 2.65

本项目应用了动态发展理念，把可变性和可发展性贯穿医院设计的始终。一反常规，将主楼设置于南侧，裙房设置在北侧，这样裙房可以与保留的门诊医技裙房及二期门诊医技裙房连成整体，形成门诊医技大平面，相较于零碎的空间更有利于医疗功能的更新。这个理念对医院改扩建项目有很好的借鉴意义，在二期实施后更有示范性。

The project adopts the concept of dynamic development, with variability and developability running throughout the design of the hospital. In an unconventional way, the main building is placed on the south side and the podium is placed on the north side, so that the podium can be integrated with the reserved outpatient medical technology podium and the outpatient medical technology podium in the second phase, forming a large plane of outpatient medical technology buildings, which is more conducive to the update of medical functions in contrast with a piecemeal space. This concept has a good reference for the reconstruction and expansion of the hospital and it is more exemplary after the implementation of the second phase.

山东省立医院儿科综合楼及辅助保障设施建设项目
Pediatric Complex and Auxiliary Supporting Facilities Construction Project of Shandong Provincial Hospital

设计师：钱宁亚、杜强、高岩、马向群、周建昌、任立全、
　　　　霍亭、孙瑞锋、张元伟、赵强、石颖、刘波
项目地点：山东 济南
建筑面积：73 383 m²
地上建筑面积：48 292 m²
地下建筑面积：25 091 m²
用地面积：1.5930 hm²
容积率：3.03

　　新建儿科综合楼工程建设规模为73 383 m²（其中地上建筑面积48 292 m²，地下建筑面积25 091 m²），共设置床位600张，停车位452个。地下车库部分2层，非车库部分3层，地上19层。连接改造原有部分建筑面积5 275 m²（地上5层）。建设完成后，医院将重新统一整合儿科医疗资源，集儿科门（急）诊、病房、医技、手术室、ICU等功能于一体。

Designers: Ningya Qian, Qiang Du, Yan Gao, Xiangqun Ma, Jianchang Zhou, Liquan Ren,
　　　　　 Ting Huo, Ruifeng Sun, Yuanwei Zhang, Qiang Zhao, Ying Shi, Bo Liu
Location: Ji'nan, Shandong
Building Area: 73 383 m²
Ground Area: 48 292 m²
Underground Area: 25 091 m²
Site Area: 1.5930 hm²
Plot Ratio: 3.03

　　The newly-built Pediatric Complex covers a total building area of 73 383 m² (including an above-ground building area of 48 292 m² and an underground building area of 25 091 m²), with a total of 600 beds and 452 parking berths. There are 2 floors for the underground garage, 3 floors for the non-garage parts and 19 floors above ground. The original part of the building has been renovated and the building area is 5 275 m² (5 floors above ground). After the completion of the construction, the hospital will re-integrate the pediatric medical resources, combining the functions of the pediatric department (emergency), wards, medical technology building, operating rooms, and ICU.

烟台天马中心
Yantai Tianma Center Plaza

设计师：孙萍、李继开、李长君、葛序尧、王佃友、左廷荣、吴晨光
项目地点：山东 烟台
建筑面积：210 276 m²
用地面积：2.75 hm²
容积率：6.27

Designers: Ping Sun, Jikai Li, Changjun Li, Xurao Ge, Dianyou Wang, Tingrong Zuo, Chenguang Wu
Location: Yantai, Shandong
Building Area: 210 276 m²
Site Area: 2.75 hm²
Plot Ratio: 6.27

项目设计从城市角度出发，结合周边现状，综合考虑交通组织、功能分区、布局结构等内在联系。建筑群体排布突出空间的通透性，避免对北侧道路、建筑及西侧广场产生连续遮挡，同时高低搭配，起伏有致，营造出优美的城市天际线轮廓。建筑形象立意的灵感来自于电子信息产业里的印刷电路板。电子信息产业作为烟台经济技术开发区的主导产业之一，其重要意义不言而喻。通过印刷电路板提炼出来的建筑形象，个性鲜明同时又彰显烟台经济技术开发区信息化的时代特色。

From the perspective of the city, the design combines with the surrounding conditions to make a comprehensive consideration of the internal links of traffic organization, functional division, and layout structure. The arrangement of the building groups highlights the transparency of the space and avoids continuous occlusion to the roads and buildings on the north side roads and the square on the west side; at the same time, the high-low collocations and undulations create a beautiful city skyline outline. The architectural image is inspired by printed circuit boards in the electronic information industry. As one of the leading industries of Yantai Economic and Technological Development Zone, the electronic information industry is of great significance. The architectural image extracted through the printed circuit boards has a distinct personality and highlights the characteristics of the informatization of the Yantai Economic and Technological Development Zone.

潍坊滨海经济开发区综合商务中心
Comprehensive Business Center in Binhai Economic Development Zone, Weifang

设计师: 韩少龙、王韬、卢慧、李向东、康恺、孙庆、
　　　　公晓丽、潘学良、姜长川
项目地点: 山东 潍坊
建筑面积: 105 381 m²
用地面积: 5.13 hm²
容积率: 1.55

Designers: Shaolong Han, Tao Wang, Hui Lu, Xiangdong Li, Kai Kang, Qing Sun, Xiaoli Gong, Xueliang Pan, Changchuan Jiang
Location: Weifang, Shandong
Building Area: 105 381 m²
Site Area: 5.13 hm²
Plot Ratio: 1.55

　　潍坊滨海经济开发区综合商务中心项目位于滨海中央商务区主要景观轴线的节点上,以开放、包容、自由的姿态有机地嵌入周围的基地环境之中,体现了为民服务的精神及市政服务高效、透明的特性。

　　主体建筑呈三维曲面形状。从四层往上每层都有呈弧线形状的出挑,每层出挑长度均不相同,呈现三维方向的不同造型。由玻璃、铝板、石材组合成的幕墙现代感强。整个建筑动感强烈,又兼具柔和的感觉,是对风吹海水的波动体现,亦是与滨海场所环境有机契合。

　　The project is located at the node of the main landscape axis of Binhai Central Business District. It is embedded in the surrounding base environment in an open, inclusive and free manner, reflecting the efficient and transparent characteristics of serving the people and municipal services.

　　The main building has a three-dimensional curved shape. Each floor from the fourth floor to the top is in the shape of an arc with different lengths, showing different shapes in three dimensions. The curtain wall made of glass, aluminum plate and stone has a strong sense of modernity. The entire building is dynamic and yet has a soft feeling, which is like the wave in the sea and is also an organic fit into the coastal environment.

永州市一宫两馆四中心
One Palace, Two Halls and Four Centers in Yongzhou

设计师：乔永学、刘京东、李好明、秦旭、张寅、唐明
项目地点：湖南 永州
建筑面积：131 508 m²
用地面积：7.42 hm²
容积率：1.21

Designers: Yongxue Qiao, Jingdong Liu, Haoming Li, Xu Qin, Yan Zhang, Ming Tang
Location: Yongzhou, Hunan
Building Area: 131 508 m²
Site Area: 7.42 hm²
Plot Ratio: 1.21

　　项目位于永州市新城区公共文化服务区的核心地块，与北侧的两中心、景观湖、水系、绿化及东侧的文化设施共同构建了永州大道主轴线上的放大节点，为永州生态新城的点睛之地，牵领整个区域的文化格调，延续北侧两中心及自然景观带给城市的活力，是创造丰富公共生活、打造场所精神的关键所在。复合化的功能分布创造了多层次的公共空间和漫游体验，创造了一个贴近市民生活的场所，这里有市民待人接物的"客厅"，有市民聆听知识的讲堂，有市民阅读的"书房"，有市民休闲娱乐的"阳台"，成为市民家的延伸。它犹如一颗城市之心，是整个市民公共文化服务区乃至永州的新的活力之源。

　　The project is located in the core area of the public cultural service area in the New Town District, Yongzhou City. Together with the two centers, the landscape lake, the water system, the greening on the north side and the cultural facilities on the east side, the project builds the enlarged node on the main axis of Yongzhou Avenue, which is the punch line in Yongzhou Ecological New City. It will lead the cultural style of the entire region, and make a continuation of the vitality of the two centers and natural landscapes on the north side, which is the key to creating a rich public life and creating a site spirit. The composite functional distribution creates a multi-level public space and roaming experience, as well as a place close to the people's life, in which there is a living room for the citizens to receive people, a lecture hall for citizens to listen to the speech, a study room for citizens to read, and a balcony for citizens to enjoy entertainment. It has become an extension of the home of the citizens. It is like a heart of the city, a new source of vitality for the entire public cultural service area and even Yongzhou.

山西省建筑设计研究院有限公司
Shanxi Architectural Design and Research Institute Co., Ltd.

山西省建筑设计研究院有限公司（简称SXIAD），成立于1953年，是新中国第一批大型综合建筑设计单位。2019年，公司由山西省建筑设计研究院转企改制为山西省建筑设计研究院有限公司，现隶属于山西建设投资集团。

公司业务范围包括建筑设计、城乡规划、工程勘察、工程测量、工程咨询、工程造价、建筑工程消防文件审查、工程总承包等。

公司现有资质：建筑行业（建筑工程）甲级、市政行业（给水工程）甲级、工程咨询单位甲级资信、工程勘察专业类（岩土工程）甲级、城乡规划编制资质乙级、市政行业（桥梁工程、排水工程、道路工程）专业乙级、风景园林工程设计专项乙级、工程测量乙级。

公司现有职工650余人，其中正高级工程师43人，高级工程师178人，具有国家相关执业注册资格人员180余人，享受国务院特殊津贴专家4人，山西省学术技术带头人8人，三晋英才14人，公司拥有一个学术造诣较高、省内外知名的优秀专家团队。

公司立足山西，面向全国，高质量完成大批城市公共建筑、工业建筑和居住建筑项目，设计项目涵盖教育、办公、科研、商业、金融、文娱、医疗、体育、交通、住宅、工业等各个领域，工程遍及全国20多个省、自治区、直辖市及世界10多个国家和地区，获国家及部、省级优秀设计奖160余项。

公司先后获得"全国先进工程勘察设计企业""全国勘察设计行业创优型企业""全国建筑业技术创新先进企业""当代中国建筑设计百家名院""全国勘察设计行业诚信单位""中国最具业主满意度设计机构""山西省勘察设计行业优秀企业""山西省十佳勘察设计院"等荣誉。

Shanxi Architectural Design and Research Institute Co., Ltd. (SXIAD for short), established in 1953, is the first batch of large-scale comprehensive architectural design units in New China. In 2019, the company was transformed from Shanxi Architectural Design and Research Institute to Shanxi Architectural Design and Research Institute Co., Ltd., which is now affiliated to Shanxi Construction Investment Group.

SXIAD's business scope includes: architectural design, urban and rural planning, engineering survey, engineering surveying, engineering consulting, engineering cost, review of construction engineering fire protection documents, general contracting, etc. The company's existing qualifications: construction industry (construction engineering) grade A, municipal industry (water supply engineering) grade A, engineering consulting unit grade A credit, engineering survey professional category (geotechnical engineering) grade A, urban and rural planning preparation grade B, municipal Industry (Bridge Engineering, Drainage Engineering, Road Engineering) Professional Grade B, Landscape Architecture Engineering Design Special Grade B, Engineering Survey Grade B.

SXIAD has more than 650 employees, including 43 senior engineers, 178 senior engineers, more than 180 people with relevant national practice registration qualifications, 4 experts enjoying special allowances from the State Council, 8 academic and technical leaders in Shanxi Province, and 14 Sanjin talents. The company has a group of outstanding experts with high academic attainments and well-known inside and outside the province.

Based in Shanxi and facing the whole country, SXIAD has completed a large number of urban public buildings, industrial buildings and residential buildings with high quality. The design projects cover education, office, scientific research, commerce, finance, entertainment, medical care, sports, transportation, housing, industry and other fields. Covering more than 20 provinces, autonomous regions, municipalities directly under the Central Government and more than 10 countries and regions in the world, it has won more than 160 national, ministerial and provincial excellent design awards.

SXIAD has successively won the "National Advanced Engineering Survey and Design Enterprise", "National Excellent Enterprise in Survey and Design Industry", "National Advanced Enterprise of Technology Innovation in Construction Industry", "Contemporary China's 100 Famous Architectural Design Institutes", "National Survey and Design Industry Industry integrity unit", "China's most owner-satisfied design agency", "Shanxi province survey and design industry outstanding enterprise", "Shanxi province top ten survey and design institutes" and other honors.

地址：山西省太原市杏花岭区府东街5号	邮编：030013	电话：0351-3285361	邮箱：ssjyzhb@163.com	网址：www.sxiad.com
Add: No.5, Fudong Street, Xinghualing District, Taiyuan, Shanxi	PC: 030013	Tel: 0351-3285361	Email: ssjyzhb@163.com	Web: www.sxiad.com

潇河新城项目
Project of Xiaohe New City

潇河新城酒店项目
Project of Xiaohe New City Hotel

项目地点：山西 太原
建筑面积：250 900 m²

Location: Taiyuan, Shanxi
Building Area: 250 900 m²

潇河新城酒店项目位于山西转型综合改革示范区潇河产业园区。项目包括1号酒店和2号、3号酒店，客房数1 639套，其中五星级酒店客房数367套，四星级酒店客房数1 272套。

项目作为潇河国际会议、会展中心的配套酒店，使用者可从"空中花园""空中大堂""城市阳台"远眺潇河景色，其建筑群与会展南侧组团围合形成内院，给繁忙城市中的往来者一片宁静的休憩地。

Located in Xiaohe Industrial Park, Shanxi Transformation and Comprehensive Reform Demonstration Zone, the project consists of No.1 Hotel, No.2 Hotel and No.3 Hotel with 1 639 guest rooms, including 367 guest rooms of five-star hotel and 1 272 guest rooms of four-star hotel.

As a supporting hotel of Xiaohe International Conference and Exhibition Center, the project allows users to overlook the view of the Xiaohe River from the "sky garden", "sky lobby" and "city balcony". The building group and the groups on the south side of the exhibition center are enclosed to form an inner courtyard, giving visitors a quiet rest place in the bustling city.

潇河国际会展中心项目
Project of Xiaohe International Convention and Exhibition Center

项目地点：山西 太原
建筑面积：281 000 m²

Location: Taiyuan, Shanxi
Building Area: 281 000 m²

潇河国际会展中心项目位于山西转型综合改革示范区潇河产业园区。项目由北、中、南3个组团组成，净展面积约100 000 m²，可提供约5 000个国际标准展位。项目建成后，将成为一座可承接国际化、高端化、专业化大型会展的国际型展馆。

巍巍太行，黄河汤汤。项目在笼山络野之地，以盘盘山脊化作屋脊，黄河翻腾起的浪花作为连绵屋顶，诠释表里山河意象。

Located in Xiaohe Industrial Park, Shanxi Transformation and Comprehensive Reform Demonstration Zone, the project consists of three groups: north group, middle group and south group, with a net exhibition area of about 100 000 m², which can provide about 5 000 international standard booths. After its completion, the project will become an international exhibition hall that can undertake international, high-end and professional large-scale exhibitions.

The project is surrounded by mountains, and the image of the mountains and rivers is interpreted by turning the mountain ridge into the house ridge and using the waves of the Yellow River as the continuous roof.

潇河国际会议中心项目
Project of Xiaohe International Conference Center

项目地点：山西 太原
建筑面积：102 000 m²

Location: Taiyuan, Shanxi
Building Area: 102 000 m²

潇河国际会议中心项目位于山西转型综合改革示范区潇河产业园区。项目中，丰富的通高中庭是组织建筑内部空间的核心，连接了1 800座剧场式会议厅、3 800 m²的国际会议厅、3 000 m²的宴会厅三大主要功能空间。

项目造型"檐如新月、长桥凌空"。通过大尺度的悬挑屋檐，彰显建筑的气势；以纯净的玻璃幕墙，体现了建筑的科技感。

Located in Xiaohe Industrial Park, Shanxi Transformation and Comprehensive Reform Demonstration Zone, the project takes the rich atrium as the core of organizing the architectural internal space, connecting the three main functional spaces of 1 800-seat theater-style conference hall, 3 800 m² international conference hall and 3 000 m² banquet hall.

The large-scale cantilevered eaves highlight the magnificence of the building; The pure glass curtain reflects the sense of science and technology of the building.

太原武宿国际机场三期改扩建工程航站区工程
Construction of the Terminal Area of the Third Phase Renovation and Expansion Project of Taiyuan Wusu International Airport

项目地点：山西 太原

建筑面积：总建筑面积约670 000 m²（其中航站楼约400 000 m²，交通换乘中心约50 000 m²，停车楼约130 000 m²，轨道交通预留工程约90 000 m²）

Location: Taiyuan, Shanxi

Building Area: The building area is about 670 000 m². The terminal covers an area of about 400 000 m². The traffic transfer center covers an area of 50 000 m². The parking building covers an area of 130 000 m². The rail transit reserve project covers an area of about 90 000 m².

太原武宿国际机场为国内省会级干线机场，是山西省最大的国际航空口岸。该机场战略性质定位为"区域枢纽机场"，是山西省唯一枢纽机场及一类对外开放口岸，同时还是首都机场的备降机场之一。本次改扩建以终端年4 000万人次旅客吞吐量为标准，建设60余万平方米的航站楼、陆侧交通枢纽设施及新的进出场快速通道。

T3航站楼整体建筑造型参考山西古建筑样式，提炼"轴、脊、翼、叠、瓦"等传统元素为建筑特征，打造外雄内秀的建筑空间及形象。T3航站楼、综合交通中心和停车楼通过连廊相连，形成"一进一天地、一轴一山河"的整体布局。

Taiyuan Wusu International Airport, as the domestic provincial level trunk airport, is the largest international aviation port in Shanxi Province. It is strategically positioned as a "regional hub airport". It is also the only hub airport and A-Class port opened to outside in Shanxi Province, as well as one of the alternate airports of the Capital Airport. This renovation and expansion project is based on the terminal annual passenger throughput of 40 million to build the termimal with an area of more than 600 000 m², land side transportation hub facilities, and a new inbound and outbound express channel.

The overall architectural modeling of T3 terminal refers to the ancient architectural style of Shanxi Province, and extracts traditional elements such as "axis, ridge, wing, fold and tile" as architectural features to create an architectural space and image that looks magnificent externally and elegant internally. The T3 terminal, the comprehensive transportation center and the parking building are connected through the corridor, forming the overall layout of "different landscapes with every move and every axis".

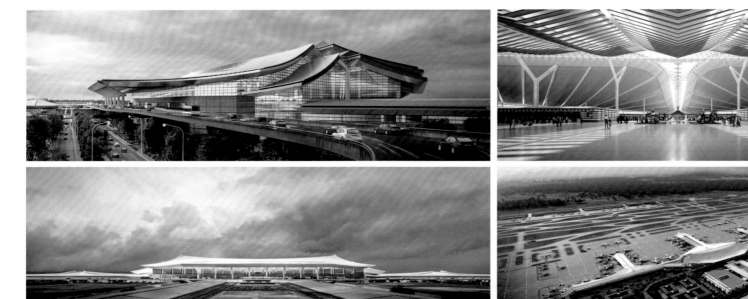

新源智慧建设运行总部A座
Building A of Xinyuan Intelligent Construction and Operation Headquarters

项目地点：山西 太原
建筑面积：14 353 m²
设计目标：全省首例"绿色建筑三星级、近零能耗建筑、AAA级装配式建筑"

Location: Taiyuan, Shanxi
Building Area: 14 353 m²
Design Goal: The province's first "three-star green building, near-zero energy consumption building, AAA-level prefabricated building"

设计定位
Design Positioning

新源智慧建设运行总部A座位于山西转型综合改革示范区潇河产业园区，项目立足"绿色""示范"这两个关键点，满足了山西·潇河新城建设、管理、展示、运营等功能要求，全面融入近零能耗、装配式建筑技术设计体系，全过程实施绿色建设，构建了三星级绿色建筑的低碳技术、创新体系、智慧运营管理系统，促进潇河新城的建设向高质量高标准发展。

Located in Xiaohe Industrial Park, Shanxi Transformation and Comprehensive Reform Demonstration Zone, the project is based on the two key factors of "green and demonstration" to meet the functional requirements of construction, management, display and operation of Shanxi · Xiaohe New City. The project fully integrates into the near-zero energy consumption and prefabricated building technology design system, and implements green construction in the whole process. The low-carbon technology, innovation system and intelligent operation and management system of three-star green buildings should be constructed to promote the construction of Xiaohe New City with high quality and high standards.

创新亮点
Innovation Highlights

近零能耗建筑关键技术

采用连续外保温、连续气密性、被动外窗体系、无热桥设计、带热回收功能的通风系统，应用太阳能光电系统、中深层无干扰地热系统等清洁能源技术，达到恒温、恒湿、恒氧、恒静、恒洁的五恒空间。目前项目已取得国家近零能耗建筑设计阶段的认证，其建筑能耗水平较国家《公共建筑节能设计标准》要求的65%基础上再降低61.78%，整体建筑节能水平达到85%以上。

绿色建筑三星级品质标准

在满足近零能耗与AAA级装配式创新要求的基础上，通过提升建筑节能环保、健康舒适、智慧互联、开放共享、绿色低碳等方面性能，达到了绿色建筑三星级的标准要求，综合得分为86.2分。

该项目获得"近零能耗建筑"设计认证，是全省首个集"近零能耗建筑、AAA级装配式建筑、三星级绿色建筑"三位一体的高科技智慧建筑。通过多项技术创新，为践行习近平总书记提出的"双碳"目标，积极引领我省建筑领域的绿色发展，起到了很好的示范带头作用。

AAA级装配式钢结构建造体系

该体系包含结构系统、外围护系统、内装系统、设备与管线系统4部分，建筑装配率达到91%以上。基于BIM技术的应用，装配建造过程运用信息管理技术平台、编码标识系统、模拟预拼装技术、虚拟仿真技术实现柱跨模数化、节点标准化、构件通用化、拼装数字化、生产自动化、方案最优化，保证结构安全的同时，提高装配效率与施工精度。

其中外围护系统创新采用近零能耗建筑装配式复合墙体系统，该系统集成了装配、防火、保温、气密性、超低能耗等多种功能，其应用尚属全国首例。

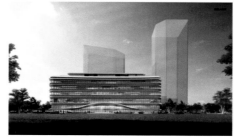 

Key Architectural Technologies of Near-zero Energy Consumption

The project adopts continuous external insulation, continuous air tightness, passive external window system, no thermal bridge design, ventilation system with heat recovery function, with the application of solar photovoltaic system, medium and deep non-interference geothermal system and other clean energy technology, to create a space with constant temperature, constant humidity, constant oxygen, constant static, constant clean. At present, the project has obtained the certification of the national near-zero energy consumption in the architectural design stage, and its building energy consumption level is 61.78%, lower than the 65% required by the national "Public Building Energy Saving Design Standard", and the overall building energy saving level has reached more than 85%.

Three-star quality standard for green buildings

On the basis of meeting the requirements of near-zero energy consumption and AAA level prefabricated innovation, the project has reached the three-star standard requirements of green building, with a comprehensive score of 86.2, by improving energy conservation, environmental protection, health and comfort, intelligent interconnection, open sharing, green and low-carbon. The project has obtained the design certification of "building with near-zero energy consumption", and is the first high-tech intelligent building in the province that integrates "near-zero energy consumption building, AAA level prefabricated building, three-star green building". Through a number of technological innovations, the project plays a good demonstration and leading role in practicing the "double carbon" goal proposed by the General Secretary Xi Jinping and actively leading the green development of the construction field in our province.

AAA level prefabricated steel structure construction system

The system consists of four parts: structural system, peripheral protection system, internal system, equipment and pipeline system, and the building assembly rate reaches more than 91%. Based on the application of BIM technology, information management technology platform, coding marking system, simulation pre-assembly technology and virtual simulation technology are used in the assembly and construction process to realize column span modularization, node standardization, component generalization, assembly digitization, production automation and program optimization, so as to ensure structural safety and improve assembly efficiency and construction accuracy.

The peripheral protection system creatively adopts the building prefabricated composite wall system with near-zero energy consumption, which has various integrated functions such as assembly, fire prevention, heat preservation, air tightness and ultra-low energy consumption. The application of this technology is the first case in China.

太忻一体化展示馆
Taiyuan-Xinzhou Integrated Exhibition Hall

项目地点：山西 太原　　Location: Taiyuan, Shanxi
建筑面积：26 500 m²　　Building Area: 26 500 m²

太忻一体化展示馆位于大盂产业新城起步区。

1. 坚持国际视野，高标准设计

向国际先进城市对标，向国内发达地区看齐，高标准规划，高水平设计，确保在控制时间和成本的基础上建造出高品质高端化的标志性建筑。

2. 坚持零碳园区，智能化设计

利用近零能耗和装配式建造新技术、新材料、新产品，实行以实效节能为导向的建筑能耗限额设计，实现建筑超低能耗，打造建筑节能技术创新示范工程。

3. 坚持先行先试，实用化设计

实现技术研发、成果转化、技术交流等功能集聚发展，引领我省建筑产业绿色、装配式、智能建造融合发展，打造我省建筑产业创新发展新高地。

Taiyuan-Xinzhou Integrated Exhibition Hall is located in the starting area of Dayu Industrial New Town.

1. Adhere to international vision and design with high standard

High-standard planning and high-level design are adopted to keep up with the international advanced cities and the domestic developed areas, so as to build high-quality and high-end landmark buildings on the basis of controlling time and cost.

2. Adhere to zero-carbon park and intelligent design

New technologies, new materials and new products are created by using near-zero energy consumption and prefabricated system. Building energy consumption quota design oriented by effective energy conservation is implemented to achieve ultra-low energy consumption in buildings, creating a demonstration project for building energy conservation technology innovation.

3. Adhere to practical design

The project aims to realize the centralized development of technology research and development, achievement transformation, technology exchange, to lead the integrated development of green, prefabricated and intelligent construction in the construction industry of our province, and to create new heights of innovation and development in the construction industry of our province.

规划从城市规划、景观建筑学的整体观念出发，综合考虑多方面因素，强调使用功能和交通的合理组织，注重反映"高起点、高科技、高引领"的思想，致力于创造一座品质卓越、功能明晰、个性鲜明的亮点示范工程；同时规划设计符合"大盂产业新城"先进的规划建设思路，体现"创新、协调、绿色、开放、共享"的发展理念；紧密围绕"双碳"的关键要素，通过精细化设计、精细化建设、精细化运营，实现山西特色文化元素与现代设计的兼容并蓄。

Starting from the overall concept of urban planning and landscape architecture, the planning takes various factors into comprehensive consideration, emphasizing functions of use and reasonable organization of traffic. It pays attention to reflect the idea of "high starting point, high technology, high leading", striving to create a demonstration project with outstanding quality, clear function, and distinct personality. At the same time, the planning and design conforms to the advanced planning and construction idea of "Dayu Industrial New Town", reflecting the development concept of "innovation, coordination, green, open and sharing"; It revolves around the key elements of "double carbon" to realize the integration of Shanxi characteristic cultural elements and modern design through fine design, fine construction and fine operation.

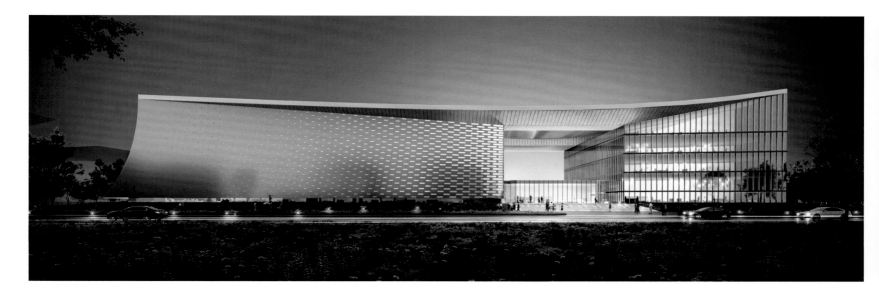

中美清洁能源研发中心
China-USA Clean Energy Research and Development Center

项目地点：山西 太原
建筑面积：47 350 m²

Location: Taiyuan, Shanxi
Building Area: 47 350 m²

设计理念：模块化的研发合院有机生长，共同形成了"邻里单元"的空间格局，拥有强烈的可识别性，成为创新城一处重要的地标。山西格盟中美清洁能源研发中心，作为首批入驻山西转型综合改革示范区科技创新城的重点企业，对山西创新城的启动发展起到了良好的示范作用，被列为2015－2016年度山西省重点工程。项目位于创新城核心区，目前，4号楼已通过竣工验收，并正式投入使用。

技术难点：4号楼长154 m，宽64.2 m，属于超长结构，为保证建筑功能及立面完整性，结构设计采取相应措施，在不设缝的前提下，消除了由温度变化对结构产生的不利影响。该项目为绿色二星级建筑，运营期间，各种生活实验污水经有效处理，可达标排放。

技术创新：邻里单元中建筑单体布局紧凑，可以阻挡冬季寒冷的北风；地块南侧分散布置体量较小的模块，可以引入夏季东南风，创造出冬暖夏凉的区域内部环境。建筑内部上下贯通的生态庭院，可加速内部空气流速，起到通风降温的作用。中央庭院的营造大大改善了建筑通风采光效果，屋顶花园也有着隔热降温的作用。

Design concept: Modular research and development courtyards grow organically, forming a spatial pattern of "neighborhood units" with strong identification and becoming an important landmark of Innovation City. As one of the first key enterprises to enter the Science and Technology Innovation City of Shanxi Transformation and Comprehensive Reform Demonstration Zone, Shanxi Gemeng China-USA Clean Energy Research and Development Center has played a good demonstration role in the start-up and development of Shanxi Innovation City, and has been listed as the key project of Shanxi Province in the year 2015-2016. The project is located in the core area of Innovation City. At present, Building 4# has passed the completion acceptance and officially been put into use.

Technical Difficulties: Building 4# is 154 m long and 64.2 m wide, belonging to the ultra-long structure. In order to ensure the architectural function and the integrity of the facade, the design of the structure takes corresponding measures to eliminate the adverse effects of temperature changes on the structure., under the premise of no cracks. This project is a green two-star level building. During operation, all kinds of experimental sewage can be discharged up to the standard after effective treatment.

Technological innovation: The compact layout of the single buildings in the neighborhood unit can block the cold north wind in winter, while the scattered smaller modules on the south side of the plot can introduce the southeast wind in summer, creating an internal environment of the region that is warm in winter and cool in summer.

The ecological courtyard through the top and bottom of the building can accelerate the internal air flow rate and play a role in ventilation and cooling. The central courtyard not only greatly improves the ventilation and lighting effect of the building, but also the roof garden has the function of heat insulation and cooling.

太原工人文化宫大修改造工程B段
Renovation Project B of Taiyuan Workers Culture Palace

1. 设计概况

太原工人文化宫,又称为"南宫",位于太原市迎泽区迎泽大街243号。项目北临迎泽大街,南侧为南宫南街,西侧为南宫西路,东侧为南宫东路;与北侧财贸大楼隔街相望,东侧距离迎泽公园500 m左右;南侧为古玩街、教学楼等。太原工人文化宫建造于1956年—1958年,采用仿苏联建筑风格,为砖混加内框架结构建筑,是当时太原市重要的公共建筑,迄今已经有60多年的历史,历经历史沧桑而屹立于太原市中心,是太原市重要的历史文化建筑遗存,承载着几代太原人的历史记忆。

2. 设计理念

第一:赋予老建筑新生命。以保护保留建筑为前提,对历史建筑部分进行结构加固和外立面的复原修缮,对内部的设备系统、消防系统和室内环境更新,给予建筑新的生命力。

利用扩建部分,将车行、贵宾、货运流线独立互不干扰。新建部分沿着南宫南街实现对原建筑的立面进行完善,与原建筑立面统一协调;通过材质的变化,增加其可识别性。新建建筑与保留建筑通过增设灰空间、退让内庭院而和谐有序地衔接在一起。

第二:延续历史文脉。大修改造坚持以"保护"为原则,以"再生"为理念,以"文化"为载体。保留历史信息和文化符号,营造太原市独一无二的城市节点。前厅室内空间完整保留历史风貌,顶面为花卉纹饰的装饰浮雕及装饰线脚,柱子、柱帽及柱础均为历史信息,门套、窗套、扶手栏杆、楼梯踏面、墙面灯饰均为历史保留式样,地面为水磨石地面。东西展廊、展厅室内空间完整保留历史风貌,顶面为花卉纹饰的装饰浮雕及装饰线脚。展览沿着流线布置展板,地面为石材装饰地面。

第三:超越历史风格,传承创新。新建观众厅采用大跨度框架结构,柱网规整,采用现代的舞台设备,内部空间更加完善,提升其品质及使用率,增加新的业态,完善配套,提升室内空间的吸引力。

第四:主立面采用古典主义"三段式",建筑整体庄严大气,比例协调,气势恢宏。建筑主体中轴对称,平面规整,主立面中间高两边低,回廊沿东西走向展开,衬托主楼的高耸。建筑虽然历经岁月的洗礼,其位置、边界没有较大的变化,建筑立面仍延续初建时期的风貌。

3. 技术创新

第一:总平面布局设计,保留原有"山"字形,对原有观众厅进行扩容和品质提升,拆除原有观众厅及东西小院内杂乱无章的建设。

第二:音乐厅室内采用扩散体吸声墙体。

第三:室内休息厅屋面采用可开启的天窗。

1.Design overview

Taiyuan Workers Culture Palace, also known as the "Southern Palace", is located at No. 243 Yingze Street, Yingze District, Taiyuan City, with Yingze Street in the north, Nangong South Street in the south, Nangong West Road in the west, and Nangong East Road in the east. It is across the street from the finance and trade building in the north, about 500 m away from Yingze Park; The south side of it is antique street, teaching building and so on. Taiyuan Workers Culture Palace was built in the year 1956-1958, adopting the Soviet architectural style which uses brick mixed with internal frame structure. The Southern Palace was an important public building in Taiyuan at that time, enjoying a history of more than 60 years and still stands in the center of Taiyuan after vicissitudes of history. It is an important historical and cultural architectural relic of Taiyuan and carries the historical memory of several generations of Taiyuan people.

2. Design Concept

First: Give old buildings new life. On the premise of preserving the buildings, structural reinforcement and restoration of the facade of the historic buildings are carried out, and the internal equipment system, fire protection system and indoor environment are updated to give new vitality.

Make use of the extension part to make traffic for vehicles, VIP and freight lines independent of each other. The newly-built part along the Nangong South Street realizes the improvement of the facade of the original building and harmonizes with the facade of the original building; The material changes increase its identification. The new building and the reserved building are connected with each other harmoniously and orderly by adding gray space and giving way to the inner courtyard.

Second: Continue the historical context. Renovation adheres to the principle of "protection", the concept of "regeneration", and the carrier of "culture" to retain historical information and cultural symbols, creating a unique urban node of our city. The interior space of the vestibule retains the historical style completely, and the top surface is decorated with decorative embossments and decorative architraves with floral patterns. Column, column cap and column base are historical information. Door covers, window covers, handrails, stair treads, wall lighting are all historical preservation patterns. The floor is terrazzo floor. The interior space of the east and west galleries and exhibition halls retains the historical style completely. The top surface is decorated with decorative embossments and decorative architraves with floral patterns. The exhibition is arranged along the streamline, and the ground is decorated with stone.

Third: Surpass the historical style with inheritance and innovation. The new auditorium adopts the large-span frame structure with regular column network. Modern stage equipment is used and the internal space is more perfect. The project improves its quality and utilization rate, increases new business forms and improve supporting facilities to make interior space more attractive.

Fourth: The main facade adopts the classicist "ternary form", which makes the overall building solemn and magnificent with coordinated proportion. The main body of the building has symmetrical axis and regular plane. The main facade is high in the middle and low on both sides, and the corridor is spread out from east to west, which sets off the towering main building. Although the building has undergone the baptism of time, its position and boundary have not changed greatly and its facade still continues the original style.

3.Technological Innovation

First: The general layout design retains the original "mountain"-shaped building while expanding the original audience hall and improving the quality. Therefore, the original audience hall and the construction of the east and west courtyard were dismantled.

Second: Diffuser sound absorption wall is used in the concert hall.

Third: The roof of the indoor lounge adopts the skylight that can be opened.

20世纪60年代　　　　　　　　　　　　　20世纪70年代　　　　　　　　　　　　20世纪80~90年代

山西省省情（方志）馆
Shanxi Provincial Information (Local Chronicles) Museum

项目地点：山西 太原　　Location: Taiyuan, Shanxi
建筑面积：16 500 m²　　Building Area: 16 500 m²

1. "器"——博容承载
方志作为全面记载某一时期某一地域自然、社会、政治、经济、文化的书籍文献，其涉及内容丰富全面，囊括所有行业，无所不容。因此方志馆成为承载省情（方志）内涵属性的载体，为传承地域文化做出积极的推动作用。该馆建筑融合了多元文化元素，表述了文字语言无法企及的文化内涵，是器物、制度和观念三层文化的集中体现。

2. "院"——院落围固
设计在总体布局方面，结合所处位置的独特性，在形制、布局上与山西地方文化相呼应，传统街巷、中式园林在此得以传承，精致的院落、丛丛绿竹、幽静的水面、点点睡莲，伴着和幽微风以及水面泛起的链滑，共同传承着山西传统的文化渊源。设计中汲取传统民居中院落格局、单坡屋面、砖雕艺术、园固陈设等元素的精华并将其有机融合，彰显出方案设计的地域性与人文性。

在空间处理上将功能确定的空间（展厅、编纂等）与功能不确定的空间（过厅、中庭、联系走廊、室外平台、内庭院、广场）以多样的方式相互关联而成网状。共享厅、灰空间、庭院、广场等多层次空间可以获得不同的体验，有助于科研人员的身心健康。

3. "格"——书香品格
设计中通过对山西地域元素及中式园林的空间布局研究，结合地方志的专业内添属性分析，将该馆定位为书院式的博物馆。设计中通过多种设计手法，包括理念生成、空间布局、细部推敲等，以中国传统书院为原型，融合现代博物馆的特质，塑造该省情（方志）馆的书香人文气息，提升其内在品格气质。

1. "Utensils" With a big load-bearing capacity
Local chronicles makes a comprehensive record of books and documents on nature, society, politics, economics, culture during a certain period in a certain region. The content involved is rich and comprehensive, covering all industries. Therefore, Local Chronicles Museum becomes the carrier of the connotation of the provincial information (local chronicles) and plays a positive role in promoting the inheritance of regional culture. The building of the museum integrates multi-cultural elements, expresses the cultural connotation that words cannot reach, and is the concentrated embodiment of the three-layer culture of utensils, institutions and ideas.

2. "Courtyard" Courtyard house shapes Chinese garden culture
In terms of the overall layout, combined with the uniqueness of the location, the design echoes the local culture of Shanxi in the aspects of the shape and the layout - traditional streets and Chinese gardens are inherited here - exquisite courtyards, clusters of green bamboo, quiet water surface, water lilies, accompanied by the gentle wind and the ripple on the water surface, which jointly inherit the traditional cultural origins of Shanxi. In the design, the elements of traditional dwellings - courtyard pattern, single-slope roof, brick carving art, garden solid furnishings, etc., are absorbed and integrated into the design, reflecting the regionalism and humanity of the scheme design.

In spatial processing, spaces with certain functions (exhibition hall, editing space, etc.) and spaces with uncertain functions (lobby, atrium, contact corridor, outdoor platform, inner courtyard, square) are interrelated in various ways, forming a network form: multi-level spaces such as shared hall, gray space, courtyard, square, etc. can enable people to obtain different experiences, which is conducive to the physical and mental health of researchers.

3. "Character" Literary character
The design studies the regional elements of Shanxi and the spatial layout of Chinese gardens, combined with the analysis of the professional and internal attributes of local chronicles. The museum is positioned as an academy museum. Through a variety of design techniques - concept generation, space layout, detailed elaboration, etc., the design takes the traditional Chinese academy as the prototype, integrates the characteristics of modern museums, and shapes the literary culture of the provincial information (local chronicles) museum to improve its internal character and temperament.

山西科技创新城科技创新综合服务平台（一期）项目

Science and Technology Innovation Comprehensive Service Platform (Phase I) Project of Shanxi Science and Technology Innovation City

项目地点：山西 太原　　　　Location: Taiyuan, Shanxi
建筑面积：175 308.55 m²　　Building Area: 175 308.55 m²

项目位于山西省科技创新城中心区的北侧，城内两大轴线——科技创新轴和智慧生活轴的交会处。根据项目功能的特点，设计采取品字形布局，围合出中央的庭院，如同代表着幸运和成功的三叶草，同心相向，具有强烈的凝聚力。北面建筑统领整个建筑群，集中布置科技金融服务平台及科技资源服务平台，南面将创业孵化服务平台一分为二，在基地西南面布置创业孵化平台的创新孵化器功能，东南面布置创业孵化服务平台的研发设计与检验检测及预孵化功能。中央为生态下沉式庭院，结合地下商业及餐饮活动空间布置配套服务功能，形成人们聚会、休息、交流的理想场所。3座单体围绕组团中心的下沉广场呈放射状布局，整个建筑群体既统一完整又相对独立。以围合式建筑形态为单元，在"三叶草"的每一个单元内形成一个内部共享空间，形成自己半私密的中庭；而三片"叶子"又围绕一个下沉广场布置，各单元共同围合出公共的中心绿地。整个建筑群布局紧凑，北部中间布置大体量单体可以阻挡冬季寒冷的北风，南部分散布置较小体量单体可以引入夏季东南风，创造一个自然生态、冬暖夏凉的内部环境。多个下沉广场以及建筑内部上下贯通的生态庭院，可加速内部空气流通，起到通风降温的作用。屋顶玻璃天窗可开启，确保室内较好通风采光效果，并设置百叶窗遮阳。建筑立面实体墙体占70%，采用生态材料，保温隔热。局部采用呼吸式幕墙，根据季节自然调节内部空气质量。建筑采用多项生态技术，如太阳能集热板、太阳能光电玻璃等；室内采用节能照明、中水回用系统和雨水收集系统等。节省运营成本，使该建筑成为绿色低碳的典范作品。

The project is located in the north of the central area of Science and Technology Innovation City of Shanxi Province, at the intersection of two major axes in the city - science and technology innovation axis and intelligent life axis. According to the characteristics of the project function, the "品"-shaped layout is carried out to enclose the central courtyard, just as the clover representing luck, success and strong cohesion. The building on the north side dominates the entire building group, with the technological finance service platform and science and technology resource service platform centrally arranged. The business incubation service platform is divided into two parts in the south. The innovation incubator function of the business incubation platform is arranged in the southwest of the base, and the R&D, design, inspection, testing and pre-incubation function of the business incubation service platform are arranged in the southeast. In the center is an ecological sunken courtyard, which combines the underground commercial and catering activity space to arrange supporting service functions, forming an ideal place for people to gather, rest and communicate. The three single buildings are arranged radially around the sunken square in the center of the cluster. The whole building group is unified, complete and relatively independent. Taking the enclosed building form as a unit, each unit of "clover" forms an internal shared space, forming its own semi-private atrium; The three "leaves" are arranged around a sunken square, and each unit encloses the central public green space. The whole building group is compact in layout. The central large units in the north can block the cold north wind in winter, while the small scattered units in the south can introduce the southeast wind in summer, creating an internal environment of natural ecology that is warm in winter and cool in summer. Multiple sunken squares and ecological courtyards running up and down inside the building can accelerate the internal air circulation and play a role in ventilation and cooling. The glass skylight on the roof can be opened to ensure better indoor ventilation and lighting effect, and louvers are set to shade the sun. The solid wall of the building facade accounts for 70%, and ecological materials are used for thermal insulation. Breathing curtain wall is adopted locally to adjust the internal air quality naturally. The building uses a number of ecological technologies, such as solar panels, solar photovoltaic glass, indoor energy-saving lighting, reclaimed water reuse system, rainwater collection and so on to save operating costs, making the building a green and low-carbon example.

中安创芯半导体及电子设备制造项目
Zhongan Chuangxin Semiconductor and Electronic Equipment Manufacturing Project

项目地点：山西 太原 Location: Taiyuan, Shanxi
建筑面积：419 936.55 m² Building Area: 419 936.55 m²

中安创芯半导体及电子设备制造项目位于潇河产业园，项目主要以建设半导体及电子设备制造厂房、综合配套楼为基础，围绕新一代信息技术、智能仪器仪表制造、高端装备制造、新材料等示范区主导的新兴产业，利用山西转型综合改革示范区优惠政策，为企业搭建政、产、学、研、用、资、介多方位公共服务平台，提供仓储、物流、测试、人才、金融等配套服务，实现产业经济快速发展。

打造产业特色鲜明、集群优势明显、功能布局完整的"人性化、弹性化、集约化、现代化的数字产城"。要以构筑高端要素聚集能力为主要目标，包括城市服务功能的配给以及宜居环境的营造。打造一个"吸引力中心"和宜居宜业的空间环境同样是产业园区发展的重要基础。

规划结构：一核、两轴、三区产业聚集发展。其中，一核：精品厂房单元均围绕一个完整的生态内核设置，形成中小企产业园区的标志性形象中心和精神内核。两轴：东西生活轴、南北景观主轴T形交叉构成园区主体架构，在基地内部实现建筑与景观的融合。三区产业聚集发展：分区域规划设置中心生活区、南部厂房、北部厂区，结合8种形态厂房（规模、层数、户型不同）吸引不同类型中小企业入驻。

The project is located in Xiaohe Industrial Park. Based on the construction of semiconductor and electronic equipment manufacturing plants and comprehensive supporting buildings, the project focuses on emerging industries dominated by demonstration zones such as new generation information technology, intelligent instrument manufacturing, high-end equipment manufacturing and new materials, and takes advantage of preferential policies of Shanxi Transformation and Comprehensive Reform Demonstration Zone to build a multi-directional public service platform of government, production, learning, research, use, resources, and media for enterprises, to provide warehousing, logistics, testing, talent, finance and other supporting services, and to achieve rapid development of industrial economy.

It aims to build a "humanized, flexible, intensive and modern digital city" with distinct industrial characteristics, obvious cluster advantages and complete functional layout. The main goal should be to build the aggregation ability of high-end elements, including the allocation of urban service functions and the construction of livable environment. Creating an "attraction center" and a livable space environment is also an important basis for the development of industrial parks.

Planning structure: one core, two axes, industrial cluster development in three regions. One core: The boutique plant units are set around a complete ecological core, forming the symbolic image center and spiritual core of the industrial park of small and medium-sized enterprises. Two axes: East-west life axis together with north-south landscape axis constitute the main structure of the park in the T-shaped form, realizing the integration of architecture and landscape inside the base. Industrial cluster development in three regions: Central living area, southern plant and northern plant area are planned and set up in different regions, combined with the eight types of plant (scale, number of floors and house type) to attract different types of small and medium-sized enterprises to settle in.

山西

太原市建筑设计研究院
TAIYUAN INSTITUTE OF ARCHITECTURE DESIGN & RESEARCH

扫描查看更多信息

太原市建筑设计研究院成立于1958年，是山西省最早的市属国有甲级建筑设计院，可承担建筑咨询、规划编制、建筑设计、市政设计、岩土设计、景观设计、装饰设计，以及绿色建筑、装配式建筑、BIM、人防、消防、幕墙、建筑智能化系统等专项设计服务；并可从事消防设计审查、绿色建筑评估、建筑抗震加固、建设工程质量检测、消防安全检测、工程总承包（EPC）和工程监理及项目管理服务等，形成山西省内唯一全过程一站式的甲级设计院服务体系。

多年来，我院相继设计完成省、市重点及大中型项目数百项，获得国家、省、市优秀工程设计奖数十项，并于2010年荣获中国建筑学会"当代中国建筑设计百家名院"荣誉称号，2013年被全国总工会授予"全国保障性安居工程建设劳动竞赛先进单位"荣誉称号，连续多年荣获"太原市先进单位"和"太原市文明（和谐）单位"荣誉称号。

Taiyuan Institute of Architecture Design and Research was founded in 1958. It is the earliest municipal state-owned Grade A architectural design institute. It can undertake architectural consulting, planning, architectural design, municipal design, geotechnical engineering design, landscape design, decoration design, and different special design services like green building design, prefabricated building design, BIM, civil air defense design, fire protection design, curtain wall design, and building intelligent system design. It can be engaged in fire design review, green building evaluation, building seismic reinforcement, construction engineering quality testing, fire safety testing, EPC, engineering supervision and project management services. So it forms the only first-class design institute who has the whole process and one-stop service system in Shanxi Province.

Over the years, it has accomplished more than 100 provincial and municipal key projects, large and medium-sized projects. More than 10 projects won national, provincial and municipal excellent engineering design awards. The institute has received a series of awards, such as one of "100 famous institutes of Chinese contemporary architectural design" granted by Architectural Society of China in 2010, "Advanced unit of national labor competition for affordable housing project construction" granted by Federation of Trade Unions in 2013, "Advanced Unit of Taiyuan" and "Civilized unit of Taiyuan" granted for many consecutive years.

地址：山西省太原市新建路80号创享天地
电话：0351-2020980
邮箱：tyjyds@163.com
网址：www.ty-jz.com

Add: Chuangxiang Tiandi, No.80, Xinjian Road, Taiyuan, Shanxi
Tel: 0351-2020980
Email: tyjyds@163.com
Web: www.ty-jz.com

钟楼街项目
Zhonglou Street

设计师：蒲净、张晨、梁向宏、张文进、原建伟、郭家樑、周康、吴鹏、王君、郝丽媛、薛博、赵彦贞、王伟、张旻
项目地点：山西 太原
建筑面积：162 000 m²
用地面积：11 hm²
容积率：1.5

Designers: Jing Pu, Chen Zhang, Xianghong Liang, Wenjin Zhang, Jianwei Yuan, Jialiang Guo, Kang Zhou, Peng Wu, Jun Wang, Liyuan Hao, Bo Xue, Yanzhen Zhao, Wei Wang, Min Zhang
Location: Taiyuan, Shanxi
Building Area: 162 000 m²
Site Area: 11 hm²
Plot Ratio: 1.5

钟楼街自宋代成街，历经千年，成为太原府城区域内商业、文化、司法的核心区域。该街区保护更新充分挖掘片区的建筑价值、文化价值、经济价值，以传承历史文脉，恢复街道肌理，重塑城市记忆，体现时代风貌为指导思想。抢救性修复传统建筑17处，保护老字号14处，恢复传统街巷11条，更新市政基础设施，改善历史地段风貌环境，更新建筑空间，升级商业业态。

Zhonglou street has been a street since the Song Dynasty. After thousands of years, it has become the core area of commerce, culture and justice in Taiyuan. Under the guiding ideology of the inheritance of historical context, the restoration of street texture, the reconstruction of the memory of this city and the reflection of the scenes of the time, the protection and renewal of this street fully excavates its architectural value, cultural value and economic value. 17 traditional buildings were repaired. 14 time-honored brands were protected. 11 traditional streets and lanes were restored. Municipal infrastructure was updated. The style and environment of historical sections were improved. And the building space was updated to upgrade the commercial format.

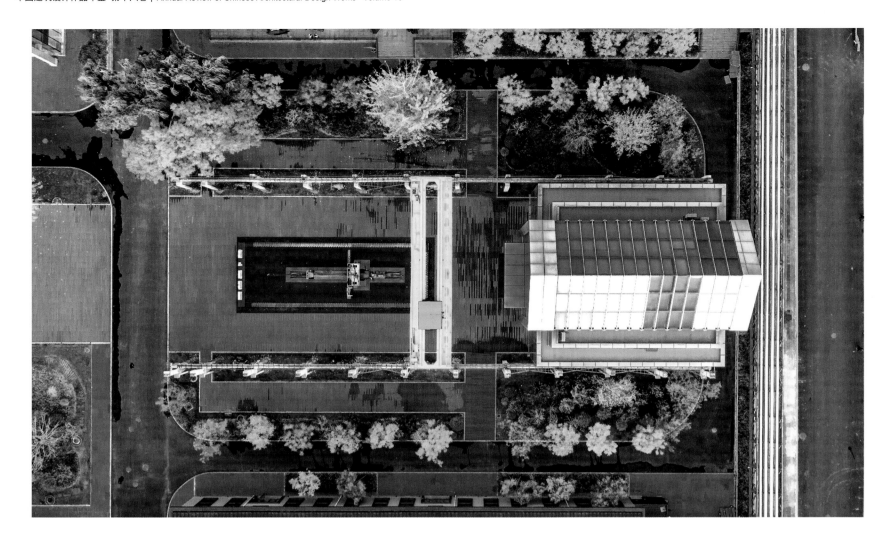

第一实验室（太原第一机床厂改造项目）
NO.1 Lab (Taiyuan NO.1 Machine Tool Works Transformation Project)

项目地点：山西 太原	Location: Taiyuan, Shanxi
竣工时间：2021年9月	Completion Time: September 2021
建筑面积：76 727 m²	Building Area: 76 727 m²
用地面积：99 900 m²	Site Area: 99 900 m²
设计团队：李志强、王原江、王君、刘香静、孙欣	Design Team: Zhiqiang Li, Yuanjiang Wang, Jun Wang, Xiangjing Liu, Xin Sun

项目位于山西省太原市小店区南内环街2号，区位优越，交通便利，总用地面积约99 900 m²。

第一机床厂于1952年建厂，是新中国成立初期成立的第一代国有企业，经过近70年的岁月洗礼，逐渐退出历史舞台。第一实验室项目在几近废弃的第一机床厂基础上，改造利用旧有厂房、办公建筑等，使之成为量子光学、碳晶薄膜、第四代半导体科研等高科技研发、生产场所。在改造提升建筑和景观风貌的同时，设计注重对工业遗迹的保护利用，留住了老工业的历史记忆，在老旧工业区、厂房保护与更新方面做出了新探索和努力。

The project is located at No.2, Nanneihuan Street, Xiaodian District, Taiyuan City, Shanxi Province, with superior location and convenient transportation. The total land area is about 99 900 m².

The First Machine Tool Factory was established in 1952. It is the first generation of state-owned enterprises established in the early days of the founding of the People's Republic of China. After nearly 70 years of baptism, it gradually withdrew from the historical stage. On the basis of nearly abandoned First Machine Tool Factory, the First Laboratory project will transform and utilize the old buildings such as workshops and offices to make it a place for high-tech research, development and production of quantum optics, carbon crystal films, the fourth generation semiconductor research and development. While upgrading the architecture and landscape, the design focuses on the protection and utilization of industrial relics, retaining the historical memory of old industries, and making new exploration and efforts in the protection and renewal of old industrial areas and plants.

解放路沿线城市风貌更新
The Urban Landscape Renewal Along Jiefang Road

项目地点：山西 太原
竣工时间：2020年10月
设计团队：李志强、周康、陈潇媛、刘剑刚、李海钊、郭家樑、
张华鹍、吴鹏

Location: Taiyuan, Shanxi
Completion Time: October 2020
Design Team: Zhiqiang Li, Kang Zhou, Xiaoyuan Chen, Jiangang Liu, Haizhao Li, Jialiang Guo, Huakun Zhang, Peng Wu

太原解放路在原大南门街、活牛市街、麻市街等狭窄街道的基础上建成,是市内南北干道之一。随着城市的发展,解放路成为太原市承载历史文化、市民记忆的南北向主要干道,也是城市发展历程展廊。本次更新以重塑名城历史文脉,提升商业文化活力,呈现景观文化廊道,提升街道环境风貌为原则,通过建筑、景观、城市家具等整体改造提升,营造温馨、有品质的生活街区。

Taiyuan Jiefang Road is built on the basis of the original narrow streets such as Dananmen Street, Huoniushi Street and Mashi Street, and is one of the north-south arterial roads in the city. With the development of the city, Jiefang Road has become a northsouth main road carrying history, culture and the memory of citizens in Taiyuan, and also a gallery of the city's development process. This renewal is based on the principle of reshaping the historical context of the famous city, enhancing the vitality of commercial culture, presenting the landscape cultural corridor, and improving the street environment. Through the overall transformation and improvement of buildings, landscapes, urban furniture, etc., it will create a warm and high-quality street for street life.

同济大学建筑设计研究院 (集团) 有限公司
TONGJI ARCHITECTURAL DESIGN (GROUP) CO.,LTD.

同济大学建筑设计研究院(集团)有限公司(TJAD)的前身是成立于1958年的同济大学建筑设计研究院,依托百年学府同济大学的深厚底蕴,TJAD确立明确的发展方向,经过半个多世纪的积累和进取,用高品质的设计在建筑设计行业中取得一席之地,全力打造建筑研究、设计与实践三个层面的有机结合。

未来,TJAD将坚持"专注"是设计创新的基础,持续地专精于高端精品项目,不断强化质量管理体系,完善人才体系建设;不放松同济人的精神核心——"进取",本着终身学习的精神,坚守传承与创新的理念,对建筑设计新领域不断探索与追求。

The predecessor of Tongji Architectural Design (Group) Co., Ltd. (TJAD) is the Architectural Design and Research Institute of Tongji University, which was founded in 1958. Relying on the profound heritage of Tongji University, a century old institution, TJAD has established a clear development direction.

After more than half a century of accumulation of experience and progress, TJAD has achieved a place in the architectural design industry with high-quality design, and strives to create an organic combination of architectural research, design and practice.

In the future, TJAD will adhere to "dedication" as the basis of design innovation, continue to specialize in high-end outstanding projects, and constantly strengthen the quality management system and improve the construction of the talent system. TJAD will not relax the spirit core of Tongji people - "forge ahead". In the spirit of lifelong learning, we will persevere with the concept of inheritance and innovation, and continue to explore and pursue new areas of architectural design.

地址:上海市四平路1230号
邮编:200092
电话:021-65987788
传真:021-65985121
网址:www.tjad.cn

Add: No.1230 Siping Road, Shanghai, China
P.C: 200092
Tel: 021-65987788
Fax: 021-65985121
Web: www.tjad.cn

西安丝路国际会展中心
Xi'an Silk Road International Convention and Exhibition Center

西安丝路国际会展中心位于浐灞生态区欧亚经济综合园区核心区,以欧亚经济论坛为依托,围绕国家"一带一路"倡议,延续历史文脉,承载时代需求,打造新丝绸之路沿线的西安新地标。其中会议中心包括会议、宴会等功能,展览中心(一期)包括登录厅、超大展厅、标准展厅等功能。建筑群以古典对称、大气庄重的建筑语汇表达了对丝绸之路以及中国传统建筑文脉的诠释与传承。

Xi'an Silk Road International Convention and Exhibition Center is located in the core area of Chanba Ecological Zone Eurasian Economic Complex. The conference center includes conference and banquet functions; the exhibition center (Phase I) includes the login hall, super-sized exhibition halls, standard exhibition halls and other functions. The building complex expresses the interpretation and inheritance of the Silk Road and the traditional Chinese architectural culture with classical symmetry and dignified architectural vocabulary.

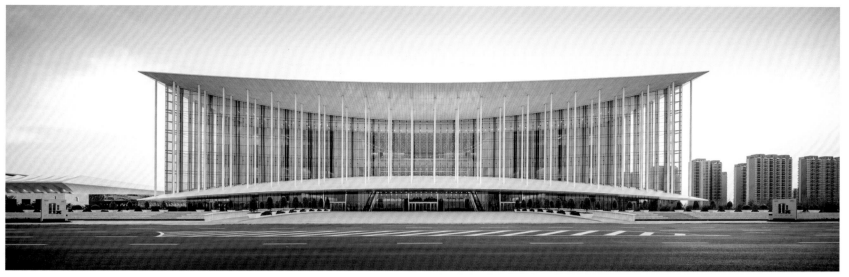

西安丝路国际会议中心
Xi'an Silk Road International Convention Center

设计师：汤朔宁、邱东晴、余雪悦、胡军锋、董天翔、赵洪刚、华轶亮、陈晓峰
项目地点：陕西 西安
建筑面积：207 112 m²
合作设计：德国gmp国际建筑设计有限公司

Designers: Shuoning Tang, Dongqing Qiu, Xueyue Yu, Junfeng Hu,
　　　　　 Tianxiang Dong, Honggang Zhao, Yiliang Hua, Xiaofeng Chen
Location: Xi'an, Shaanxi
Building Area: 207 112 m²
Collaborator: German GMP International Architectural Design Co., Ltd.

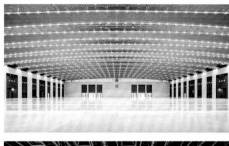

西安丝路国际展览中心（一期）
Xi'an Silk Road International Exhibition Center (Phase I)

设计师：汤朔宁、邱东晴、林大卫、徐烨、张泽震、孙宏楠、刘依朋、李姗姗、
　　　　刘洋、李阳夫、魏娜、罗益飞
项目地点：陕西 西安
建筑面积：486 678 m²
合作设计：德国gmp国际建筑设计有限公司

Designers: Shuoning Tang, Dongqing Qiu, Dawei Lin, Ye Xu,
　　　　　 Zezhen Zhang, Hongnan Sun, Yipeng Liu, Shanshan Li,
　　　　　 Yang Liu, Yangfu Li, Na Wei, Yifei Luo
Location: Xi'an, Shaanxi
Building Area: 486 678 m²
Collaborator: German GMP International Architectural Design Co., Ltd.

浦东美术馆
Museum of Art Pudong

设计师：任力之、王玉妹、孙倩、吴杰、吴睿、包恺、铁云、邹昊阳、张丽萍
项目地点：上海
建筑面积：40 590 m²
用地面积：1.3 hm²
合作设计：法国让·努维尔事务所

Designers: Lizhi Ren, Yumei Wang, Qian Sun, Jie Wu, Rui Wu, Kai Bao, Yun Tie, Haoyang Zou, Liping Zhang
Location: Shanghai
Building Area: 40 590 m²
Site Area: 1.3 hm²
Collaborator: Ateliers Jean Nouvel

浦东美术馆以含蓄平和的姿态展露于陆家嘴建筑群，它的建成为上海再添一座全新的国际性文化地标建筑，成为面向所有人的艺术"领地"。建筑立面和地面主要选材均为山东白麻大理石，通过不同的打磨工艺，构建出不同的质感，营造出建筑与周边地景的和谐统一感。隔江而望，建筑以超大玻璃幕墙面向外滩，使西立面可作为镜子反射外滩建筑群，或作为巨幅展示屏使美术馆成为一件艺术装置、一起艺术事件。

Museum of Art Pudong is located in the Lujiazui complex, adding a new international cultural landmark to Shanghai and becoming an art "territory" for all. The building facade and floor are mainly made of Shandong white marble, which is polished in different ways to create different textures, creating a sense of harmony and unity between the building and the surrounding landscape. Looking across the river, the building's large glass curtain wall faces the Bund, so that the west facade can be used as a mirror to reflect the Bund buildings, or as a giant display screen to make the museum an art installation or an art event.

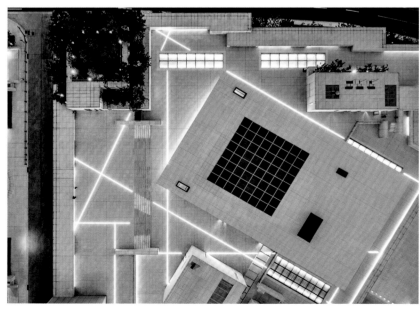

宁波院士中心
Ningbo Academician Center

设计师：吴志强、邹子敬、陈康诠、赖君恒、马天冬、马忠、刘培培、
赵昕未、高蟒、杨竞、韩佩菁、刘聪、赵明哲
项目地点：浙江 宁波
建筑面积：24 055 m²
用地面积：3.5 hm²

Designers: Zhiqiang Wu, Zijing Zou, Kangquan Chen, Junheng Lai,
Tiandong Ma, Zhong Ma, Peipei Liu, Xinwei Zhao, Mang Gao,
Jing Yang, Peijing Han, Cong Liu, Mingzhe Zhao
Location: Ningbo, Zhejiang
Building Area: 24 055 m²
Site Area: 3.5 hm²

宁波院士中心选址于东钱湖西畔陶公山南麓。项目利用历史名校原宁波师范学院的旧址建筑空间改扩建而成，以"山、水、村、厅、廊、宅"为理念，将自然景观与人文艺术完美融合，打造一座满足院士科研创新、交流研讨等多种使用需求和高品质、高标准、高定位的国际一流院士中心，成为带动整个东钱湖地区乃至宁波市科技、文化、经济发展的引爆点。

Ningbo Academician Center is located at the southern foot of Taogong Mountain on the west side of Dongqian Lake. The project makes use of the former Ningbo Normal College, a historic school, to expand and renovate the building space. With the concept of "Mountain-Water-Village, Hall-Corridor-House", it is a perfect integration of natural landscape and humanistic art to create an international first-class academician center with high quality, high standard and high positioning to meet the multiple functions of academicians' scientific research and innovation, communication and seminar, which becomes the trigger point to drive the development of science and technology, culture and economy of the whole Dongqian Lake area and even Ningbo.

中国扬州运河大剧院
The Grand Canal Theatre

设计师：张洛先、王文胜、周峻、李恒、马溪茵、叶雯、奚秀文
项目地点：江苏 扬州
建筑面积：14 4700 m²

Designers: Luoxian Zhang, Wensheng Wang, Jun Zhou, Heng Li, Xiyin Ma, Wen Ye, Xiuwen Xi
Location: Yangzhou, Jiangsu
Building Area: 144 700 m²

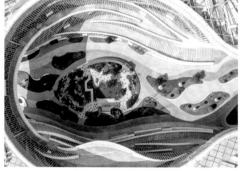

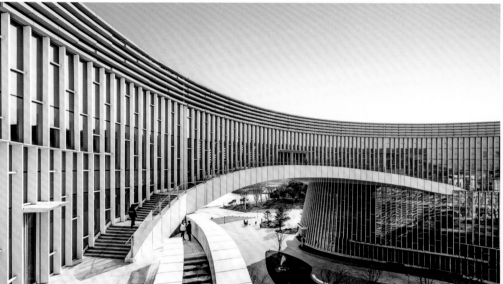

扬州运河大剧院位于明月湖畔,是一座融观演、展示、商业、休闲、园林于一体的文化综合体。大剧院包含1 600座的歌剧厅、800座的戏剧厅、500座的曲艺剧院、300座的多功能小剧场。运河大剧院的设计基于场地、建筑与城市关系的研究,探索出一种面向全民开放、筑景相融的文化建筑模式,让运河大剧院成为扬州这座公园城市的一个鲜活的样板。

Located on the shore of Lake Mingyue, The Grand Canal Theatre is a cultural complex that integrates performance viewing, exhibition, business, leisure and garden. It features a 1 600-seat opera hall, a 800-seat drama hall, a 500-seat Chinese opera hall and a 300-seat Multi-function hall. The design is based on the study of the relationship between the site, the building and the city, exploring a cultural architecture model that is open to all people and built to blend with the landscape, making the Canal Grand Theatre a vivid model of the park city of Yangzhou.

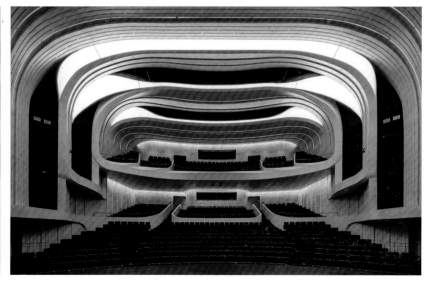

上海

BEING
Studio Architects
彼印建筑工作室

扫描查看更多信息

2020年1月，彼印建筑工作室（Being Studio Architects）成立。

彼印建筑工作室专注于文化旅居空间的营建，设计建构自然而不造作的空间，回应当下，诉诸实践。那当下的回应，会成为彼时的印迹，以示存在的意义——存在。

自2005年起，蔡沪军于多处山水酒店的营建中，感悟自然、建构、人三者的链接。蔡沪军的设计空间具有能拨动心弦、令人驻足的质素。此种质素，摒弃无谓的装饰干扰，容纳自然的光，令光在其中变化。因其无言，直抵心灵。

蔡沪军近年着意于古琴、笛箫的研习，感悟到音乐是空气中开出的花，建筑则是时间结出的果。这时间之果，即彼印的存在，具有沉着的力量，自然而不造作，让居于其中的意念得片刻歇息，看山色明暗，见湖光潋滟，听琴箫意韵，无思无虑，无想无欲。

In Jan 2020 BEING STUDIO was founded.

BEING studio focuses on the modeling of cultural residential space, designing natural and artificial spaces. Respond to the present and resort to practice.

The response at the moment will become the imprint at that time, the meaning of being as being – BEING.

Since 2005, James had realized the link between nature construction and people from the designing of many hotels. The space created by James has the ability to pluck the heartstrings and make people to stay .This character can eliminate unnecessary decoration, contain natural light, and make it change in it .Because of silence, it directly reaches the spirit.

James has devoted himself to the study of Guqin Dizi and Xiao in recent years, and realized that happiness is a flower in the sky, architecture is the fruit of time. The fruit of time is BEING, which with calming force, naturally without affectation. Being in it can take a short rest, watch the light and shade of mountains and lakes, listen to the melody of Guqin and Xiao .No thoughts, no desire.

地址：上海市黄浦区南昌路83弄3号楼
电话：021-54488568
邮箱：961411531@qq.com，1835453610@qq.com

Add: Building 3, Lane 83, Nanchang Road, Huangpu District, Shanghai
Tel: 021-54488568
Email: 961411531@qq.com, 1835453610@qq.com

黄山雨润涵月楼酒店
Huangshan Yurun Han Yue Lou Hotel

项目业主：黄山松柏高尔夫乡村俱乐部有限公司
项目地点：安徽 黄山
建筑功能：酒店建筑
建筑面积：17 790 m²
用地面积：167 424 m²
设计时间：2007年
项目状态：建成
主创设计师：蔡沪军

Project Owner: Huangshan Songbai Golf and Country Club Co., Ltd.
Location: Huangshan, Anhui
Building Function: Hotel Building
Building Area: 17 790 m²
Site Area: 167 424 m²
Design Time: 2007
Project Status: Completed
Designer: Hujun Cai

项目秉承黄山地区徽文化文脉，依托丰富的自然资源，建立一个拥有自然山水景观的高端度假场所，包含99套庭院客房，属一级旅馆建筑。设计采用园林式布局，庭院室内外空间开合，依山就势，景观层叠递进。

客房区域依山形水势各具特点，区域间以水域区隔，水域的连通与融合，一方面延续了徽州民居的水文化，另一方面形成了组团的自然分隔。以自然村落的聚合形态为灵感，结合村落元素（牌坊、亭台、水巷）与错落的单元布局，营造了具有徽州聚落特征的现代村落式度假酒店。

Adhering to the cultural context of Huangshan area and relying on rich natural resources, the project establishes a high-end resort with natural landscape. It consists of 99 courtyard rooms, belonging to the first-class hotel building. The design adopts the garden-style layout. The indoor and outdoor space of the courtyard is designed according to the terrain features, with stacked and progressive landscape.

Each guest room area has its own characteristics in accordance with the terrain. The guest room areas are separated by the water area. The connectivity and integration of the water area on the one hand continues the water culture of Huizhou folk houses, and on the other hand forms the natural separation of groups. Inspired by the converging form of natural villages, the design adopts the layout combining the village elements, such as archways, pavilions, water alleys and scattered units to create a modern village-style resort hotel with Huizhou settlement characteristics.

九华山涵月楼度假酒店
Jiuhuashan Han Yue Lou Hotel

项目业主：雨润集团	Project Owner: Yurun Group
项目地点：安徽 池州	Location: Chizhou, Anhui
建筑功能：酒店建筑	Building Function: Hotel Building
建筑面积：83 044 m²	Building Area: 83 044 m²
用地面积：207 610 m²	Site Area: 207 610 m²
设计时间：2009年	Design Time: 2009
项目状态：建成	Project Status: Completed
主创设计师：蔡沪军	Chief Designer: Hujun Cai

项目依托丰富的九华山自然景观和深厚的佛教文化资源，在布局上融合了九华山的吉祥象征——莲叶。

宛转离合，契合九华山的人文情怀。

设计于基地中创造了一片水域，映照九华山的雄伟与佛像的庄严。

酒店主体以谦逊的姿态，融入山水，体现设计师对自然与人文的独特理解。

Relying on the rich natural landscape of Mount Jiuhuashan and profound Buddhist cultural resources, the project integrates the auspicious symbol of Mount Jiuhuashan "Lotus Land of Buddha" in the layout - the lotus leaf shape of the landscape, which corresponds to the humanistic feelings of Mount Jiuhuashan. The design creates a water area in the base, reflecting the majesty of Mount Jiuhuashan and the dignity of Buddhist statues. The main body of the hotel is humbly integrated into the landscape, reflecting the designer's unique understanding of nature and humanity.

安吉柏翠姚良度假酒店
Anji Bocui Yaoliang Boutique Hotel

项目业主：安吉柏翠度假酒店有限公司		Project Owner: Anji Baicui Resort Hotel Co., Ltd.
项目地点：浙江 湖州		Location: Huzhou, Zhejiang
建筑功能：酒店建筑		Building Function: Hotel Building
建筑面积：7 679 m²		Building Area: 7 679 m²
用地面积：6 658 m²		Site Area: 6 658 m²
设计时间：2016年—2017年		Design Time: 2016-2017
项目状态：建成		Project Status: Completed
主创设计师：蔡沪军		Chief Designer: Hujun Cai

项目位于浙江省安吉县梅溪镇姚良村村口，原址为一所村小学，虽然学校早已搬迁，但稻田中的两层小楼却承载着那一代人的记忆。小楼旁的文化礼堂与古枫树围合起村口的精神空间。

设计顺应地形，延续原小学及礼堂的尺度，由西向东展开，依从地形，自然呈现出贴合大地、水平延绵的感受。南侧一层为公共空间，北侧两层为客房区间，中间为院落。大堂与餐厅层位于西侧高地，大尺度的泳池空间位于大堂北侧的松园下方，休闲空间向西延展到庭院三面的书吧，巧妙地过渡了自然高差。由此，接待空间与休闲空间在不同高差上向外部敞开，与原有的田园无缝连接。

The project is located at the entrance of Yaoliang Village, Meixi Town, Anji County, Zhejiang Province. It was originally a primary school in the village. Although the school has already been relocated, the two-story building in the rice field bears the memory of that generation. The cultural auditorium beside the small building and the old maple trees enclose a spiritual space at the entrance to the village.

In line with the terrain features, the design keeps the same scale as the original primary school and the auditorium and extends the building from west to east, so as to make the project naturally present a feeling of clinging to the earth and extending along the horizon. The public space is arranged on the south side of the first floor and guest rooms on the north side of the second floor, with a courtyard in the central area in accordance with the terrain. The lobby and restaurant are situated on the highland on the west side, while the swimming pool is set below the pine garden on the north side of the lobby based on a natural elevation difference. The recreational space extends westward to the book bar surrounded by courtyards on three sides, tactically smoothing the natural elevation difference. As a result, the reception area and the recreational area open up toward the outside at different levels, seamlessly connecting with the original farmland.

崇明长兴岛开心农场酒店
Chongming Changxing Island Happy Farm Hotel

项目业主：上海住联实业（集团）有限公司
项目地点：上海
建筑功能：酒店建筑
建筑面积：14 994 m²
用地面积：8 289 m²
设计时间：2020年
项目状态：在建
主创设计师：蔡沪军

Project Owner: Shanghai Zhulian Industrial (Group) Co., Ltd.
Location: Shanghai
Building Function: Hotel Building
Building Area: 14 994 m²
Site Area: 8 289 m²
Design Time: 2020
Project Status: Under Construction
Chief Designer: Hujun Cai

项目结合崇明岛当地文脉特色及建筑特征，构建三个层次上的建造模式和控制体系，运用光线把控建筑的空间特质、尺度造型以及对细节处材质肌理、色彩构造的协调，充分体现崇明独具特色的江南海岛特质。

In combination with the local cultural characteristics and architectural features of Chongming Island, the project constructs the construction mode and control system at three levels, and uses light to control the spatial characteristics and scale modeling of the buildings as well as the coordination of the material texture and color structure of the details, fully reflecting the unique characteristics of Chongming Island.

希尔顿格芮酒店二期
Hilton Curio Hotel Phase II

项目业主：中海地产	Project Owner: China Overseas Real Estate
项目地点：江西 庐山	Location: Lushan, Jiangxi
建筑功能：酒店建筑	Building Function: Hotel Building
建筑面积：7 888 m²	Building Area: 7 888 m²
用地面积：12 600 m²	Site Area: 12 600 m²
设计时间：2020年	Design Time: 2020
项目状态：在建	Project Status: Under Construction
主创设计师：蔡沪军	Chief Designer: Hujun Cai

建筑依岛形而建，将客房分为两组，中间留出视觉通廊，公共区域均向湖面打开，将西海气象万千的景致纳入所有空间。接待、会议、餐厅等主要功能空间以盒子的形式，结合景观资源，穿插于建筑之间；庭院、观景长廊等禅意空间与自然山水相映，营造和谐宁静的空间感受；丰富的室外休闲观景平台，增加室内外连通互动。

Built in accordance with the shape of the island, the rooms are divided into two groups with visual corridors, and the public areas open to the lake, incorporating the spectacular views of the West Sea into the spaces. The main functional spaces such as reception, conference and restaurant are interspersed between the buildings in the form of boxes, combined with landscape resources; The courtyard, viewing gallery and other Zen spaces are set against the natural landscape, creating a harmonious and quiet space feeling; The rich outdoor leisure viewing platform increases the interaction between indoor space and outdoor space.

前卫14队研学拓展培训基地
Qianwei Team 14 Research Training Base

项目业主：上海前卫旅游发展有限公司
项目地点：上海
建筑功能：教育基地
建筑面积：6 000 m²
用地面积：250 000 m²
设计时间：2020年
项目状态：在建
主创设计师：蔡沪军

Project Owner: Shanghai Qianwei Tourism Development Co., Ltd.
Location: Shanghai
Building Function: Education Base
Building Area: 6 000 m²
Site Area: 250 000 m²
Design Time: 2020
Project Status: Under Construction
Chief Designer: Hujun Cai

基地位于上海崇明区长兴岛郊野公园东北侧，以公园为依托，为公园提供多功能的配套及服务。基地东侧自然景观良好，南侧与东侧有杉林景观，西侧草地为拓训活动场地。研学基地约90~100间房，可接待200~230人。

Located in the northeast of Changxing Island Country Park, Chongming District, Shanghai, the base relies on the park to provide multi-functional facilities and services for the park. There is a good natural landscape on the east side of the base, fir forest landscape on the south side and the east side, and a training venue on the west side of the grassland. The research base has about 90-100 rooms and can accommodate 200-230 people.

西安云山湖酒店
Xi'an Yunshan Lake Hotel

本项目位于秦岭山脉中一处僻静的峡谷，自西安曲江驱车前往仅需40分钟左右。周边群山环绕，云雾缭绕，村落蜿蜒，宛如江南山水。投资方旨在打造一处面向都市人群避世度假的"云山湖慢生活度假区"，包含儿童乐园、温泉酒店、商业街和高端地产。本酒店为其中的高端配套项目。

在多次踏勘场地后，设计方决定尽量保留场地中的竹林、石溪和村居小屋。并将各个区域现存原始建筑作为核心，网格化布置，使得新旧建筑充分融合，同时最大限度地保留原有建筑中鲜明的地域特色和文化属性。

通过对东方文化概念——"道法自然、天人合一""静居静隐"的探究，在设计中体现东方美学的三个重要特点：诗性的思维方式、象征的符号体系、意象的艺术世界。成功打造远离尘俗、寄情山水、在隐逸中寄托精神和领悟禅意的秘境。

设计方相信，顶级的野奢体验，不在于华丽的建筑外表，而是强调一种紧密的"人与自然的关联"，以达古语"天人合一"的超脱境界。

The project is located in a secluded gorge in the Qinling Mountains. It only takes about 40 minutes to drive from here to Qujiang in Xi'an. Surrounded by green and mist-shrouded mountains as well as winding villages, it represents the scenery in the south of the Yangtze River. The investors aim to create a "Yunshan Lake Slow Life Resort" for city dwellers, which consists of a children's playground, a spa hotel, a shopping mall and high-end real estate. This hotel is one of the high-end supporting projects.

After many visits to the site, the designer decided to preserve the bamboo forest, stone stream and village hut as much as possible. Besides, the existing original buildings in each area are taken as the core texture with grid layout, making the old and new buildings fully integrated; At the same time, the distinctive regional characteristics and cultural attributes of the original buildings are preserved to the greatest extent.

Through the exploration of Oriental culture concept - "following nature's course with harmony between man and nature" and "retreating in static implicit", the design reflects three important characteristics of the Oriental aesthetics : poetic way of thinking; the symbolic system of symbols; the art world of imagery. The project successfully builds the secret environment away from the mundane, in which people can abandon themselves to nature in pursuit of spiritual ballast and Buddhist mood.

The designer believes that the top wild luxury experience does not lie in the gorgeous architectural appearance, but emphasizes a close "connection between man and nature", in order to achieve the transcendent state of the ancient saying "the unity of nature and man".

项目业主：西安荣禾投资有限公司
建设地点：陕西 西安
建筑功能：酒店
建筑面积：16 139 m²
用地面积：63 723 m²
设计时间：2021年
项目状态：方案完成
主创设计师：蔡沪军

Project Owner: Xi'an Ronghe Investment Co., Ltd.
Location: Xi'an, Shaanxi
Building Function: Hotel
Building Area: 16 139 m²
Site Area: 63 723 m²
Design Time: 2021
Project Status: Scheme Completed
Chief Designer: Hujun Cai

宜兴凤凰山康养谷
Yixing Fenghuang Mountain Health Care Valley

项目业主：山西宏源集团有限公司	Project Owner: Shanxi Hongyuan Group Co., Ltd.
项目地点：江苏 宜兴	Location: Yixing, Jiangsu
建筑功能：康养	Building Function: Health Care
建筑面积：243 940 m²	Building Area: 243 940 m²
用地面积：173 483 m²	Site Area: 173 483 m²
设计时间：2021年	Design Time: 2021
项目状态：概念方案完成	Project Status: Conceptual Scheme Completed
主创设计师：蔡沪军	Chief Designer: Hujun Cai

本项目位于江苏宜兴市凤凰山。基地北侧凤凰山蜿蜒起伏；南侧笔架山、龙山山形优美，植被茂密，为基地提供良好的景观资源；东侧不远为浩瀚的太湖；基地内部平坦微坡、水系纵横，实乃建设康养项目的风水宝地。

投资方旨在打造有"村落感"的康养社区，将客群定位于高净值人群中的"有活力的长者"，以封闭式管理方式，倡导康养生活方式，体现生活品质与生命价值。

设计方在研究了山西传统村落中以行业为核心的聚落空间以及江南村落中逐水而居的水街空间形态后，运用从局部到整体的设计方法，创造出聚落空间自然成长的空间形态，并以"水森林""清净地""自在心"为核心理念，打造出符合现代人生活方式的传统村落空间环境。

The project is located in Fenghuang Mountain, Yixing City, Jiangsu Province. The winding and undulating Phoenix Mountain on the north side, as well as the beautiful shape and dense vegetation of Bijia Mountain and Longshan Mountain on the south side provide good landscape resources for the base. Not far to the east is the vast Taihu Lake. The inside area of the base is flat with gentle slope and water system, which is actually an appropriate place for the health care project.

The investor aims to build a health care community with a "sense of village", positioning the customers as "dynamic elderly" among the high net worth individuals, advocating a health lifestyle with a closed management mode and reflecting the quality of life and the value of life.

After studying the industry-centered settlement space in Shanxi traditional villages and the spatial form of water street in Jiangnan villages, the designer adopts the design method from part to whole to create the spatial form of natural growth of settlement space. And with the core concept of "water forest", "clean place" and "free heart", the design creates a traditional village space environment in line with modern life style.

上海

TOPSUM | 上海砼森建筑规划设计有限公司
Shanghai TOPSUM Architectural Planning and Design Co., Ltd.

上海砼森建筑规划设计有限公司(简称砼森建筑)隶属于砼森国际设计集团。该集团是一个综合性的工程设计集团,是由上海砼森建筑规划设计有限公司(同时具备建筑设计甲级资质、规划乙级资质、园林景观乙级资质)、上海砼森结构设计事务所和上海砼森装配式建筑设计中心共同组成的综合性工程设计集团。集团自2006年进入中国市场,在上海设立总公司后,在淄博、济南分别设立了分支机构,在广东设立了联络处。

设计团队的项目范围包括城市规划、城市设计、建筑设计、景观设计、室内设计、房地产咨询等多个领域,并在城市综合体、集群商业、大型办公建筑、文化纪念、场馆建筑、教育建筑、医疗建筑、居住区规划、高档住宅社区、绿色建筑、工业建筑以及特种建筑、钢结构设计、建筑装配化等方面形成了初步的专业优势。

我们集合了一批涵盖各个设计专业领域、富有经验且极富创新精神的优秀专家、建筑师和工程师。伴随着在中国近20年的创作实践,砼森建筑逐步形成了特有的风格和符合中国特色的工作方法,并在公司内部形成了一套严谨的工作流程。与此同时,公司特别强调发掘设计过程的内在逻辑,积极推行整体化的设计和成本控制,将城市设计、建筑设计、景观设计与室内设计视为完整的设计整体加以对待,为使用者提供最适合的解决方案。从项目前期策划研究直至最终建成并投入使用,开发建设的每个设计环节都是我们关注的目标。

砼森建筑特别注重建筑技术领域的研究探索,除新能源与绿色建筑研究中心和装配式建筑设计研究中心外,集团内还设置有专业的建筑声学和建筑光环境研究部门。专业严密的逻辑分析和严谨的流程控制保证了设计品质和产品的价值,并使我们在作品中所倾注的理性与激情都能得以最大限度的实现。

梦想改变世界,砼森建筑愿与您携手开创未来!

Shanghai TOPSUM Architectural Planning and Design Co., Ltd. belongs to Concrete Forest International Design Group, which is a comprehensive engineering design group with Shanghai Concrete Forest Architectural Planning and Design Co., Ltd. (with Class A qualification for architectural design, Class B qualification for planning and Class B qualification for landscape architecture); Shanghai Concrete Forest Structural Design Firm (Class A Qualification of Structural Firm); A comprehensive engineering design group jointly formed by Shanghai Concrete Forest Fabricated Building Design Center. Since entering the Chinese market in 2006, the Group has set up its head office in Shanghai, China, branches in Zibo and Jinan, and liaison offices in Guangdong.

The personal performance of the design team covers urban planning, urban design, architectural design, landscape design, interior design, real estate consulting and other fields, and in urban complexes, cluster commerce, large office buildings, cultural commemoration, venue buildings, education buildings, medical buildings, residential area planning, high-end residential communities, green buildings, industrial buildings and special buildings, steel structure design Preliminary professional advantages have been formed in building assembly and other aspects.

We have gathered a group of experienced and innovative excellent experts, architects and engineers covering various design professional fields. With nearly 20 years of creative experience in China, TOPSUM has gradually formed a unique style and working methods with Chinese characteristics, and has formed a set of rigorous work processes within the company. At the same time, the company especially emphasizes to explore the internal logic of the design process, actively promote integrated design and cost control, treat urban design, architectural design, landscape design and interior design as a complete design, and provide users with the most appropriate solutions. From the preliminary planning and research of the project to the final completion and use, each design link of the development and construction is our focus.

TOPSUM pays special attention to the research and exploration in the field of architectural technology. In addition to the Research Center for New Energy and Green Buildings and the Research Center for Fabricated Building Design, the Group also has a professional research department for architectural acoustics and architectural light environment. Professional and strict logical analysis and strict process control ensure the design quality and product value, and enable us to maximize the rationality and passion we put into our works.

Dream of changing the world, TOPSUM is willing to work with you to create the future!

地址:上海市淞沪路303号创智天地广场三期1101-1108室
邮编:200433
电话:021-33623936
传真:021-33626289
网址:www.topsumchina.com
邮箱:topsumsd@163.com

Add: Room 1101-1108, Phase III, Chuangzhi Tiandi Square, 303 Songhu Road, Shanghai
P.C: 200433
Tel: 021-33623936
Fax: 021-33626289
Web: www.topsumchina.com
Email: topsumsd@163.com

淄博市城市馆、美术馆建设项目
Construction Project of Zibo City Museum, Art Museum

项目业主:淄博市自然资源和规划局		Project Owner: Zibo Natural Resources and Planning Bureau
项目地点:山东 淄博		Location: Zibo, Shandong
建筑功能:展厅、办公		Building Function: Exhibition Hall and Office
建筑面积:79 920 m²		Building Area: 79 920 m²
用地面积:37 880 m²		Site Area: 37 880 m²
设计时间:2022年7月		Design Time: July 2022
项目状态:方案		Project Status: Scheme
设计团队:谭东、陈志文、庞珍珍、张树清、宋丽、王庆坤、王小强、刘奇、闫鹏		Design Team: Dong Tan, Zhiwen Chen, Zhenzhen Pang, Shuqing Zhang, Li Song, Qingkun Wang, Xiaoqiang Wang, Qi Liu, Peng Yan

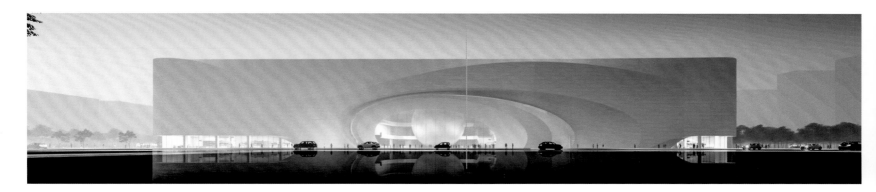

项目位于淄博新区,东临新环西路,西临上海路,南侧为世贸璀璨珑府项目,北临联通路,总用地面积3.8 hm²。

淄博是一座工艺美术之城,博山的琉璃、陶瓷、内画;周村的丝绸、印染、彩灯享誉四方。这座城市本身就是一件艺术作品,把城市馆、美术馆装入同一个巨大的建筑中,就仿佛把一个微缩的城市连同它的珍藏收纳在一个宝盒之中。这个宝盒拥有一个雕塑般的外壳,内部除了大量展品还镶嵌着一颗"琉璃宝珠"。宝珠由白色半透明的琉璃包裹,在琉璃之后隐藏全彩泛光照明,夜幕降临后,整座建筑都将透射出流光幻彩的美妙光华。

两馆作为淄博市最重要的公共建筑之一,造型上既要与政务中心简洁稳重的建筑风格协调,又要保留文化建筑的性格。为此,方案设计确立了平面规则、形态简洁的总体设计策略,仅在细节和重点部位加以变化。立面使用山东本地的白色石材,且保持平面规则,只将表面沿曲线折叠,使平整的立面转化成如周村丝绸般柔美飘逸的流动曲面。

The project is located in Zibo New District, with Xinhuan West Road in the east, Shanghai Road in the west, Shimao Shine Longfu Project in the south, and Liantong Road in the north, with a total land area of 3.8 hm².

Zibo is a city of arts and crafts. Glass, ceramics, and interior painting in Boshan as well as silk, printing and dyeing, and colorful lanterns in Zhoucun are well known throughout the world. The city itself is a work of art. Putting the city museum and art museum into the same huge building is like storing a miniature city and its treasures in a "treasure box". The treasure box has a sculptural shell and a "glass pearl" in addition to a large number of exhibits. The glass pearl, wrapped with white translucent glass, hides the full color lighting behind the glass. After night falls, the whole building will give off a wonderful light.

As one of the most important public buildings in Zibo City, the two buildings should not only coordinate with the simple and stable architectural style of the government affairs center, but also retain the character of cultural buildings. To this end, the scheme design establishes the overall design strategy of plane rules and simple form, only with changes in the details and key parts. The facade is made of white stone native to Shandong, and the plan rules are maintained. Only the surface is folded along the curve, transforming the flat facade into a flowing curved surface as soft and elegant as silk in Zhoucun.

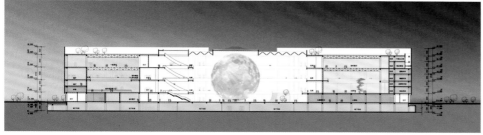

淄博市"一馆两中心"项目
Zibo City "One Pavilion and Two Centers" Project

项目业主：淄博市自然资源和规划局	Project Owner: Zibo Natural Resources and Planning Bureau
项目地点：山东 淄博	Location: Zibo, Shandong
建筑功能：展厅、办公	Building Function: Exhibition Hall and Office
建筑面积：134 450 m²	Building Area: 134 450 m²
用地面积：66 027 m²	Site Area: 66 027 m²
设计时间：2022年7月	Design Time: July 2022
项目状态：方案	Project Status: Scheme
设计团队：谭东、陈志文、庞珍珍、张树清、宋丽、王庆坤、余林梦、孙小琳、杨国其、徐亚东、冯莹烨	Design Team: Dong Tan, Zhiwen Chen, Zhenzhen Pang, Shuqing Zhang, Li Song, Qingkun Wang, Linmeng Yu, Xiaolin Sun, Guoqi Yang, Yadong Xu, Yingye Feng

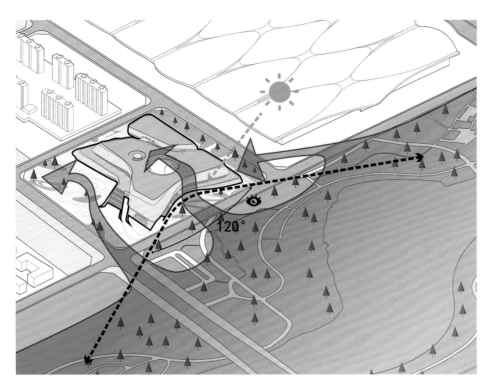

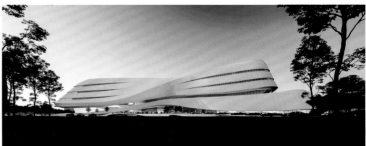

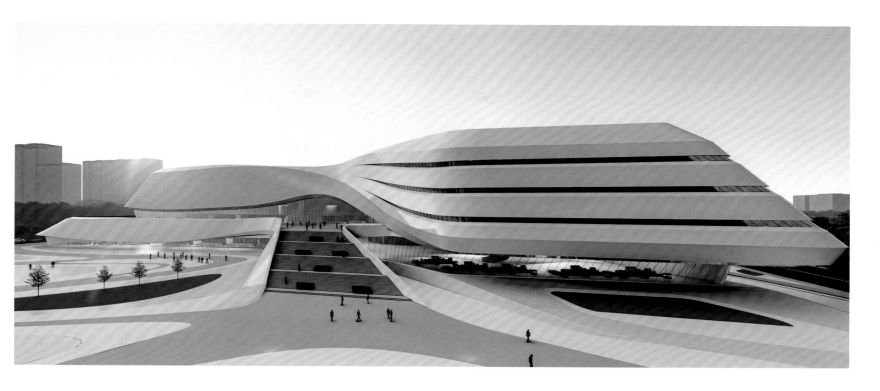

  本项目位于淄博经济开发区,淄博市国际会展中心(拟建)以南,海岱大道以北,孝妇河以西,总用地面积为66 027 m²。项目由3部分组成,包括淄博科技馆、青少年活动中心和妇女儿童事业发展中心,地上总建筑面积约为95 000 m²。

  "一馆两中心"项目作为市级重大公共服务设施落位于孝妇河西岸,是贯彻淄博市拥河发展策略的重要节点,本方案充分考虑了与北侧国际会展中心的联动关系,以及与东侧黄土崖湿地公园的景观融合,在交通组织、空间规划、景观设计、建筑造型等方面,都充分考虑了与上位规划和城市设计的衔接。

  在本方案中,一馆两中心被融合在同一座建筑之中,并根据各自不同的功能和性质,分别规划于不同区域。3大功能板块除共享一套物业管理和消防系统以外,还将部分同类功能设施合并建设。

This project is located in Zibo Economic Development Zone, to the south of Zibo International Convention and Exhibition Center (planned), north of Haidai Avenue and west of Xiaofu River, with a total land area of about 66 027 m². The project consists of three parts, including Zibo Science and Technology Museum, Youth Activity Center and Women and Children Career Development Center, with a total above ground floor area of about 95 000 m².

As a major municipal public service facility, the project of "One Museum and Two Centers" is located on the west bank of Xiaofu River, which is an important node to implement the development strategy of depending on river in Zibo City. The project fully considers the linkage relationship with the International Convention and Exhibition Center in the north and the landscape integration with the Huangtuya Wetland Park in the east. It also gives full consideration to the connection with the upper planning and urban design in terms of transportation organization, spatial planning, landscape design and architectural modeling.

In this scheme, One Museum and Two Centers are integrated into the same building and planned in different areas according to their different functions and properties. In addition to sharing a set of property management and fire protection system, the three functional blocks will also combine some similar functional facilities.

淄博市环理工大学创业创新带
Entrepreneurship and Innovation Belt Surrounding Shandong University of Technology in Zibo

项目业主：山东齐赢产业投资发展有限公司	Project Owner: Shandong Qiying Industrial Investment Development Co., Ltd
项目地点：山东 淄博	Location: Zibo, Shandong
建筑功能：办公、酒店、公寓、商业	Building Functions: Office, Hotel, Apartment and Commerce
建筑面积：290 000 m²	Building Area: 290 000 m²
用地面积：42 000 m²	Site Area: 42 000 m²
设计时间：2022年2月至今	Design Time: February 2022 to Now
项目状态：拟建	Project Status: Proposed
设计团队：谭东、张树清、宋丽、王庆坤、陈志文、庞珍珍、毕于强、王小强、刘奇、闫鹏	Design Team: Dong Tan, Shuqing Zhang, Li Song, Qingkun Wang, Zhiwen Chen, Zhenzhen Pang, Yuqiang Bi, Xiaoqiang Wang, Qi Liu, Peng Yan

环理工大学创业创新带项目位于新村路以南，西九路以西，西十路以东，总占地面积42 000 m²，主要规划理工大学创新中心、张店会客厅、校友经济创业园、技术转移转化中心、学术交流中心、创意街区等板块。

本次规划项目共涉及4个地块，分别分布在北京路与新村路交叉处东西两侧以及重庆路和新村路交叉处东侧。1号、2号、3号地块较规则，整体呈长面宽窄进深特性，因此建筑方案整体布局临新村路东西向依次排开；4号地块完整，南北向和东西向尺寸富裕，建筑整体可围合布局。

项目利用口袋公园、城市绿廊及文化广场等公共文化场所和开放城市空间，紧密联系东西3大区域，在城市天际线设计上，由于地块西侧为新城中心，且可连接于此，地块整体建筑高度呈西高东低的形态，1号地块设计一栋高度为150 m的塔楼，作为园区地标建筑，延续城市界面，重点打造沿新村路的城市形象。

The Entrepreneurship and Innovation Belt Surrounding Shandong University of Technology in Zibo is located in the south of Xincun Road, west of Xijiu Road and east of Xishi Road, covering a total area of 42 000 m². It is mainly planned to construct buildings such as the Innovation Center of Shandong University of Technology, Zhangdian Meeting Room, Alumni Economic Pioneer Park, Technology Transfer and Transformation Center, Academic Exchange Center and creative blocks.

This planning project involves four plots, which are distributed on the east and west sides of the intersection of Beijing Road and Xincun Road and on the east side of the intersection of Chongqing Road and Xincun Road. Plots 1#, 2# and 3# are regular. Therefore, the overall layout of the building scheme is arranged in the east-west direction along Xincun Road. Plot 4# is complete, with rich dimensions from north to south and east to west, and the whole building can take the enclosed layout.

The project makes use of public cultural places such as pocket parks, urban green corridors and cultural squares as well as open urban spaces to closely connect the three eastern and western regions; In terms of urban skyline design, as the west side of the plot is the center of the new city, the height of the overall building is high in the west and low in the east. A 150 m tower is designed for Plot No.1, which is a landmark building of the park to continue the urban interface and focus on creating a city image along Xincun Road.

淄博美达菲国际学校项目
The MacDuffie School Project in Zibo

项目业主：淄博美达菲国际学校
项目地点：山东 淄博
建筑功能：学校
建筑面积：67 482 m²
用地面积：65 309 m²
设计时间：2021年
项目状态：已建成
设计团队：谭东、陈志文、庞珍珍、王小强、周江、董雪、张益飞、佘睿杰

Project Owner: Zibo Meidafi International School
Location: Zibo, Shandong
Building Function: School
Building Area: 67 482 m²
Site Area: 65 309 m²
Design Time: 2021
Project Status: Completed
Design Team: Dong Tan, Zhiwen Chen, Zhenzhen Pang, Xiaoqiang Wang, Jiang Zhou, Xue Dong, Yifei Zhang, Ruijie She

该项目为美达菲国际学校山东省淄博市校区，该校区涵盖了幼儿园以及九年一贯制学校。我们在设计中，不但充分考虑了不同年龄段的学生，在空间中的玩、学、食等基本活动展开的舒适度，以多角度的人体工程学展现了设计的人性化，而且运用了低饱和度的多样化颜色，在给予充足人工照明的前提下，潜移默化地帮助少年儿童进行身心健康的学习生活。

项目中大量运用了自然界中最常见的弧形，尽可能地抹平了边角，在保护学生的同时也增添了空间中的趣味性。各个不同的学制场所，通过相似的颜色、近似的元素、环保的材料，形成了一个有区别、有个性又相互联系的有机整体。

The project is The MacDuffie School in Zibo City, Shandong Province. Zibo campus covers kindergartens and nine-year education schools. The design not only fully considers the students of different ages, with multi-angle ergonomics to show the humanization of the design and ensure the comfort level in playing, learning, eating and other basic activities, but also adopts diverse colors with low saturation to have a subtle influence on children's physical and mental health in their study and life, under the premise of sufficient artificial lighting.

In the project, the most common shape of arc in nature is extensively used to smooth out the corners as much as possible, which not only protects the students but also adds interest to the space. Different school places form a distinct, personalized yet interrelated organic whole, through similar colors, similar elements, environmental protection materials.

淄博火车站北城市设计项目
Urban Design Project of the North Square of Zibo Railway Station

项目业主：淄博市房屋建设综合开发有限公司
项目地点：山东 淄博
建筑功能：住宅、商业、办公、酒店
建筑面积：25.15 m²
用地面积：56 752 m²
设计时间：2019年至今
项目状态：在建
设计团队：谭东、张树清、宋丽、陈志文、庞珍珍、王庆坤、毕于强、余林梦、孙小琳

Project Owner: Zibo Housing Construction Comprehensive Development Co., Ltd.
Location: Zibo, Shandong
Building Functions: Residential, Commercial, Office and Hotel
Building Area: 251 500 m²
Site Area: 56 752 m²
Design Time: 2019 to Now
Project Status: Under Construction
Design Team: Dong Tan, Shuqing Zhang, Li Song, Zhiwen Chen, Zhenzhen Pang, Qingkun Wang, Yuqiang Bi, Linmeng Yu, Xiaolin Sun

火车站北广场片区位于淄博市老城区核心地段，这里曾经是全市最繁华的交通、商业、文化和工业中心。但是，随着淄博南站、淄博高铁北站陆续建成并投入使用，位于胶济线北侧的淄博火车站所承担的客流量未来将减少3/4。另外，由于出行方式的改变，淄博火车站周边区域早已经呈现出萧条的迹象，传统的核心已成为城市边缘，并面临着加速衰败的窘迫局面。老城区正在加速老龄化、交通空心化，并导致商业衰败、基础设施闲置、文化娱乐活动消失、周边城区衰弱、环境恶化等一系列问题。过去最有活力的老城区，已经丧失了引人驻足的魅力。

2021年初，淄博市政府决心加快老城区城市更新的步伐，为火车站北广场片区的新生提供了契机。传统的交通枢纽未来应该如何转型？怎样才能改变老城区拥挤脏乱的破败面貌？老城区需要什么样的公共空间？如何才能使老城区焕发新的活力？对这些问题的回答，就是本次城市设计需要探索的方向。

设计理念：本设计试图赋予老城更新以另一种意义，弱化其本已弱化的交通职能，将过去以交通为主导的核心功能区转变为服务于人居环境的绿色新城区。将车站、公园绿化、商业服务、生活配套融为一体；将铁路沿线棚户区转化成绿色开放空间带（铁路公园）；将年久失修的历史建筑保护、转化为热闹的特色文化商业街；将混乱的平面交通转化为分层的立体交通网络；将空旷的硬质广场转化为绿色的休憩空间；以车站地区改造为起点，改善城市环境、提升街区品质、吸引居民回流、培育新产业、培育新商业、提升土地价值，重塑城区活力。

The North Square area of the Railway Station is located in the core of the old city of Zibo, which used to be the most prosperous transportation, commercial, cultural and industrial center of the city. However, as Zibo South Railway Station and Zibo North High-speed Railway Station have been successively completed and put into use, the passenger flow of Zibo Railway Station on the north side of Jiaozhou-Jinan Railway Line will be reduced by three quarters in the future. In addition, due to the change of travel mode, the area around Zibo Railway Station has already shown signs of depression, and the traditional core has become the edge of the city, and is facing the distress situation of accelerated decline. The old city is aging at an accelerated pace while the traffic is hollowing out, leading to a series of problems such as commercial decline, empty infrastructure, disappearance of cultural and recreational activities, weakening of the surrounding urban areas, environmental deterioration and so on. The old city, once the most vibrant one, has lost its charm.

In early 2021, the Zibo Municipal government decided to accelerate the pace of urban renewal of the old city, which provided an opportunity for the renewal of the North Square area of the Railway Station. How should traditional transportation hubs be transformed in the future? How should we change the dilapidated appearance of the crowded and dirty old city? What kind of public space does the old city need? How can the old city be revitalized? The answer to these questions is the direction that this urban design needs to explore.

Design concept: This design attempts to give another meaning to the renewal of the old city, weakening its already weakened function of transportation, and transforming the former functional area with transportation as the leading core into a green new urban area serving the living environment. It aims to integrate the station, the park greening, commercial services and life supporting facilities, to transform shantytowns along the railway into green open space belts (railway parks), to protect the historical buildings in disrepair and transform them into lively cultural and commercial streets, to transform the chaotic plane traffic into a layered three-dimensional network, to transform the empty hard square into a green rest space, to renovate the crowded and dilapidated shanty towns along the route into beautiful homes surrounded by greenery. With the transformation of the station area as the starting point, the design strives to improve the city environment, enhance the quality of the block, attract residents to reruen, cultivate new industries and new businesses, enhance the land value, and rebuilt the vitality of the city.

上海鼎实建筑设计集团有限公司
Shanghai DE-SIGN Architectural Design Group Co., Ltd.

上海鼎实建筑设计集团有限公司是一家聚焦价值创造的全程一体化建筑及规划设计集团公司。

我们擅长用多维的角度审视城市建筑的核心价值,为开发商、零售商、产业方及政府部门提供富有创意的专业设计解决方案。我们在建筑全周期的每个阶段中关注价值增长的核心点,在最关键的环节上勇于创新,并最终实现物业资产价值的高效提升。我们相信设计可以为空间创造价值,可以为生活创造意义,可以为体验创造记忆,在致力于打造富有创意的城市场景的同时,我们也严谨地控制设计全过程当中的技术与管理质量,并且成为国内最早参与总包、BIM等先进设计及管理技术落地实践的创新企业。在中国地产行业10余年间的不懈耕耘中,我们服务的项目遍及全国28个省市,服务面积超过5 000余万平方米。

Shanghai DE-SIGN Architectural Design Group Co., Ltd. is an integrated architecture and planning and design group focusing on value creation.

We are good at examining the core value of urban architecture from a multi-dimensional perspective, providing creative professional design solutions for developers, retailers, industry and government departments. We pay attention to the core points of value growth in each stage of the whole construction cycle, and dare to innovate in the most critical links, and ultimately achieve the efficient improvement of property asset value. We believe that design can create value for space, meaning for life and memory for experience. While committed to creating creative urban scenes, we also strictly control the technology and management quality in the whole process of design, and become the first innovative enterprise to participate in the implementation of advanced design and management technologies such as general contracting and Bim in China. In more than ten years of unremitting efforts in China's real estate, we serve projects throughout 28 provinces and cities, with a service area of more than 50 million m².

地址:上海市静安区西藏北路199号光明地产大厦7-8楼
邮编:200070
电话:021-65153550
邮箱:xx@mrd.sg
网址:www.groupmrd.com

Add: 7-8 F, Guangming Real Estate Building, No.199 North Xizang Road, Jing'an District, Shanghai, China
P.C: 200070
Tel: 021-65153550
Email: xx@mrd.sg
Web: www.groupmrd.com

鑫创科技园展示中心
Xinchuang Science Park Exhibition Center

设计师:谢璇、孙月书、郑化龙、盘广湖、詹建锋	Designers: Xuan Xie, Yueshu Sun, Hualong Zheng, Guanghu Pan, Jianfeng Zhan
项目地点:广东 佛山	Location: Foshan, Guangdong
建筑面积:7 000 m²	Building Area: 7 000 m²
用地面积:15.3 hm²	Site Area: 15.3 hm²
容积率:2.4	Plot Ratio: 2.4

展示中心是整个鑫创科技园园区的核心,是以展示、展陈、发展、招商等为主的现代化复合型展示中心,是园区发展的重要驱动力,同时兼有办公、休闲、餐饮等辅助配套功能,为企业提供全方位的服务支持的同时,也是知识信息共享及传递的重要载体。建筑形象采用现代手法,融入科技主题,将芯片概念抽象为建筑形体,并在入口处采用高科技技术,体现建筑的科技感。

项目定位西江新城创新型增长极,智慧城市生态化示范谷,人工智能(AI)科创企业成长集聚区。

项目采用一轴两中心的规划结构,西侧组团为产业办公组团,东侧为产业配套组团。

The exhibition center is the core of the whole park. It is a modern composite exhibition center focusing on exhibition, development and investment attraction. It is an important driving force for the development of the park. At the same time, it also has auxiliary supporting functions such as office, leisure and catering. It not only provides comprehensive service support for enterprises, but also an important carrier for knowledge and information sharing and transmission. The architectural image adopts modern techniques, integrates the theme of science and technology, abstracts the chip concept into the architectural form, and adopts high-tech technology at the entrance to reflect the sense of science and technology of the building

The project is positioned as the innovative growth pole of Xijiang new town, the ecological demonstration valley of smart city and the growth cluster of AI science and innovation enterprises.

The project adopts the planning structure of one axis and two centers. The West Group is the industrial office group and the East Group is the industrial supporting group.

深圳龙岗万达广场
Shenzhen Longgang Wanda Plaza

设计师: 谢璇、孙月书、周慧、郑化龙
项目地点: 广东 深圳
建筑面积: 304 416.48 m²
用地面积: 80 700 m²
容积率: 2.69

Designers: Xuan Xie, Yueshu Sun, Hui Zhou, Hualong Zheng
Location: Shenzhen, Guangdong
Building Area: 304 416.48 m²
Site Area: 80 700 m²
Plot Patio: 2.69

深圳龙岗万达广场总用地面积为80 700 m²,建筑面积为304 416.48 m²。其立面设计理念立根于深圳当地传统文化"一夜鱼龙舞",即"凤箫声动,玉壶光转,一夜鱼龙舞"。"龙"即"兴隆","飞"即"一飞冲天"。翻转的立面形态暗喻龙鳞的舞动,时隐时现高雅的金色使龙鳞熠熠生辉,代表着深圳龙岗万达广场必定光彩夺目,一飞冲天。

The total land area of Shenzhen Longgang Wanda Plaza is 80 700 m², and the building area is 304 416.48 m². Its facade design concept is rooted in the local traditional culture of Shenzhen, "fish and dragon dance in one night". That is, "the sound of the phoenix flute moves, the light of the jade pot turns, and the fish and dragon dance overnight". Dragon means prosperity, and flying means soaring into the sky. The inverted facade shape implies the dancing of the dragon scale. The elegant golden color makes the dragon scale glitter, representing that Wanda Plaza in Longgang, Shenzhen, must be dazzling and soaring.

武汉绿地天河国际会展城
Wuhan GEC International Exhibition

设计师：谢璇、李卓、韩力、于永飞	Designers: Xuan Xie, Zhuo Li, Li Han, Yongfei Yu
项目地点：湖北 武汉	Location: Wuhan, Hubei
建筑面积：6 372 m²	Building Area: 6 372 m²
用地面积：1.44 hm²	Site Area: 1.44 hm²
容积率：0.45	Plot Ratio: 0.45

项目地处会展、居住、文化体育3大板块交会处，是绿地国际会展城项目的重要门户，代表着对武汉"后疫情时代"的崛起与发展的期待。建筑秉承"中式形韵"的建筑美学思想，以现代建筑形式体现了绿地集团在地性和未来性相结合的开发理念。

The project is located at the intersection of exhibition, residence, culture and sports. It is an important image portal for the overall project of Greenland International Convention and Exhibition City and represents the expectation of the rise and development of Wuhan in the "post epidemic era". The architecture inherits the architectural aesthetic thought of "Chinese style shape and rhyme", and reflects the development concept of combining the local and future of Greenland space city building in the form of modern architecture.

合肥银泰中心二期项目
Introduction of Hefei Intime Center Phase II Project

设计师：谢璇、何海洋	Designers: Xuan Xie, Haiyang He
项目地点：安徽 合肥	Location: Hefei, Anhui
建筑面积：104 559.50 m²	Building Area: 104 559.50 m²
用地面积：1.1 hm²	Site Area: 1.1 hm²
容积率：6.2	Plot Ratio: 6.2

合肥银泰中心二期项目位于合肥市旧城中心,背靠合肥最繁华的淮河路步行街,为项目带来了便利的客户优势以及配套优势。银泰中心二期与一期相互配合及错位经营,定位为年轻、时尚及有活力的新商场,业态类型丰富,致力打造合肥市最高端商业旗舰综合体。

Hefei Intime Center Phase II project Located in the center of the old city of Hefei, it is backed by the most prosperous pedestrian street of Huaihe Road in Hefei, which brings convenience to customers and supporting advantages to the project. Intime Center Phase II and Phase I cooperate with each other and operate out of alignment. It is positioned as a young, fashionable and dynamic new shopping mall with a variety of business types, committed to building the most high-end business flagship complex in Hefei.

上海五角场万达改造
Renovation of Shanghai Wujiaochang Wanda Plaza

设计师：谢璇、孙月书、郑化龙、徐洪鹏
项目地点：上海
建筑面积：94 000 m²

Designers: Xuan Xie, Yueshu Sun, Hualong Zheng, Hongpeng Xu
Location: Shanghai
Building Area: 94 000 m²

上海五角场万达位于上海杨浦区，是新市中心的核心区域，与徐家汇同为上海商圈影响力最大的城市副中心之一。

改造后的五角场万达将创建一个以零售、休闲、娱乐等为一体的新城市区域生活中心。设计充分挖掘地块的最佳商业价值，合理布置商业空间，提高整个地块和周边地区的商业活力，结合商业室内外步行街与中庭的设计，体现公共空间的共享性、舒适性与休闲功能。

Located in Yangpu District of Shanghai, Shanghai Wujiaochang Wanda is the core area of the new city center. Together with Xujiahui, it is the most influential urban sub center of Shanghai business district.

The reconstructed Wujiaochang Wanda will create a new urban regional life center integrating retail, leisure and entertainment, fully tap the best commercial value of the plot, reasonably arrange the commercial space, improve the commercial vitality of the whole plot and surrounding areas, and combine the design of indoor and outdoor pedestrian streets and atriums to reflect the sharing, comfort and leisure of public space.

万萃城二期(1)M1组团项目
M1 Group Project of Wancui City Phase II

设计师：谢璇、何海洋、詹建锋 Designers: Xuan Xie, Haiyang He, Jianfeng Zhan
项目地点：重庆 Location: Chongqing
建筑面积：314 642.88 m² Building Area: 314 642.88 m²
用地面积：6.32 hm² Site Area: 6.32 hm²
容积率：3.5 Plot Ratio: 3.5

本案将充分挖掘项目潜力，通过整合设计将项目从商业综合体提升至万州城市旅游"名片"，将商业体验与万州文化相结合，用现代手法重新演绎流水、码头、戏台等万州传统文化特色，结合城市脉络，追溯万州起源，打造一处具有立体、人文、生态属性的新城市客厅。项目包含7栋高层住宅、9栋多层商业街区（老字号文化主题）及1栋购物中心。

This case will fully exert potential project, the project by integrating design from the commercial complex to wanzhou city tourism business "CARDS", combining business experience and wanzhou culture and use modern methods to deduce water, wharf, wanzhou stage such as traditional culture characteristics, combining with the urban context, traces the origin of wanzhou, build a three-dimensional, cultural and ecological property of new urban living room. It contains 7 high-rise residential buildings, 9 multi-storey commercial blocks (time-honored cultural theme) and 1 shopping mall.

绿地济南CBD项目
Greenland Jinan CBD Project

设计师：谢璇、何海洋、詹建锋
项目地点：山东 济南
建筑面积：180 815.03 m²
用地面积：58 346 m²
容积率：2.21

Designers: Xuan Xie, Haiyang He, Jianfeng Zhan
Location: Ji'nan, Shandong
Building Area: 180 815.03 m²
Site Area: 58 346 m²
Plot Ratio: 2.21

项目位于济南市历下区奥体中心核心位置,由商业、办公、公寓组成,设计充分遵循打造济南区域级金融城市的上位规划原则,建设紧凑的特色街区,依托济南城市人文特色,将"山、水、城"作为整个项目的设计理念,以超高层组群为"山",以流动、穿梭于商业空间的人流为"水",以"街、里、园"特色空间为"城"。

The project is located at the core of the Olympic Sports Center in Lixia District, Jinan City, and is composed of commerce, offices and apartments. The design fully follows the upper planning principle of building a regional financial city in Jinan. Build compact characteristic streets, rely on Jinan's urban cultural characteristics, take "mountain, water and city" as the design concept of the whole project, take super high-rise groups as "mountains", take flowing and shuttling in commercial space as "water", and take "street, interior and garden" characteristic space as "city".

四川

中国建筑西南设计研究院有限公司
CHINA SOUTHWEST ARCHITECTURAL DESIGN AND RESEARCH INSTITUTE CORP.LTD

中国建筑西南设计研究院有限公司成立于1950年,是中国同行业中成立时间最早、专业最全、规模最大的国有甲级建筑设计院之一,目前隶属于世界500强第9位、全球最大的建筑投资央企——中国建筑集团公司,现有员工6 000余人。建院70余年来,中建西南院设计完成了万余项工程设计任务,项目遍及全国各省、市、自治区及全球20多个国家和地区,获省部级及以上设计咨询类奖项1 900余项,其中国家优秀设计金质奖5项、银质奖4项、铜质奖5项。在科研方面,获省部级及以上科技类奖项140余项,其中国家科技进步一等奖2项、二等奖2项,是我国拥有独立涉外经营权并参与众多国外设计任务经营的大型建筑设计院之一。2019年,中建西南院入选国务院国资委创建世界一流企业的试点单位。目前正按照中建集团"一创五强"战略目标,坚持"技术、管理、投资"的发展模式,全面对标世界顶级设计机构,加快建设世界一流设计企业。

Established in 1950, China Southwest Architectural Design and Research Institute Corp. Ltd ("CSWADI") is one of the earliest class-A state-owned architectural design institutes with largest scale and full disciplines. It is controlled by China State Construction Engineering Corporation, a world top 500 enterprise. CSWADI has nearly 6 000 employees in total. In its 70 years'history, CSWADI has undertaken the design of nearly 10 000 projects across all provinces, municipalities and autonomous regions in China and over 20 countries and regions in the world. CSWADI has won more than 1 700 national, ministerial and provincial awards for its excellence in engineering design and scientific research, and 5 national golden prizes, 4 silver prizes and 5 bronze prize for excellent design. In the field of scientific research, more than 140 Ministerial and Provincial Prizes for Progress in Science and Technology have been awarded to CSWADI, including 2 first-class, and 2 second-class National Prizes for Progress in Science and Technology, which is one of the large-scale architectural design institutes with independent overseas operation rights and participating in many foreign design tasks. In 2019, CSWADI was selected as a pilot site for establishing a world-class enterprise by the State-owned Assets Supervision and Administration Commission of the State Council. According to the CSCEC's strategic targets of "One Creation and Five Strong Points", insisting the development mode of "technology, management, investment". CSWADI will take world top-level design firms as the benchmark to improve its strength for becoming world first-calss design enterprise.

地址:四川省成都市天府大道北段866号 Add: No.866 North Section, Tianfu Avenue, Chengdu, Sichuan
邮编:610042 P.C: 610042
网址:https://xnjz.cscec.com/ Web: https://xnjz.cscec.com/

成都金融城双子塔
Chengdu Financial City Twin Towers

设计师:钱方、刘艺、张宗腾
项目地点:四川 成都
建筑面积:164 000 m²
用地面积:2.32 hm²
容积率:7.06

Designers: Fang Qian, Yi Liu, Zongteng Zhang
Location: Chengdu, Sichan
Building Area: 164 000 m²
Site Area: 2.32 hm²
Plot Ratio: 7.06

项目系成都金融城园区内的策划加建工程。通过增建双子塔,完善园区功能,强调该区域城市东西轴线,丰富了城市空间感知维度。项目以现代、简约、优雅的方式,融入金融城园区花园般的环境中。在总体布局、体量塑造、表皮策略上,分别采取向心、同构、双层表皮的方法,实现和而不同的目标。项目建成后给该城市片区带来了巨大的公共活力。

This project is aplanned addition project in Chengdu Financial City. By adding the Twin Towers, the functions of the park are improved, meanwhile, the east-west axis of thecity in the region is emphasized, and the perception dimension of urban spaceis enriched. The Twin Towers have been blended into the garden-like environment of the Financial City in a modern, simple and elegant way. Centripetal, isomorphic and double-layer methods are adopted to answer the important issues, which are the overall layout, volume shaping and skin strategy. As a result, harmonization and identity have been perfectly balanced in this project. Great public vitality is the biggest gift that the Twin Towers provide for this area.

成都城市音乐厅
Chengdu City Concert Hall

设计师：刘艺、王永炜	Designers: Yi Liu, Yongwei Wang
项目地点：四川 成都	Location: Chengdu, Sichuan
建筑面积：102 619 m²	Building Area: 102 619 m²
用地面积：1.98 hm²	Site Area: 1.98 hm²
容积率：1.92	Plot Ratio: 1.92

成都城市音乐厅是国内少有的在城市核心区新建的大型综合观演建筑。包含1 600座的歌剧厅、1 400座的音乐厅、400座的戏剧厅、300座的室内小型音乐厅以及2 000人露天音乐厅广场。设计在充分体现对"地域和艺术文化"传承的同时，追崇人性化的思考方式，以一种谦和开放的态度，将人的行为模式贯穿于设计的始终，造就了一座真正为人而设计的建筑。

Chengdu City Concert Hall, one of the few new constructions located in the heart of the city, serves as a large-scale integrated performance project. It includes 1 600-seat Opera Hall, 1 400-seat Concert Hall, 400-seat Drama Performance Hall, 300-seat Recital Hall and a capacity of nearly 2 000 people open-air concert hall with views to the city. The ambition of the design is to express not only indigenous culture, but also humanistic based methodology: the environment/behavior leading design strategy, picturing a human-centered design.

南沙青少年宫
Nansha Youth Palace

设计师：刘艺、王珏	Designers: Yi Liu, Jue Wang
项目地点：广东 广州	Location: Guangzhou, Guangdong
建筑面积：56 028.86 m²	Building Area: 56 028.86 m²
用地面积：30 036 m²	Site Area: 30 036 m²
容积率：1.3	Plot Ratio: 1.3

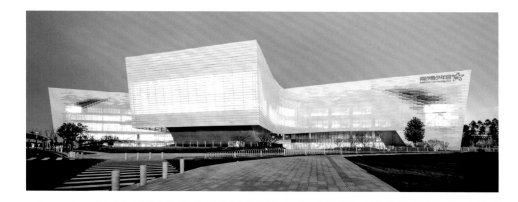

南沙青少年宫位于广州市南沙区凤凰大道，坐落于南沙自贸区门户口岸。建筑功能区包含1 000座的儿童剧场、文化交流中心、图书馆、科技互动展厅、报告厅以及各类教学培训用房。

青少年宫是城市的"儿童之家"与"第二课堂"，设计师为青少年宫师生提供了探索新型教育空间设计的契机。南沙青少年宫开放式功能布局与首层架空的建筑形式，用建筑内外的游走路径串联起来的空间序列；工程中所运用的结构技术和BIM参数化设计，共同体现"技"与"艺"的有机结合。

Nansha Youth Palace is located in Phoenix Avenue, Nansha District, Guangzhou, at the gateway port of Nansha Free Trade Zone. The building features 1 000 children's theaters, cultural exchange centers, libraries, interactive exhibition halls for science and technology, lecture halls, and various teaching and training rooms.

As the city's "children's home" and "second classroom", the designers provide opportunities for teachers and students of the Youth Palace to explore the design of new educational spaces. The open functional layout of Nansha Youth Palace and the architectural concept of the first floor overhead, as well as the spatial sequence connected by the wandering paths inside and outside the building, and the structural technology and BIM parametric design application used in the project, reflect the organic combination of the designer's "technique" and "art".

西昌邛泸景区游客中心
Xichang Qionglu Scenic Spot Tourist Center

设计师：郑勇、肖迪佳	Designers: Yong Zheng, Dijia Xiao
项目地点：四川 西昌	Location: Xichang, Sichuan
建筑面积：2 980 m²	Building Area: 2 980 m²
用地面积：2.958 hm²	Site Area: 2.958 hm²
容积率：0.1	Plot Ratio: 0.1

西昌邛泸景区游客中心项目力图在建筑与湿地环境之间建立有力的联系，通过地景化建筑处理，让部分屋顶从地面逐渐升起，自然地融入整个湿地景观中。在建筑内部精心布置的一系列坡道、天桥、通道、片墙，形成了各功能空间之间必要的交通联系，也引导游客去探索游玩，而当游客行走来到建筑坡顶上，更是能高点远眺，饱览湿地及邛海景致。

Xichang Qionglu Scenic Spot Tourist Center sets up a strong connection between the architecture and wet land. With the method of landscape architecture design, the roof stretches out of the ground and integrates into natural environment. A series of ramps, bridges, pathways and walls are arranged orderly in the building, which organize the traffic connection of different function spaces and inspire tourists to discover the building for fun. When people reach the top of the building, they can enjoy a distant view of the whole wet land.

青岛胶东国际机场
Qingdao Jiaodong International Airport

设计师：陈荣锋、潘磊	Designers: Rongfeng Chen, Lei Pan
项目地点：山东 青岛	Location: Qingdao, Shandong
建筑面积：547 731 m²	Building Area: 54 773 1 m²
用地面积：253.5 hm²	Site Area: 253.5 hm²
容积率：0.2	Plot Ratio: 0.2

青岛胶东国际机场定位于"面向日韩具有门户功能的区域性枢纽机场,环渤海地区国际航空货运枢纽",新机场设计目标年为2025年,满足年旅客吞吐量3 500万人次。

新机场航站楼概念设计按照"规划导引、安全第一、功能齐备、便捷舒适、环保节能、协调美观、质优价公"的原则,确定以"齐"字状为总体布局,以富有张力的连续曲面将极具向心力的5个指廊与大厅融为整体,实现大集中与单元式的合理平衡。

Qingdao Jiaodong International Airport is positioned as "a regional hub airport as a gateway for Japan and South Korea and an international air cargo hub around the Circum-Bohai-Sea Region." The new airport is designed to meet the annual passenger throughput of 35 million by 2025.

The conceptual design of the terminal of the new airport follows the principles of "planning guided, safety first, complete functions, convenience and comfort, environmental protection and energy saving, coordination and beauty, high quality and reasonable price". The overall layout of the new airport is designed to be as the shape of Chinese character "齐", and the five extremely centripetal airside concourses together with the hall have been integrated into a whole with a continuous curved surface full of tension to achieve a reasonable balance between large concentration and separated sections.

成都露天音乐公园
Chengdu Open Air Music Park

设计师：佘念
项目地点：四川 成都
舞台面积：15 816.09 m² (公园总建筑面积：47 342.53 m²)
用地面积：31.77 hm²
容积率：0.09

Designer: Nian She
Location: Chengdu, Sichuan
Stage Area: 15 816.19 m² (Total building area of the park: 47 342.53 m²)
Site Area: 31.77 hm²
Plot Ratio: 0.09

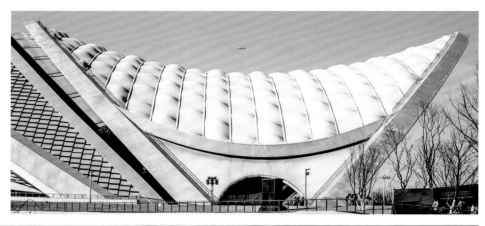

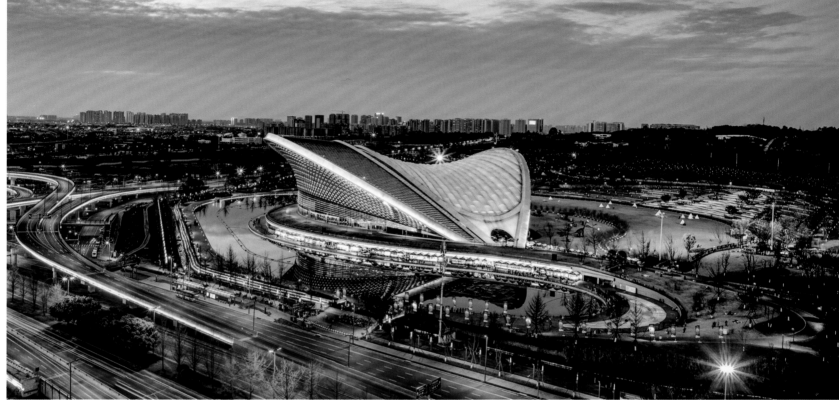

成都露天音乐公园是以音乐文化为主题的城市公园，也是成都市环市一周的天府绿道的重要节点。作为露天音乐圣地，公园能承办超大型露天音乐节，主舞台能举办同时容纳4万人的露天音乐会或1万人的半室内音乐会。"凤凰于飞、翙翙其羽"，自2019年5月开园至今，主舞台已举办多场盛大音乐活动，成为成都市建设国际音乐之都的新地标。

Chengdu Open Air Music Park is an urban park with the theme of music, which is also an important node on the Tianfu Greenway around the Chengdu city. As a mecca for open-air music, the entire area can host large scale open-air music festivals. The main stage can hold an open-air concert for 40 000 people or a semi-indoor concert for 10 000 people at the same time. "Phoenix Flying, Its Feathers Flying", since the opening of the park in May 2019, many grand music activities have been held on the main stage, which has become a new landmark for Chengdu to build an international music city.

中建科技成都绿色建筑产业园研发中心
Research & Development Center in CCSTC Green Industrial Park

设计师：李峰、佘龙	Designers: Feng Li, Long She
项目地点：四川 成都	Location: Chengdu, Sichuan
建筑面积：4 409.69 m²	Building Area: 4 409.69 m²
用地面积：10.85 hm²	Site Area: 10.85 hm²
容积率：1.28	Plot Ratio: 1.28

中建科技成都绿色建筑产业园研发中心是中建未来零碳建筑的一次探索，将装配式建筑技术、近零能耗建筑技术、绿色建筑技术及智慧建筑技术等4大技术体系相融合，是全国首例装配式混凝土结构近零能耗被动式公共建筑。

项目为中德合作被动式低能耗建筑和多项国家"十三五"重点研发计划示范工程。

The project of Research & Development Center in CCSTC Green Industrial Park is a discussion on the future zero-carbon building. The project integrates four technology systems: prefabricated building technology, nearly zero energy building technology, green building technology and smart building technology, which makes the first prefabricated concrete structure nearly zero energy consumption passive public building in China.

The project is a Sino-german Passive Low Energy Building Demonstration Project and a number of national "13th Five-Year" Key Project Demonstration Projects.

四川

四川省大卫建筑设计有限公司
Sichuan David Architectural Design Co., Ltd.

四川省大卫建筑设计有限公司是由中华人民共和国住房和城乡建设部批准成立的综合性建筑工程甲级设计企业，业务涉及建筑工程、市政行业、风景园林、旅游规划、工程造价咨询、工程勘察、文物保护勘察等。公司有一支由国家注册设计师组成的专业精英团队，其思维活跃、意识前卫、注重实际、诚信敬业。大卫人致力于营建诗意栖息的艺术空间，在设计领域长期坚持生态和文化的注入，建立起国际化的管理模式和先进的设计理念，不断创作出优秀的作品。30年多的经营，大卫团队创造出骄人业绩：

——2012年 全球人居环境规划设计奖（联合国环境规划署）
——2013年 中国经济发展最具发展潜力企业（新华社）
——2014年 全国人居经典建筑规划设计方案竞赛双金奖（中国建筑学会）
——2015年 中国建筑设计金拱奖（中国建筑文化中心）
——2016年 人居生态国际建筑规划设计方案竞赛金奖（中国建筑学会）
——2018年 美国建筑师协会（AIA）首届中国建筑设计奖
——2019年 亚洲美丽乡村营造优胜奖（亚洲人居环境协会）
——2020年 亚洲景观规划设计范例奖（亚洲景观设计学会）
——2020年 第19届中国精瑞人居奖（中华人民共和国科学技术部）
——2022年 亚洲都市景观奖（联合国人居署）

大卫设计与时俱进，积极践行绿色建筑推广，2014年成为BRE-Global设在中国成都的国际绿色建筑评估机构。大卫设计还发起成立四川智慧计算机辅助设计技术研究院。

四川省大卫建筑设计有限公司扎根本土，放眼全球。倡导"文化最具竞争力"，弘扬民族建筑文化，响应国家"一带一路"倡议，与国际先进技术团队合作，聚集了国内外BIM研发高精尖人才，并在国家低碳绿色建筑国际联合研究中心支持下，建立了"中国西南地区绿色建筑示范推广基地"。与国际知名设计师广泛合作，在非洲、中东和东南亚等地拓展业务，荣获四川省勘察设计"一带一路"工作先进单位。更多年轻的海归设计人才为大卫品牌所吸引而汇聚，推动大卫设计进一步致力于将先进的绿色生态、节能环保等高新技术及全新理念融入设计中，为当今中国城市与建筑的创新及可持续发展作出积极贡献。

Sichuan David Architectural Design Co., Ltd. was established in 1984 and restructured in 2002 restructuring. The company was approved by the Ministry of Housing and Urban-Rural Development of the People's Republic of China (MOHURD); and has obtained the Class-A Integrated Qualification for professional contracting of architectural engineering. The business range covers construction engineering, municipal industry, landscape architecture, tourism planning, construction cost consultation, engineering investigation, cultural relics preservation investigation etc.

The company possess on elite team involving a group of national registered designers, who are active thinking and forward-looking consciousness, with integrity, professionalism, pragmatic. David Architecture Design devoted to the construction of the art space of a poetic habitat for contemporary human, and insisted in integrating ecology and culture in the field of design in the long term, incorporating international management mode and advanced design concept, to constantly emerge out excellent design works. In the aftermath of the over 30 years of operation, David team has made remarkable achievements, as following:

—2012 Global Human Settlement Award on Planning and Design (United Nations Environment Programme)
—2013 The most potential enterprise with best investment opportunity in China's economic development (XINHUA NEWS)
—2014 Gold Awards on Planning and Architecture of National Habitat Classic Scheme Competition (Architectural Society of China)
—2015 Gold Arch Award - Annual Review of Chinese Architectural Design Works (China Architectural Culture Centre)
—2016 Gold Awards on Eco - habitat Architecture Planning & Design International Competition (Architectural Society of China)
—2018 AIA Shanghai Design Awards - Architecture (the American Institute of Architects)
—2019 Asian Townscape Awards (Asian Habitat Society)
—2020 Asian Landscape Planning and Design Example Prize (Asian Landscape Architecture Society)
—2020 19th Jingrui Habitat Award (Ministry of Science and Technology, PRC)
—2022 Asian Townscape Awards(UN-Habitat)

Sichuan David Architectural Design Co., Ltd. stands not only at localization, but also in globalization, advocating "Culture is the most competitive", to carry forward the national building culture, responding to the guidance of national strategy of the Belt and the Road. The company works with the international advanced technology team to set up overseas working organizations in the US and Israel; and the extensive cooperation with internationally renowned designers to expand business in places such as, Africa, Middle East and Southeast Asia; and given the honour of Pioneer Enterprise of the "one belt, one road" of Sichuan province's reconnaissance designing.

David Architecture Design keeps pace with the times and actively practices the promotion of green buildings; in 2014 became an established new branch in Chengdu, China of BRE Global, an international approvals body offering certification of sustainability assessment for green buildings. David Architecture Design also initiated to establish the Sichuan Provincial BIM Research Institute (SCBIM), and for the Research and development of BIM has gathered of highly talented professionals here and abroad; also under the support of the International Joint Research Centre for national low-carbon green building, created the Demonstration and Extension Base of Green building in Southwest China.

In recent years, more and more overseas returned design talents have come together attracted by david's cultural brand to promote David design to further commit to the advanced green ecology, incorporating Hi-tech like environmental protection & energy conservation and new concepts into the design, for the positive contribution of the Innovation and sustainable development of urban and architecture in today's China.

地址：四川成都天府大道天府二街138号蜀都中心3号楼6层	Add: 6 Floor, No.3 Building, Shudu Center, Tianfu 2nd Street No.138, Wuhou, Chengdu, Sichuan
邮编：610041	P.C: 610041
电话：028-65292530	Tel: 028-65292530
传真：028-65292530	Fax: 028-65292530
邮箱：275010918@qq.com	Email: 275010918@qq.com
网址：www.scdavid.cn	Web: www.scdavid.cn

红楼梦酒坊遗址馆
Hongloumeng Distillery Site Museum

项目地点：四川 宜宾
建筑面积：3 000 m²
用地面积：450 m²
容积率：6.66

Location: Yibin, Sichuan
Building Area: 3 000 m²
Site Area: 450 m²
Plot Ratio: 6.66

糟房头酿酒作坊遗址，位于四川省宜宾市叙州区喜捷镇红楼梦村，是一处明清时期白酒作坊遗址，是目前在四川地区发现的年代最早、保存最完整的一处白酒酿造作坊遗址。

设计遵循保护性修复原则，建筑特征以"居""坊""竹""景"元素为主。方案将考古成果转化为生产力，使传统焕发新生，打造一个集遗址保护、文物展览、酒文化参观、非物质文化教育、产品生产展示、品评洽谈、旅游观光功能为一体的华夏非物质文化遗产国际孵化基地、工业旅游景点。

The site of Zaofangtou brewing workshop located in Hongloumeng village, Xijie Town, Xuzhou District, Yibin City, Sichuan Province, is the site of baijiu (Chinese liquor) brewing workshop in the Ming and Qing Dynasties. It is the earliest and most complete site of Baijiu brewing workshop found in Sichuan Province.

The design follows the protective principle, and the architectural features combine the design elements of residence, workshop, bamboo and scenery. The Design scheme converts the archaeological achievements into productive forces, brings new vitality to the tradition, and creates an international incubation base of Chinese intangible cultural heritage and industrial tourist attraction, which integrates various of function including site protection, cultural relics exhibition, wine culture activity, intangible culture education, product production & exhibition, product evaluation & negotiation and tourism.

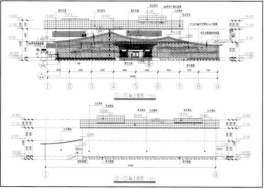

皓月·江屿城
Haoyue · Jiangyucheng

项目地点：四川 江油	Location: Jiangyou, Sichuan
建筑面积：413 222.71 m²	Building Area: 413 222.71 m²
用地面积：121 347.02 m²	Site Area: 121 347.02 m²
容积率：3.4	Plot Ratio: 3.4

皓月·江屿城临江夜景透视图

江屿城大门入口透视图

皓月·江屿城内庭半鸟瞰图

皓月·江屿城入户门厅透视图

项目位于四川江油古城腹地,与明月岛公园隔河相望,东临邀月路,整个广场南北向长300 m,东西向最宽处达90 m,整体呈带状空间。

设计在提高土地经济效益的同时,因势利导,力求提高环境质量品质,创造具有"皓月当空"意境的江屿环境,形成独具地域特色的诗意空间。本次设计将城市性格、人居状态、国际前沿理念连接起来,充分利用临涪江之美和简洁现代的建筑造型创造出李白故里——江油古城的当代人居新空间。

The project is located in the hinterland of Jiangyou Ancient City, Sichuan Province, with the Mingyuedao Park across the river, and yaoyue Road to the east. The whole public square is about 300 m long in the north-south direction and 90 m wide in the east-west direction, and is a belt like space.

While improving the economic benefits of land, we should make good use of the situation, strive to improve the quality of the environment, forming a poetic space with unique regional characteristics.

This design will be the city's character, human settlements connected, make full use of the beauty of Linfujiang, concise modern architectural modeling.

This design connects the urban character, living conditions of the residents and international frontier conception, expressing beauty of Fujiang River, with concise and contemporary architectural modeling to create a new contemporary residential space in Jiangyou ancient city, the hometown of Li Bai.

中国川西林盘非遗（马椅子）工坊更新
Update of the Workshop of Intangible Cultural Heritage (Ma's chair) in Western Sichuan, China

项目地点：四川 都江堰　　Location: Dujiangyan, Sichuan
建筑面积：426.6 m²　　　Building Area: 426.6 m²

项目位于青城山麓,中国长江上游文明最重要的起源地——芒城古城址的核心区。这个由工坊主人自己动手构建的全手工竹建筑聚落,凭借尊重当地自然人文、活力创新、助力乡村振兴和非遗保护等设计亮点,赢得了全球人居环境规划设计奖,让世界瞩目。

项目由3座不同造型的竹建筑(料坊、工坊和茶坊)构成,最大的亮点莫过于光影可以"自由飘拂"的工坊,设计灵感来自唐代李白《蜀道难》:"蚕丛及鱼凫,开国何茫然",仿佛在幽僻川西林盘中的蜀地蚕茧,暗喻古蜀国始祖蚕丛劝蚕所用的金茧之美。

The project lies in the foothills of the Mount Qingcheng, is the core area of the site of Mangcheng Ancient City, the most important origin of civilization in the Upper Reaches of the Yangtze River in China. This fully hand-made bamboo architectural settlement built by the workshop owner, by virtue of the design highlights such as respecting the local nature and culture, vitality and innovation, support of rural revitalization and intangible cultural heritage protection, won the Award of Global Human Settlements Award on Planning and Design and attracted the international attention.

The project consists of the three bamboo architectures with different styles (material warehouse, workshop and Tea house). The great highlight of this project is the effects of sunlight and shade of the workshop, this design inspiration comes from the poem "The Difficulty of the Shu Road," by Li Bai, a great poet in the Tang Dynasty of China, which writes thus: "Can Cong and Yu Fu, their founding of a state - how hazily distant !". It is like a silkworm cocoon in the peaceful and secluded Linpan, implying the beauty of the golden cocoon used by the ancestor.

中国两弹城
Liangdancheng of China

项目地点：四川 梓潼　　Location: Zitong, Sichuan
建筑面积：200 000 m²　Building Area: 200 000 m²
用地面积：2 000 000 m²　Site Area: 2 000 000 m²
容积率：0.1　　　　　　Plot Ratio: 0.1

中国两弹城位于四川梓潼，是原中国工程物理研究院旧址，建于20世纪60年代后期，是我国第二个核武器研制基地。在第三次全国文物普查中被列入"全国100大文物""全国红色旅游经典景区"。

两弹城规划范围为院部旧址、长卿山森林公园等区域。核心区为院部旧址，规划结构为"一心四区"。大卫设计以"修旧如旧、原貌恢复、原状展陈、弘扬精神"的规划原则，开展其修缮与展陈工作，让人们更深切地感受中国核武器探索的峥嵘岁月，追寻跨越60余年的精神传承。

Liangdancheng of China in Zitong County of Mianyang, southwest China's Sichuan Province, Built in the late 60's, it is the former site of China Academy of Engineering Physics (CAEP), is the headquarters of China's second nuclear weapon research and development base, in the Third National Relic Survey was listed "in the top 100 cultural relics in China", also, it has been listed as "National Red tourism classic scenic spot".

The planning range of Liangdancheng covers the former site of the CAEP, Changqingshan Forest Park and other areas. The core planning area is the previous site of the CAEP, with the planning structure of one center to four districts.

Based on the planning principle of "repairing the old as the old, restoration of original appearance, original state displaying and carrying forward the spirit", the design carries out its repair and exhibition work, so that people can feel the eventful years of Chinese exploration of nuclear weapons much more deeply and pursue the spiritual inheritance spanning over 60 years.

四川华西建筑设计院有限公司
Sichuan Huaxi Architectural Design Institute Co., Ltd

四川华西建筑设计院有限公司（简称华西设计）是一家提供综合性规划及建筑设计解决方案的企业，主营业务包括建筑设计、规划设计、装配式建筑设计、BIM设计、顾问咨询、设计研发、设计优化、精细化审图、绿建咨询等，涵盖居住、商业、办公、酒店、文化教育、体育、医疗、TOD、工业园区等不同业态及建筑类型，拥有建筑行业甲级设计资质和丰富的实践经验。

多年来，凭借雄厚的实力，华西设计与多地政府机构、国内一线开发企业及专家建立了长期战略合作关系，坚持以客户为导向，践行设计创造价值的理念，持续为客户提供优质的设计产品和良好的服务体验，先后获得多家知名开发企业授予的优秀设计单位、优秀战略合作单位、优秀设计质量奖等荣誉，多次获得行业权威机构授予的优秀设计等奖项。

Huaxi Design is a national enterprise providing comprehensive planning and architectural design solutions. Its main businesses include architectural design, planning and design, prefabricated architectural design, BIM, consulting, design research and development, design optimization, refined drawing review, green building consulting, etc., covering different business types and building types such as residential, commercial, office, hotel, culture and education, sports, medical, TOD, industrial parks, etc, It has Class A design qualification and rich practical experience in the construction industry.

Over the years, relying on its strong strength, Huaxi Design has established a long-term strategic cooperation relationship with many local government institutions, domestic front-line development enterprises and academic authorities, adhered to the customer orientation, practiced the concept of creating value through design, and continued to provide customers with high-quality design products and good service experience. It has been awarded excellent design units, excellent strategic cooperation units, excellent design quality awards and other honors by many well-known development enterprises, It has won many awards such as excellent design awarded by industry authorities.

地址：四川省成都市武侯区益州大道中段555号星宸国际B座21、27、28楼
电话：028-65419060
邮箱：contact@huaxidesign.com

Add: 21, 27, 28/F, Building B, Xingchen International, No.555, Middle Yizhou Avenue, Wuhou District, Chengdu, Sichuan
Tel: 028-65419060
Email: contact@huaxidesign.com

成都市蜀都小学
Chengdu Shudu Primary School

项目地点：四川 成都
建筑面积：30 000 m²
用地面积：26 000 m²

Location: Chengdu, Sichuan
Building Area: 30 000 m²
Site Area: 26 000 m²

成都市蜀都小学项目为48班规模的5层小学教学楼，位于成都市成华区八里庄路旁，总用地面积约26 000 m²，总建筑面积约30 000 m²。蜀都小学的建成，推动了成华区教育资源升级，其富有特色的建筑立面也成为城市名片。

建筑物整体呈组团状结构，功能划分明确，交通便捷流畅。包容性的设计，让不同领域的人互相交流，让不同人群的认知与需求更好地在教育空间中得到表达。

考虑到教育空间的生命周期长的特点，本项目设计简洁，可根据未来不同需求改变空间功能；建筑红白配色与成华小学本部遥相呼应，另有"C型玉龙""可园"等特色元素，展现了中华民族优秀文化的艺术韵味。

The project of Chengdu Shudu Primary School is a 5-floor teaching building for 48 classes, located on the roadside of Balizhuang, Chenghua District, Chengdu, with a total land area of about 26 000 m² and a total building area of about 30 000 m². The completion of Shudu Primary School promotes the upgrading of education resources in Chenghua District, and the distinctive building surface also becomes the symbol of the city.

The whole building adopts a group structure with clear functional division, simple and smooth traffic, and inclusive design, allowing people in different fields to communicate with each other, so that different people's cognition and needs can be better expressed in the education space.

Considering the long life cycle of the education space, the design of this project is simple and can change the function of the space according to different needs in the future; The red and white color scheme of the building echoes the main campus of Chenghua Primary School. There are other characteristic elements such as "C-shaped Jade Dragon" and "Ke Garden", conveying the outstanding cultural and artistic charm of the Chinese nation.

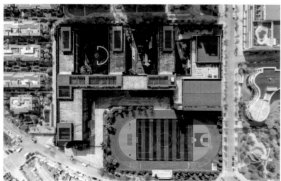

成都时代天境
Chengdu Times Paradise

项目地点：四川 成都
Location: Chengdu, Sichuan

时代天境项目是由3幢商业、办公复合业态的建筑构成的开放共享的都市街区。设计师从成都的特质切入，融合传统街巷空间与现代化城市脉络，打造未来城市文化区与社交生活的目的地。

设计师从不同尺度积极介入城市环境，赋予街区丰富空间体验，构建未来多样化生活场景，探索高密度环境下城市复合街区新范式。

项目从成都传统川西坡顶民居中汲取灵感，演化成形态各异的坡顶空间，再融合开放式街区布局，重现成都传统街巷肌理，高低区结合，营造一个层次丰富、充满艺术与活力的城市界面。

The Times Paradise project consists of three commercial and office buildings, which constitute an open and shared urban block. Starting from the characteristics of Chengdu, the designer integrates traditional street space and modern urban context to create a destination for future urban culture and social life.

Architects are actively involved in the urban environment at different scales, giving blocks rich spatial experience, constructing diversified living scenes in the future, and exploring a new paradigm of urban composite block in a high-density environment.

The project draws inspiration from the traditional western Sichuan hilltop dwellings in Chengdu and evolves into hilltop spaces of different forms. By integrating the open block layout, the project represents Chengdu's traditional street texture and combines high and low areas to create an urban interface with rich layers, art and vitality.

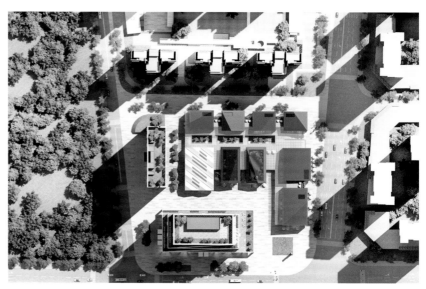

中铁二院工程集团有限公司
China Railway Eryuan Engineering Group Co., Ltd.

中铁二院工程集团有限公司(以下简称"中铁二院"),成立于1952年,隶属于世界双500强企业——中国中铁股份有限公司,是国内特大型工程综合勘察设计企业。

中铁二院业务范围涵盖规划、勘察设计、咨询、监理、产品产业化、工程总承包等基本建设全过程服务,横跨铁路、城市轨道交通、公路、市政、港口码头、民航机场、生态环境等多个领域,现有职员6 000多人,拥有全国工程勘察设计大师5人、省级工程勘察设计大师17人。

建院以来,中铁二院参与建成的铁路通车里程占全国四分之一,其中高速铁路通车里程占全国三分之一;建成的城市轨道交通通车里程占全国已运营里程的五分之一;先后设计了近5 000 km高速公路。

中铁二院以服务全球交通和市政建设为使命,以交通建设为核心,以技术创新为先导,努力成为跨行业、涵盖工程建设生命全周期、具有完整产业链的国际型工程公司。

China Railway Eryuan Engineering Group Co., Ltd. (Abbreviation:"CREEC"), founded in 1952, is a large-scale engineering comprehensive survey and design enterprise in China. CREEC is affiliated to China Railway Group Limited, one of the world's top 500 enterprises.

CREEC provides services for the whole procedure of capital construction such as planning, survey and design, consultation, supervision, product industrialization, and Engineering Procurement Construction (EPC) etc., covering fields such as railway, urban rail transit, highway, municipal administration, port and wharf, civil aviation airport, and ecological environment etc. CREEC now has more than 6 000 employees, including 5 national engineering survey and design masters and 17 provincial engineering survey and design masters.

Since the establishment of CREEC, CREEC has participated in the construction of one fourth of the railway mileage, including one third of the high-speed railway mileage; the mileage of urban rail transit completed accounts for one fifth of the total operating mileage and nearly 5 000 km of expressways have been designed in China.

CREEC, with the mission of serving global transportation and municipal construction, takes transportation construction as the core and technological innovation as the guide, strives to become an international engineering company with a complete industrial chain covering the whole life cycle of engineering construction.

地址:成都市通锦路3号 Add: No.3 Tongjin Road, Chengdu, P.R. China
邮编:610031 P.C: 610031
电话:028-87668866 Tel: 028-87668866

贵安站
Gui'an Railway Station

项目地点:贵州 贵阳 Location: Guiyang, Guizhou
设计时间:2013年—2015年 Design Time: 2013-2015
竣工时间:2021年12月 Completion Time: December 2021
建筑面积:60 580 m² Building Area: 60 580 m²
用地面积:19 840 m² Site Area: 19 840 m²

  贵安站位于贵州省贵安新区核心区中心,属于贵州省大型重点综合交通枢纽之一。交通综合体主要由3部分组成:站房工程、东广场及地下空间工程、市政接驳工程。站房工程设沪昆、西南环、贵南共7台15线,并预留远期车场扩建场地,站房为线侧下式加高架候车厅,采取上进下出客流组织方式。一期已完成站房建筑面积60 000余m²,并正在组织设计二期贵南场扩建工程12 000 m²,扩建后站房总规模达73 000 m²。站前广场及地下工程设置地下车场204 000 m²,含商业、社会停车、出租车及市政接驳工程82 800 m²,轨道交通S1和G1号线从站房中心处下穿并设置站点,车站与市政交通互联互通,可实现便捷换乘,满足乘客高效便捷的出行需求。

  贵安新区是国家级新区,如何"与城市共生、与历史共融",是贵安站设计的出发点和归宿,也是设计过程中的重点。贵安站既要有"黔"之韵的建筑传统,又要有苗族文化、侗族文化及古夜郎国文化特点。

  Gui'an railway station is located in the center of the Gui'an New district, Guizhou Province. It is one of the large-scale key comprehensive transportation hubs in Guizhou Province. The traffic complex is mainly composed of three parts; "Station building project", "East square and underground space project", "municipal connection project". The station building project is provided with 7 sets of 15 train lines in Shanghai to Kunming line, Southwest ring line and Guinan line, also a long-term parking lot expansion site is reserved. The station is a low-lying station building with an elevate waiting hall type station, and the passenger flow organization mode of upper entrance and lower exit is adopted. The construction area of the station building in phase I have been more than 60 000 m², and the second phase of gui nan chang expansion project is being organized and designed with 12 000 m². After the expansion, the total scale of the station building will reach to 73 000 m². The underground parking lot, the station ground and underground works is about 204 000 m², including 82 800 m² of commercial, public parking lot, taxi parking lot and municipal engineering. The rail transit lines S1 and G1 pass through the center of the station building and set up stations. The station and municipal transportation are interconnected and convenient to meet the needs of passengers for efficient and convenient transfer.

  Gui'an new area as a national new developed area, how Gui'an Station make to "coexists with the city and integrates with history"point is the starting point and destination of the design. Gui'an railway station should not only have the architectural tradition of "Guizhou"charm, but also include the Miao traditional culture, Dong traditional culture and the cultural characteristics of "ancient Yelang" state.

川藏铁路拉萨至林芝铁路站房设计
Design of railway station buildings of Lhasa to Nyingchiwithin the Sichuan-Tibet Railway

项目地点：西藏
设计时间：2018年—2019年
竣工时间：2021年
建筑面积：45 170 m²
用地面积：29 200 m²

Location: Tibet
Design Time: 2018-2019
Completion Time: 2021
Building Area: 45 170 m²
Site Area: 29 200 m²

川藏铁路拉萨至林芝段位于西藏自治区东南部，线路从协荣站引出，向东经贡嘎、扎囊、山南、桑日、加查、朗县、米林到达林芝。

林芝站建筑造型通过层层上升的建筑形态，体现巨大山水落差和壮丽景色。

The section of the Sichuan-Tibet Railway from Lhasa to Nyingchi is located in the southeast of the Tibet Autonomous Region. The railway starts from Xierong Station and runs through Gongga, Zhanang, Shannan, Sangri, Jiacha, Lang County and Milin till Nyingchi.

Nyingchi Station's architectural shape forms the magnificent landscape through the continuousrising architectural form, which reflects the huge height gap between the local mountain and river.

雍布拉康是西藏历史上第一座宫殿，也是西藏最早的建筑之一，依山势而建，雄伟挺拔。站房建筑造型提取雍布拉康层层叠叠上的特征，两侧实墙采用干挂预制彩色混凝土板，玻璃幕墙与铝板虚实互补，柱廊和屋面的挑檐构成入口的灰空间，突出整个建筑的雄伟气势。

室内设计提炼莲花为母题，候车厅吊顶用软膜印花灯片组合成吉祥节图案，莲花柱头、莲花图案磨砂玻璃栏板等在旅客视线重点部位点缀运用，沿用白、红、金黄的色彩，营造"雪域升红日、金莲迎盛世"的室内氛围。

Yongbrakang is the first palace in the history of Tibet as well as one of the earliest buildings in Tibet, which is built against the mountain, majestic and tall. The architectural shape of the Shannan Railway Station is extracted from the characteristics of Yongbrakangpalace. The solid walls on both sides are dry-hung precast colored concrete panels, the glass curtain wall and aluminum panels complement each other, as well as the colonnade and roof eaves form the foyer at the entrance, highlighting the grand momentum of the whole building.

The indoor space extracts lotus flower as the motif, the soft film printed lantern pieces on the ceiling of the waiting hall are combined into auspicious festival patterns, which are applied on frosted glass bars, headpiece of columns and other key parts in the sight of passengers. The white, red and golden colors are used to create a magnificent indoor atmosphere of "red sunrising in snow field and golden lotus in full bloom to welcome prosperous times".

玉磨线站房
Yumo Section

建设地点：云南玉溪至磨憨
设计时间：2018年—2020年
竣工时间：2021年底
建筑面积：40 000 m²
用地面积：50 500 m²

Location: Yuxi Mohan, Yunnan
Design Time: 2018-2020
Completion Time: The end of 2021
Building Area: 40 000 m²
Site Area: 50 500 m²

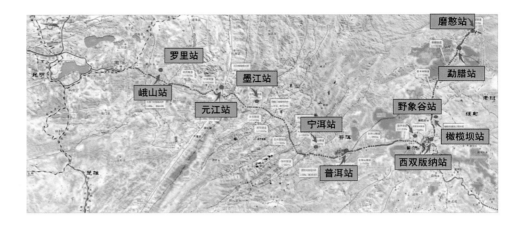

中老昆万铁路玉磨段全长508.533 km，北起玉溪，南至磨憨，与国外磨万段连通，是"一带一路"倡议的重要线路工程。

玉磨段共设11座铁路客运站房，分别为峨山站(2 500 m²)、化念站(1 500 m²)、元江站(3 000 m²)、墨江站(5 000 m²)、宁洱站(8 000 m²)、普洱站(12 000 m²)、野象谷站(2 500 m²)、西双版纳站(10 000 m²)、橄榄坝站(2 500 m²)、勐腊站(3 000 m²)、磨憨站(16 000 m²)，我院作为总体设计单位，完成了除西双版纳站、磨憨站以外的9座车站及站区设计。其创建的"精品站房、美丽站区"工程获得国铁集团好评，部分站点的站房与站区已被国铁集团指定为样板与观摩工程，在全路推广。

The Yumo section of China/Kunming-Laos/Vientiane Railway is 508.533 km long. It starts from Yuxi in the north and ends at Mohan in the south, it's connected with the foreign Mohan section. Is an important line project of the "The Belt and Road" strategy.

There are 11 railway station in Yumo section, including E'shan station(2 500 m²)，Mojiang station(5 000 m²), Ninger station (8 000 m²),Pu'er station(12 000 m²),Yexianggu station(2 500 m²), Xishuangbanna/Sipsongpanna station(10 000 m²), Ganlanba station(2 500 m²), Mengla station(3 000 m²) and Mohan station(16 000 m²). As the general design unit, our instituter has completed the nine construction of the railway station except Xishuangbanna station and Mohan station.The project of "excellent station buildings and station areas"designed by us has been praised by the State Railway Group, some of the station has been designated as model projects and promoted through the road.

重庆西站
Chongqing West Railway Station

项目地点：重庆	Location: Chongqing
设计时间：2010年－2014年	Design Time: 2010-2014
竣工时间：2018年	Completion Time: 2018
建筑面积：119 996 m²	Building Area: 119 996 m²
用地面积：388 695 m²	Site Area: 388 695 m²

重庆西站站房及铁路综合交通枢纽工程是重庆市铁路枢纽客运系统"四主"客运站之一，作为西南地区规模最大、建设最新的高铁客站之一，是我国铁路客站建设发展基础上的又一次重大实践，是铁总进一步加强科学化、系统化、精细化管理的成果，是中国铁路实现建设、设计、施工一体化管理新高度的再一次实践，"慎于思、践于行、共铸精品"的建设理念成为重庆西站实施项目建设管理的目标和宗旨。该工程是渝黔铁路重庆枢纽的新建车站，是集各种交通方式为一体的客运综合交通枢纽。整体结合"T"形站房两侧空间，通过层叠的形态构成两江汇聚的壮阔景象，体现"两江汇聚耀西部明珠"的建筑创意，充分表达出重庆这座西部中心城市的明珠形象。该工程包括站房工程、站房雨棚工程及站前平台高架桥工程等和铁路综合交通枢纽工程两部分内容。

West station of chongqing railway station building and integrated transportation hub project is the chongqing railway hub passenger transportation system is one of the "four master" bus, southwest as one of the largest scale and the construction of new high-speed rail station, railway station construction is our country development and on the basis of a major practice, is the total iron further to strengthen the management of scientific, systematic and refinement results, It is another practice for China Railway to realize the new height of integrated management of construction, design and construction. The construction concept of "town in thinking, practice in practice, and create high-quality products together" has become the goal and purpose of project construction management of Chongqing West Railway Station. The project is a new station of Chongqing junction of Chongqing-Guizhou Railway, which is a comprehensive passenger transport hub integrating various modes of transport. Combined with the space on both sides of the "T" type station, the magnificent scene of the convergence of the two rivers is formed through the overlapping form, reflecting the architectural creativity of "the convergence of the two rivers and the bright pearl of the west", and fully expressing the pearl image of Chongqing, the central city of the west. The project includes two parts: station house project, station house canopy project, station front platform viaduct project and railway integrated traffic hub project.

站房工程建筑面积119 944 m², 采取线上式布局, 为特大型站房。站房工程地上2层, 地下1层, 局部设置夹层, 总高度38.4 m。站房雨棚工程81 108 m²、站前平台高架桥工程长度为335.4 m。站场规模29站台面(15站台)31线;铁路综合交通枢纽工程地下一二期共计430 000 m², 地面集散广场74 000 m², 面宽660 m, 进深262 m。地下部分共计5层, 分别设计综合换乘大厅、公交、长途、出租车、社会车等交通方式, 满足零换乘的设计理念;地上开发共计280 000 m², 以餐饮、办公、酒店、商务功能为主。

The construction area of the station building is 119 944 m². It adopts the online layout and is an oversized station with two floors above ground and one underground floor, and mezzanine is set locally. The total height of the station building is 38.40 m. Station canopy workers 81 108 m², the length of the station platform viaduct project is 335.4 m. Station scale 29 platform surface (platform 15) 31 line; The first and second underground phase of the railway comprehensive transportation hub project is 430 000 m², the ground distribution square is 74 000 m², the surface width is 660 m, the depth is 262 m, the underground total 5 floors, the comprehensive transfer hall, bus, long-distance, taxi, social car and other transportation organization ways are designed respectively, to meet the design concept of zero transfer. A total of 280 000 m² of above-ground development, mainly catering, office, hotel and business.

天津

天津大学建筑设计规划研究总院有限公司
Tianjin University Research Institute of Architectural Design and Urban Planning Co., Ltd.

1952年天大北洋的常青藤上又发新枝——国家重点高校中最早成立的甲级设计研究院——天津大学建筑设计规划研究总院初立。经过半个多世纪的成长与发展，2020年5月天津大学建筑设计规划研究总院改制为天津大学建筑设计规划研究总院有限公司（以下简称"总院有限公司"），注册资本人民币5 000万元。

总院有限公司始终致力于全产业类型同步发展战略，业务领域涵盖各类公共与民用建筑工程设计、城市总体规划、国土空间规划、城市设计、详细规划编制与专项规划编制、居住区规划与住宅设计、古建筑保护修缮、保护规划、风景园林、旅游规划、室内设计、市政交通、加固改造、绿色生态、装配式设计、岩土设计、TOD设计、冷链物流、复杂结构优化、全过程工程咨询等。服务跨越咨询、策划、可研、设计、优化、施工把控、后期评估全过程，伴随整个项目生命周期。

总院有限公司现拥有国家相关部门授予的11项设计资质：工程设计建筑行业（建筑工程）甲级、环境工程专项（水污染防治工程）甲级、风景园林工程专项甲级（A112001898）、城乡规划编制甲级（自资规甲字21120028）、工程咨询建筑专业甲级资信（甲022021010202）、旅游规划设计甲级（旅规甲08-2011）、文物保护工程勘察设计甲级（文物设甲字0101SJ0008）、工程设计市政行业（给水工程、排水工程、道路工程、桥梁工程）专业乙级（A212001895）、工程勘察［岩土工程（设计）］专业乙级（B212001895）、电力行业（送电工程、变电工程）丙级（A212001895）、特种设备设计许可证（压力管道GB2、GC2）（TS1812033-2025）。

In 1952, a new branch on the ivy of Tianjin University (founded as Peiyang University in 1895) -- the first class-A design institute in the national key universities - Tianjin University Research Institute of Architectural Design and Planning was established. After more than half a century, it was restructured into Tianjin University Research Institute of Architectural Design and Planning Co., Ltd. (hereinafter referred to as "AATU") in May 2020, with a registered capital of RMB 50 million.

AATU has always been committed to the synchronous development of all types of industries, with its business fields covering various types of public and civil architectural engineering design, urban master planning, land and space planning, urban design, site planning and special planning, residential area planning and housing design, conservation and restoration of historic buildings, conservation planning, landscape architecture, tourism planning, interior design, municipal transportation, reinforcement and reconstruction, green ecology, prefabricated design, geotechnical design, TOD design, cold chain logistics, complex structure optimization, whole-process engineering consultation, etc. The service spans the whole process of consulting, planning, feasibility study, design, optimization, construction control, and post-assessment, embracing the whole process of a project.

AATU now has acquired eleven design qualifications from relevant state departments: Architectural Design Engineering (Architectural Engineering) Class A, Environmental Engineering Special (Water Pollution Prevention and Control Project) Class A, Landscape Architecture Engineering Special Class A (A112001898), Urban and Rural Planning Class A (Self-funded Regulation A 21120028), Architecture Engineering Consulting Class A (A 022021010202), Tourism Planning and Design Class A (Travel Regulation A 08-2011), Survey and Design of Cultural Relics Conservation Project Class A (Cultural Relics Design A 0101SJ0008), Municipal Civil Engineering (Water Supply Engineering, Drainage Engineering, Road Engineering, Bridge Engineering) Class B (A212001895), Engineering Survey (Geotechnical Engineering (Design)) Class B (B212001895)), Power Industry (Power Transmission Engineering, Substation Engineering) Class C (A212001895), Special Equipment Design License (Pressure Pipeline GB2, GC2) (TS1812033-2025),CMA Qualification of Inspection and Testing Institution(220201060079).

地址：天津市南开区鞍山西道192号1895天大建筑创意大厦
电话：022-27404753
邮箱：tdqhb@sina.com
网址：www.aatu.com.cn

Add: 1895 Tianda building, No.192, Anshan West Road, Nankai District, Tianjin
Tel: 022-27404753
Email: tdqhb@sina.com
Web: www.aatu.com.cn

武夷新区体育中心
Wuyi New District Sports Center

项目地点：福建 南平
项目规模：94 181 m²
竣工时间：2020年
所获奖项："海河杯"天津市优秀勘察设计一等奖

Location: Nanping, Fujian
Project Scale: 94 181 m²
Completion Time: 2020
Awards: "Haihe Cup" Tianjin Excellent Survey and Design First Prize

本项目为福建省2022年省运会比赛主场馆。武夷山有着独特的自然山水特色和历史人文底蕴，自然风光以武夷山为最，山脚下的武夷山庄极具闽北民居特色，曲径回廊、浅滩流水，乃"武夷风格"建筑的代表作，有着"中国历史文化名村"之称的下梅古镇，是最原汁原味的闽北民居遗存，代表了文化底蕴深厚的武夷建筑形制。

本方案从武夷特有的建筑形制出发，采取"粉墙黛瓦"的民族建筑风格，用民间折纸的设计手法构造两座武夷岩茶花造型的场馆建筑，与北部的山脉交相呼应，相得益彰。巨大的二层平台将体育场与体育馆有机地联系在一起，平台中部自由的庭院设置恰好为综合活动提供更好的室外空间。

剑指云天的火炬塔成为建筑群的点睛之笔，蓬勃向上的外观昭示着体育竞技永不服输、顽强进取的奋斗精神。建筑主立面韵律感十足，将武夷岩茶花的神韵展现得淋漓尽致。错落有致的花瓣、渐变孔洞的铝合金板、恢宏的尺度不掩细腻的情感表达。

This project is the home stadium of Fujian 2022 Provincial Games. Wuyi Mountain boasts unique natural landscape features as well as historical and cultural background, and its natural scenery is the most prominent. Wuyi Mountain Resort at the foot of the mountain is a representative work of "Wuyi style" architecture. Xiamei ancient Town, which is known as "Chinese historical and cultural relic", is the most authentic relic of folk houses in northern Fujian. It represents the architectural form of Wuyi with profound cultural background.

Starting from the unique architectural form of Wuyi, this project adopts the national architectural style of "white walls and black tiles" and uses folk origami design techniques to construct two venue buildings in the shape of Wuyiyan Camellia, which echoes and complements each other with the northern mountains. The stadium is organically linked to the gymnasium by a large second floor platform, and the free courtyard in the middle of the platform provides a better outdoor space for the integrated activity area. The torch tower like the sword pointing to the sky becomes the finishing touch of the building group. Its vigorous and upward appearance shows the indomitable and enterprising spirit of the sports competition that never gives in. The main facade of the building is full of rhythm, and the charm of Wuyiyan Camellia is vividly displayed. The well-placed petals, the aluminum alloy plates with gradual holes, and the magnificent scale do not mask the delicate expression of emotion.

长株潭绿心中央公园总体城市设计方案
Overall Urban Design of Changzhutan Green Heart Central Park

项目地点：湖南 长株潭都市圈
项目规模：604.82 m²
设计时间：2022年

Location: Changzhutan, Hunan
Project Scale: 604.82 m²
Design Time: 2022

  长株潭绿心中央公园位于国家级都市圈——长株潭都市圈的核心区位置，地跨长沙、株洲、湘潭3市，规划面积超过600 km²，其中绿心中央公园371 km²，临园片区231 km²。

  围绕"大国苑囿、无界绿湾"的规划愿景，创新演绎湖湘营城空间模式，构建山水交织、礼乐交融的空间格局。以器显礼，堪舆湖湘历史"山·水·城"格局，锻造南起金霞山-古桑洲、北至天心阁-橘子洲的虚实相间的湖湘礼轴；以境显乐，依托生长于湘江、浏阳河的藤蔓体系，打造自然灵动、美好生活的赏乐轴线。

Changzhutan Green Heart Central Park is located in the the core position of the national metropolitan region - Changsha - Zhuzhou - Xiangtan metropolitan region, with a planned area of more than 600 km², including 371 km² of Green Heart Central Park and 231 km² of the neighbouring area.

Centering on the planning vision of "garden of Great Country, boundless Green Bay", the design makes an innovative interpretation of the space pattern and constructs a space pattern combining mountains and rivers. With the instrument showing ceremony, the historical pattern of "mountain · water · city" forges an axis from Jinxia Mountain - Gusangzhou (known as Catfish Island) in the south, to Tianxin Pavilion - Orange Island in the north; With the scenery showing happiness and the design relies on the system growing in Xiangjiang River and Liuyang River to create a natural and smart axis of enjoying happy life.

河北工程大学新校区规划设计
Planning and Design for the New Campus of Hebei University of Engineering

项目地点：河北 邯郸
项目规模：规划用地约273 hm², 总建筑面积约973 100 m²
竣工时间：2021年

Location: Handan, Hebei
Project Scale: The planned land is about hm², The total construction area is about 973 100 m²
Completion Time: 2021

河北工程大学是河北省重点大学，本项目涵盖了新校区总体规划，道路、绿化景观、市政等各类专项规划及多个建筑单体设计。设计团队以邯郸的地域文化和邯郸东部新区城市设计为出发点，体现以人为本、先进、合理、实用的原则，打造具有"工科特色、现代气质"的高等学府。

教学区设计中将图书馆和基础教学楼作为校园中的核心建筑群，并呈放射状布局。学院组团围绕核心建筑群、生活区围绕整个学院组团布局，学生到学院和到核心建筑群都非常便捷。学生生活区分别与相应学院组团一一对应，最大限度地减少学生日常往返于宿舍和学院的步行距离。横向大气严谨的"学术长轴"，满足了学生老师学术讨论、休闲娱乐、文化展示等多种教学、娱乐、生活的需要，使轴线空间被充分利用。

Hebei University of Engineering is a key university in Hebei Province. This project covers the overall planning for the new campus, special planning for road, green landscape, municipal engineering as well as the design of several individual buildings. Starting from the regional culture of Handan and the urban design of the eastern New District of Handan, the design team aims to create an institution of higher education with "engineering characteristics, modern temperament" according to the people-oriented, advanced, reasonable, practical principles.

The teaching area has a radial layout, with the library and basic teaching buildings as the core buildings on the campus. The college group is around the core buildings, and the living area is around the whole college group, so it is very convenient for students to go to each college and the core buildings. The students' living areas correspond respectively to each college group to minimize the daily walking distance between the dormitory and the college. The horizontal and rigorous "academic long axis" satisfies the needs of students and teachers for various teaching and recreational life such as academic discussion, recreation, cultural display, etc., so that the axis space is fully utilized.

河北工程大学新校区图书馆
Library on the New Campus of Hebei University of Engineering

项目地点：河北 邯郸
项目规模：70 980 m²
竣工时间：2021年

Location: Handan, Hebei
Project Scale: 70 980 m²
Completion: 2021

河北工程大学新校区图书馆作为特色鲜明的大型工程类大学图书馆，具有简洁硬朗、科学严谨、富有力量感的建筑特征。通过叠加组合的建筑几何形体，展现工程类大学特有的刚强建筑"表情"与工程化空间趣味。庭院与灰空间作为建筑与环境的过渡衔接，既丰富了外立面，又使建筑室内空间与室外环境紧密相连，避免了大尺度建筑容易造成的呆板与压迫感，形成错落有致且完整大气的建筑形态。

As a large engineering university library with distinctive characteristics, the Library on the New Campus of Hebei University of Engineering has the architectural characteristics of simplicity, rigor and sense of strength. Through the overlay combination of architectural geometries, it shows the unique expression of strong architecture and the interest of engineered space of engineering universities. Courtyard and gray space, as a transitional connection between the building and the environment, not only enrich the facade, but also make the indoor space of the building closely connected with the outdoor environment, avoiding the rigidity and sense of pressure easily caused by large-scale buildings while forming a well-arranged and complete architectural form.

雄安新区启动区高教园区图书馆
Higher Education Park Library of the Start-up Area, Xiongan New Area

项目地点：河北 雄安新区
项目规模：70 000 m²
设计时间：2021年
Location: Xiong'an, Hebei
Project Scale: 70 000 m²
Design Time: 2021

　　本项目是雄安新区启动区城市级标志建筑，与南区城市级标志建筑共同组成总部商务片区的南北双核，形成片区的第一天际线。由3栋核心建筑塔楼围合形成开阔的城市广场，结合小街密路的城市肌理，以及地上地下一体化的空间设计理念，全力打造极具城市名片形象的门户建筑群。

　　图书馆作为启动区标志性建筑，设计目标强调图书馆在整个大学园片区的标志性和核心地位，大尺度的建筑体量主导广场空间，成为校园轴线重要节点。

This project is a city-level landmark building in the start-up area, which forms the north-south dual core of the headquarters business area together with the city-level landmark building in the southern area and forms the first skyline of the area. Surrounded by three core buildings and towers, it forms an open city square. Combined with the urban texture of small streets and dense roads, and the space design concept of integrating above ground and under ground, it strives to create a gateway architectural complex with the image of a city card.

　　As the landmark building of the start-up area, the library is designed to emphasize the landmark and core status of the library in the whole university area. The large-scale building volume dominates the square space and becomes an important node of the campus axis.

邢台新能源职业技术学院设计
Design of Xingtai New Energy College of Vocationand Technology

项目地点：河北 邢台
项目规模：368 100 m²
设计时间：2022年

Location: Xingtai, Hebei
Project Scale: 368 100 m²
Design Time: 2022

学院以本科职业技术教育为主，立足新能源行业，服务河北，面向全国，重点针对新能源、智能制造、环保新材料等战略性新兴行业开展学历教育与培训，助推区域产业转型和绿色发展。

项目位于河北省邢台市宁晋县宁北新区，占地面积约65.14 hm²，其中一期专科用地约39 hm²，二期本科用地约26.14 hm²。基地东侧紧邻凤鸣湖公园和宁晋县文化艺术中心，为学校提供了更好的基础条件。

规划以图书馆为核心，南侧康宁大道作为校园主入口，一侧是建筑，一侧是汪洋沟绿道景观。以图书馆为对景，沿着梧桐大道、凤仪广场形成校前文化轴；西侧形成衔接一二期的分期发展轴；北侧连接308国道形象入口和产业园区形成产学融合的产学互动轴；学习中心和教学实训楼共同形成产教环；宿舍构成犹如凤凰展翼的桃李芬芳环。"一心三轴两环"共同构成生态校园环境，美好的校园环境必将吸引更多的人才。

With focus on undergraduate vocational and technical education, the college serves Hebei and faces the whole country based on the new energy industry. It focuses on academic education and training for strategic emerging industries such as new energy, intelligent manufacturing and environmental protection new materials, so as to promote regional industrial transformation and green development. The project is located in Ningbei New District, Ningjin County, Xingtai City, Hebei Province, covering an area of 65.14 hm², including 39 hm² of the first-phase technical college land and 26.14 hm² of the second-phase undergraduate college land. The base is close to Fengming Lake Park and Ningjin County Culture and Art Center on the east side, which provides better basic conditions for the school. The plan takes the library as the core, Kangning Avenue in the south as the main entrance to the campus, with buildings on one side and Wangyanggou Greenway landscape on the other. With the library as the opposite view, the cultural axis in front of the school is formed along Wutong Avenue and Fengyi Square. On the west side, a phased development axis connecting Phase I and Phase II is formed. On the north side, an interactive axis integrating production and academy is formed connecting the image entrance of National Road 308 and the industrial park. The learning center and the teaching and training building jointly form a production-education ring. The dormitory forms a fragrant ring like the wings of a phoenix. "One core, three axes and two rings" together constitute an ecological campus, whose beautiful campus environment attracts more talents.

杨柳青大运河国家文化公园（元宝岛）
Yangliuqing Grand Canal National Cultural Park (Yuanbao Island)

项目地点：天津	Location: Tianjin
项目规模：653 500 m²	Project Scale: 653 500 m²
设计时间：2021年	Design Time: 2021

运河水畔的千年古镇杨柳青人文历史底蕴深厚，拥有以四大木版年画之首杨柳青年画为代表的年画文化及其他极具特色的民间文化艺术。在京津冀协同发展、打造国家大运河文化带的背景下，作为大运河中国北部区域的重要节点，她肩负着以文化为载体、传递文化精神的历史责任，以及作为城市中的公共空间，通过文化公园的形式与市民生活进行连接的时代使命。

元宝岛是杨柳青大运河国家文化公园的核心区，是集中展示明清天津运河文化盛景的全域、全时、全景、全要素的国际文旅目的地。

设计旨在打造一座以运河文化和杨柳青文化为主题的活态博物馆，方案从记录着元宝岛历史痕迹的老照片入手，以"再生"为核心概念，根据不同地块的特质分别通过再活化、再开发的方法，进行再组织；恢复过去这片土地上的湿地范围，恢复原景观，重现过去的院子或道路的记忆；置入与大运河和杨柳青文化相关的业态单体，恢复场地历史地文的记忆。

Yangliuqing, a millennium-old town by the canal, has a profound cultural and historical background. It boasts the New Year painting culture represented by Yangliuqing New Year wood-block paintings, the top of the four major New Year wood-block paintings, and other distinctive folk culture and art. Under the background of the coordinated development of the Beijing-Tianjin-Hebei region and the creation of the National Grand Canal Culture Belt, as an important node of the Grand Canal in northern China, it shoulders the historical responsibility of taking culture as the carrier to convey the cultural spirit, and the mission of the times to connect with citizens' life in the form of cultural park as a public space in the city.

Yuanbao Island is the core area of Yangliuqing Grand Canal National Cultural Park. It is an international cultural and tourism destination with full scope, full time, full panorama and full elements to display the spectacular cultural scenes of Tianjin Canal in Ming and Qing Dynasties. The design aims to create a living museum with the theme of canal culture and Yangliuqing culture. Starting from the old photos that record the historical traces of Yuanbao Island, the project takes "regeneration" as the core concept and reorganizes by means of reactivation and redevelopment according to the characteristics of different plots. It restores the wetland extent of the land in the past as well as the original landscape, and recreates the memory of the original yard or road. Commercial single buildings related to the Grand Canal and Yangliuqing culture are adopted to restore the historical and cultural memory of the site.

天津第二工人文化宫公园改造提升（二宫）
Renovation and Upgrading of Tianjin Second Workers' Culture Palace Park (Second Palace)

项目地点：天津	Location: Tianjin
项目规模：240 000 m²	Project Scale: 240 000 m²
竣工时间：2022年	Completion Time: 2022

第二工人文化宫是天津早期建立的公园式工人文化宫，其中剧场和图书馆为天津市文物保护单位，承载着几代人的历史记忆。设计延续场地风貌与文脉，最大限度尊重场地现状，植入全新功能，合理排布路网、广场、亭廊等景观元素，使场地焕然一新。

本次共对7个单体建筑进行改造，包括剧场、图书馆、艺体培训中心、综合球馆、篮球馆、劳模疗休养中心、足球场等，总建筑面积约38 000 m²。

改造之后的二宫公园绿化面积约95 000 m²，新增近百种植物，通过景观设计强化公园属性，结合低碳理念，最大限度拓展生态空间；通过促进碳吸收增加碳汇作用；通过城市绿地降低热岛效应。

本次改造还建设了包括橡胶健身步道、石材步道、木栈道等多级道路系统，总长约14.2 km。同时新增中式廊亭、广场、健身运动区、文化长廊等数十处景观节点。

The Second Workers' Culture Palace is a park-style workers' culture palace established in the early days of Tianjin. The theater and library are the cultural relic protection units of Tianjin, carrying the historical memory of several generations. The design keeps the style and culture of the site, respects the current situation of the site to the greatest extent, implants new functions, and reasonably arranges the landscape elements such as road network, square, and pavilion to make the site take on a new look.

This time, 7 individual buildings has been transformed, including theatre, library, art and sports training center, comprehensive arena, basketball arena, model workers' recuperation center and football field, covering a total construction area of about 38 000 m².

After the transformation, the green area of Second Palace is about 95 000 m², and nearly 100 new plants have been added. The landscape design strengthens the attributes of the park, combines the concept of low carbon, and maximizes the ecological space, so as to promote carbon absorption and reduce the heat island effect through urban green space. A multi-stage road system, including rubber fitness trails, stone trails and wooden trestles, has also been built in the renovation, with a total length of about 14.2 km. At the same time, there are dozens of new landscape nodes, such as Chinese style pavilion, square, fitness and sports area and cultural corridor.

河北工业大学北辰校区化工、海洋学院教学实验楼
Teaching Experiment Building, College of Chemical Engineering and Marine Science, Beichen Campus, Hebei University of Technology

项目地点：天津
项目规模：29 677 m²
竣工时间：2020年
所获奖项："海河杯"天津市优秀勘察设计一等奖

Location: Tianjin
Project Scale: 29 677 m²
Completion Time: 2020
Awards: "Haihe Cup" Tianjin Excellent Survey and Design First Prize

本项目为河北工业大学建设的由化工学院及海洋科学与工程学院组成的综合实验楼，建设规模为29 677 m²，主体5层，局部4层。建筑布局采用"回"字形平面，把房间分散布置，既满足了每层平面的规模要求，又避免了房间集中布置的拥挤感。根据任务书要求，把两个学院进行分区的同时，将各专业按层划分，空间使用方便，布局合理。内庭院布置景观及小品，供人员停留休息，创造出丰富的交流空间。

This project is a comprehensive experiment building consisting of the College of Chemical Engineering and the College of Marine Science and Engineering constructed by Hebei University of Technology. It covers a construction area of 29 677 m², with five-floor main building and four-floor partial buildings. The layout of the building adopts the shape plane of the Chinese character "回" and the rooms are scattered, which not only meets the scale requirements of each floor plane, but also avoids the sense of being crowded. According to the requirements of the assignment book, the design divides the two colleges while arranging each major acoding to floors, so that the space is convenient to use and the layout is reasonable. The interior courtyard is decorated with landscape sketches for people to stay and rest, creating a rich space for communication.

天津市梅江地块医疗和公共卫生服务组团新建工程
New Construction of Medical and Public Health Service Cluster in Meijiang Plot, Tianjin

项目地点：天津
项目规模：122 350 m²
设计时间：2020年
竣工时间：在建

Location: Tianjin
Project Scale: 122 350 m²
Design Time: 2020
Completion Time: Under Construction

按照天津市卫生健康委员会总体战略布局，于梅江地区建设由天津市妇女儿童保健中心、天津市急救中心、天津市口腔医院组成的医疗和公共卫生服务组团。天津市口腔医院工程总建筑面积约74 600 m²。项目门诊部配置牙椅300台，门急诊日流量约2 500人。住院部配置床位150张。天津市妇女儿童保健中心总建筑面积约34 100 m²，门诊日流量约1 600人，是一所集公共卫生管理、预防、保健、科研、教学为一体的现代化省级妇幼保健机构。天津市急救中心总建筑面积约13 650 m²，主要承担院前急救及天津市疾控指挥中心等功能，在编员工500人，救护车停车位40个，值班停车位2个，特种车辆停车位6个。

设计合理规划了三种不同性质的医疗公共卫生服务机构的功能布局及活动流线，巧妙解决三者之间复杂的内外部交通问题。同时，打造智慧医疗时代背景下，先进的医疗和公共卫生资源共享管理运行模式，实现医疗和公共卫生服务体系的新突破。

In accordance with the overall strategic layout of the Health Commission of Tianjin, a medical and public health service group consisting of Tianjin Women and Children Health Center, Tianjin Emergency Center and Tianjin Stomatological Hospital has been built in Meijiang area. The total construction area of Tianjin Stomatological Hospital project is about 74 600 m². The outpatient department of the project is equipped with 300 dental chairs, and the daily flow of outpatient and emergency care is about 2 500 people. The inpatient department is equipped with 150 beds; Tianjin Women and Children Health Center covers a total construction area of 34 100 m², with a daily flow of 1 600 patients. It is a modern provincial women and children care institution integrating public health management, prevention, health care, scientific research and teaching. Tianjin Emergency Center has a total construction area of 13 650 m², mainly responsible for emergency, Tianjin disease control command center and other functions. The project has 500 employees, 40 parking spots for ambulances, 2 parking spots for duty and 6 parking spots for special vehicles.

The design reasonably plans the functional layout and activity streamline of three kinds of medical and public health service institutions with different natures, and skillfully solve the internal and external complex traffic problems. At the same time, advanced medical and public health resource sharing management and operation mode will be created in the background of smart medicine to achieve new breakthroughs in the medical and public health service system.

黄骅市群众艺术公园和"四馆一中心"项目
Huanghua Mass Art Park and "Four Pavilions and One Center" Project

项目地点：河北 黄骅
项目规模：28 000 m²
竣工时间：2020年
所获奖项："海河杯"天津市优秀勘察设计一等奖

Location: Huanghua, Hebei
Project Scale: 28 000 m²
Completion Time: 2020
Awards: "Haihe Cup" Tianjin Excellent Survey and Design First Prize

  黄骅是一座以英雄名字命名的城市。它既有内陆城市的大气沉稳，又有滨海城市的浪漫时尚。针对黄骅这一特征，设计提出"刚柔并举·英雄港城"的整体设计理念，并将这一理念贯穿于整个规划与建筑设计之中。

  场地规划布局延续上位规划的对称格局，建筑以中轴线为核心呈扇形展开，东侧布置图书馆和科技馆，西侧布置文化艺术馆、档案馆和青少年活动中心。在东、西、南、北四个方向分别设置开放空间，呈现不同的主题：北面设置较为开阔的群众活动广场，便于容纳群众性集体活动；南面沿湖面设置形态自由、绿化丰富的生态广场，作为主要的市民休闲观景场所；场西侧设置以抽象艺术景观为主的艺术广场，与文化艺术馆和档案馆、青少年活动中心相对应；东侧设置以景观为主的科技广场，与科技馆和图书馆配套。整体规划曲中有直、静中有动，功能布置有分有合、各得其所，为城市提供了一处独具特色的公共活动空间。

  Huanghua is a city named after a hero. It has both the calm atmosphere of an inland city and the romantic fashion of a coastal city. Accordingly, the design puts forward the overall design concept of "Hardness with Softness · Hero Port City", and this concept runs through the whole planning and architectural design.

  The layout of the site keeps the symmetrical pattern of the upper plan. The building is laid out in the shape of a fan with the central axis as the core, with the library and science and technology museum on the east side, and the cultural art museum, archives and youth activity center on the west side. Open spaces are set up in four directions: east, west, south and north to present different themes. In the north, a relatively open mass activity square is set up to accommodate mass collective activities. In the south, there is an ecological square with free form and abundant green along the lake, which serves as the main recreational viewing place for citizens. On the west side of the site is an art square with abstract art landscape, which corresponds to the cultural art museum, archives and youth activity center. On the east side, a science and technology square with rational landscape is set up to match the science and technology museum and library. In the overall planning, there are curved, straight, static and dynamic elements, and the functional layout is divided and combined, and each is in its proper place, providing a unique public activity space for the city.

宝鸡大剧院
Baoji Grand Theater

项目地点：陕西 宝鸡
项目规模：41 000 m²
竣工时间：2021年
所获奖项：教育部优秀勘察设计评选一等奖、中国钢结构金奖

Location: Baoji, Shaanxi
Project Scale: 41 000 m²
Completion Time: 2021
Awards: The first prize of the Ministry of Education's excellent survey and design selection, China Steel Structure Gold Award

宝鸡大剧院位于宝鸡市金台区，占地约30 000 m²，总建筑面积约41 000 m²，地下1层，地上4层。剧院功能以演出大型歌剧、舞剧为主，同时满足国内外各类音乐剧、交响乐、地方戏曲等大型舞台类演出及大型会议的使用要求。

剧院设计按国内领先、西北一流的设计定位，完全按照国际建筑声学标准进行剧场工艺设计。大剧院由两个剧场组成，主剧场可容纳1 203人，小剧场可容纳500人。其中，主剧场舞台面积达1 600 m²，并采用国际先进的机械设备和灯光设备、阶梯式自动升降台和旋转舞台，可满足各种会议的召开和国内外不同类型的文艺演出需求。小剧场为辅助演出空间，采用机械式多功能舞台，在同一空间内满足不同的演出要求。

作为2020年陕西省"第九届艺术节"开幕式主会场，宝鸡大剧院是宝鸡市重点工程，今后将作为彰显宝鸡文化发展的新载体、新地标，也将成为宝鸡"看"世界的重要窗口。

Located in Jintai District, Baoji City, Baoji Grand Theater covers an area of about 3 000 m², with a total construction area of about 41 000 m², with one floor underground and four floors above ground. The main function of the theater is to perform large-scale operas and dance dramas. Meanwhile, it meets the requirements of large-scale stage performances and large-scale conferences, such as musicals, symphonies and local operas at home and abroad.

The design of the theater is based on the leading design orientation in China and the first-class level in Northwest China, and the process design is carried out in full accordance with the international architectural acoustic standards. The Grand Theater consists of two theaters - the main theater can hold 1 203 people, and the small theater can hold 500 people. The stage area of the main theater is 1 600 m², and the use of international advanced mechanical equipment and lighting equipment, step automatic lifting platform and rotating stage can meet the convening of various conferences and different types of domestic and foreign theatrical performances. As an auxiliary performance space, the small theater uses a mechanical multifunctional stage to meet different performance requirements in the same space.

As the main venue of the opening ceremony of the 9th Art Festival of Shaanxi Province in 2020, Baoji Grand Theater is a key project of Baoji City. In the future, it will be a new carrier and landmark highlighting the cultural development of Baoji, and also an important window for Baoji to "see" the world.

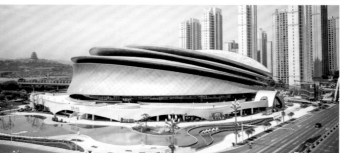

"兴文石海"世界级旅游目的地游客中心
Visitor Center of "Xingwen Shihai" World-class Tourist Destination

项目地点：四川 宜宾	Location: Yibin, Sichuan
项目规模：58 000 m²	Project Scale: 58 000 m²
竣工时间：2022年	Completion Time: 2022

世界地质公园"兴文石海"坐落于四川盆地南缘山区、山清水秀的宜宾市兴文县境内，地处川、滇、黔交界地，奇峰罗列，溶洞纵横。同时，兴文是古代僰人繁衍生息和最终消亡之地，也是四川省最大的苗族聚居地。石海自然景观文化与民族文化交相辉映，形成了独特的世界级旅游目的地。

"兴文石海"世界级旅游目的地游客中心位于兴文县中心区域，项目地块紧邻兴文高铁客运站和兴文汽车客运中心站。作为城市标志性文旅建筑和兴文城市门户形象，该项目将成为连接城市与自然、文化与环境的纽带，从而进一步推进兴文石海旅游服务、文化展示等相关产业的发展。设计希望从城市空间、地域特征、民族情感色彩三个维度进行思考，构建一个独特的、富有魅力的城市型游客中心建筑空间。

The World Geopark "Xingwen Shihai" is located in Xingwen County, Yibin City, in the mountainous area of the southern edge of Sichuan Basin with picturesque scenery. It is located at the junction of Sichuan, Yunnan and Guizhou, with a series of strange peaks and caves. Xingwen is a place where ancient Bo people live and die out, and is also the largest settlement of the Miao ethnic group in Sichuan Province. Shihai natural landscape culture and the culture of the Bo and Miao ethnic groups coexist to form a unique world-class tourist destination.

The Visitor Center of "Xingwen Shihai" World-class Tourist Destination is located in the central area of Xingwen County. The project plot is close to Xingwen high-speed railway station and Xingwen bus passenger terminal. As a landmark of cultural tourism building and the gateway image of Xingwen city, it will become a link between the city and nature, culture and environment, so as to further promote the development of Xingwen Shihai tourism services, cultural display and other related industries. The design hopes to think from three dimensions of urban space, regional characteristics and national emotion color, building a unique and charming urban tourist center architectural space.

南宫文化中心·尚小云大剧院
Nangong Culture Center - Shang Xiaoyun Grand Theater

项目地点：河北 邢台	Location: Xingtai, Hebei
项目规模：20 000 m²	Project Scale: 20 000 m²
竣工时间：2022年	Completion Time: 2022

项目坐落于河北省邢台市南宫湖畔，视野开阔、景色优美。

项目由尚小云大剧院、南宫市博物馆和南宫市艺术馆等主要建筑组成。在临湖处设置市民活动文化广场，为南宫市民提供文化活动场地，为游客游览南宫湖提供休憩场所。

体现京剧文化：以尚小云人物为主题，通过传统戏园布局体现传统京剧元素，剧院与古建式建筑外部形象相互配合，为京剧演出提供支持。

呼应湖光水色：中心广场向水面打开，临水设置开敞、灵活的亭榭廊道以及园林景观带，将远处的湖光山色引入场地内。

传统与创新结合：新中式建筑形象在提取传统建筑元素的同时满足了现代功能需求，体现南宫开拓创新的精神。

打造城市客厅：展示南宫文化，弘扬京剧国粹，打造文化高地，吸引来自全省乃至全国的客流，带动周边经济发展。

The project is located near Nangong Lake, Xingtai City, Hebei Province, with a wide vision and beautiful scenery.

The project consists of Shang Xiaoyun Grand Theater, Nangong City Museum and Nangong City Art Museum. Public Activity Culture Center is set up near the lake to provide a venue for cultural activities and a rest place for visitors to visit the lake.

Reflecting the culture of Peking Opera: Taking Shang Xiaoyun as a development opportunity, the layout of traditional opera garden reflects the elements of traditional Peking Opera. The external images of the theater and ancient buildings cooperate with each other to provide material support for the performance of Peking Opera.

Echoing the color of the lake: the central square opens to the water, and open, flexible pavilions and garden landscape belts are set near the water, bringing the distant scenery of the lake and mountain into the site.

Combination of tradition and innovation: The image of new Chinese style architecture extracts traditional architectural elements while satisfying modern functions, reflecting the pioneering and innovative spirit of Nanggong.

Creating city living room: The project displays Nangong culture, carrys forward the quintessence of Peking Opera, creates a cultural highland, attracts tourists from the whole province and even the whole country, and drives the economic development of the surrounding area.

青岛创智产业园一期工程
Qingdao Chuangzhi Industrial Park (Phase I)

项目地点: 山东 青岛
竣工时间: 2020年
建筑面积: 235 000 m²

Location: Qingdao, Shandong
Completion Time: 2020
Building Area: 235 000 m²

青岛创智产业园位于山东省青岛市西海岸新区,是一个占地面积约180 000 m²、总建筑面积304 000 m²的大型科技产业园区。园区三面环山,一面向海,内有柏果树河穿流而过,近山临海,自然环境优越。项目以西海岸及周边地区快速聚集发展的高新技术产业为支撑,吸引国际、国内中小企业设立经济行政总部,具有一定的时代性和先进性。本次设计在突出时代性的同时,旨在彰显青岛西海岸独特的地域文化,塑造令人印象深刻的独特空间魅力和建筑品质,为使用者打造充满创新创业力量的科技园区空间。

创智产业园三面环山,东南面向黄海,具有独特的地理环境特征和地域特色。本次设计从宏大的自然背景入手,在与周边环境山体及城市相协调的前提下,将城市与山水意境结合,为建筑群建造一个小环境,并将自然气息带入其中。建筑群不会破坏城市连绵起伏的天际线特色,也不会突兀地置身于周边的环境之中,而是与周边环境相协调,在空间气韵上与环绕的群山融为一体。

Qingdao Chuangzhi Industrial Park is located in the West Coast New District of Qingdao, Shandong Province. It is a large-scale science and technology industrial park covering an area of about 180 000 m² and a total construction area of 304 000 m². Surrounded by mountains on three sides and the sea on the other, with Baiguoshu River running through it, it enjoys superior natural environment. Supported by the rapid development of high-tech industries in the west coast and surrounding areas, the project has attracted international and domestic small and medium-sized enterprises to set up the economic and administrative headquarters, which has certain epochal character and advanced nature. Meanwhile, it aims to highlight the unique regional culture of the west coast of Qingdao, create impressive unique space charm and architectural quality, and create a science and technology park space full of innovation and entrepreneurship for users.

Surrounded by mountains on three sides and facing the Yellow Sea in the southeast, Chuangzhi Industrial Park has unique geographical and regional characteristics. Starting from the grand natural background, the design combines city with the artistic conception of landscape under the premise of coordinating the surrounding mountains with the city, creating a small environment for the building groups and bringing nature into it. The building groups will not destroy the characteristics of the rolling skyline of the city, and will not be abruptly placed in the surrounding environment, either. Instead, it is in harmony with the surrounding environment, and the charm of the space is integrated with the surrounding mountains.

烟台市所城里、朝阳街(一期)历史文化街区保护性改造
Yantai Suochengli, Chaoyang Street (Phase I) Historical and Cultural Block Protection Reconstruction

项目地点: 山东 烟台
竣工时间: 2021年
用地面积: 所城里历史文化街区总占地面积92 700 m², 景观62 000 m², 建筑7 855 m², 朝阳街历史文化街区总占地面积128 600 m², 景观52 000 m², 建筑27 909 m²

Location: Yantai, Shandong
Completion Time: 2021
Site Area: The total area of Suochengli Historical and Cultural District is 92 700 m², the landscape is 62 000 m², and the building is 7 855 m².
The total area of Chaoyang Street Historic and Cultural District is 128 600 m², the landscape is 52 000 m², and the building is 27 909 m²

所城里与朝阳街均位于烟台城市中心,景观优良,可达性良好。两街区相隔数百米,分别形成于明代及近代开埠时期,见证了烟台市自明代至今的历史发展,在烟台人心目中具有特殊的地位与价值。

本项目包含建筑修缮及改造、景观、照明、标识、氛围等专项设计,也包含整体街区的设计咨询工作。所城里西城墙景观带是本项目的重要节点。设计团队深入挖掘所城历史文化资源,将西城墙作为所城新篇章的"文化长卷"与"艺术长廊",移步异景的文化图卷向世人展示所城里的发展脉络、市民生活、文化教育、商业发展等故事,游走其间构成跨越时空对话的文化盛景。

Suochengli and Chaoyang Street are located in the center of Yantai city, enjoying excellent landscape and good accessibility. The two blocks, being several hundred meters apart, were formed in the Ming Dynasty and the modern port opening period respectively. They have witnessed the historical development of Yantai since the Ming Dynasty and possess special status and value in the eyes of Yantai people.

The project includes building repair and renovation and special design such as landscape, lighting, signage and atmosphere, as well as design consultation for the whole block. The western city wall landscape belt of Suocheng is an important node of the project. The design team dig deep into the city's historical and cultural resources, regarding the western city wall as the "long cultural scroll" and "art corridor" in the new chapter of the city. The cultural scroll with different scenery shows the development process of the city, life of citizens, cultural education, commercial development and so on to the world, and forms a cultural scene of dialogue with citizens across time and space.

天津市安里甘艺术中心
Tianjin Anglican Art Center

项目地点：天津

设计时间：2009年，原安里甘教堂外檐修缮工程及环境整治；
2017年，原安里甘教堂围墙复原工程；
2018年7月，原安里甘教堂副堂内檐修缮、加固及设备改造工程；
2020年11月，原安里甘教堂主堂内檐修缮工程。

竣工时间：2021年

Location: Tianjin

Design Time: In 2009, the renovation of the outer eaves of the original Anglican and the environmental improvement;
In 2017, the restoration project of the original Anglican wall;
In July 2018, the interior eaves repair, reinforcement and equipment renovation project of the former Anglican attached church;
In November 2020, the renovation project of the inner eaves of the main church of the former Anglican.

Completion Time: 2021

在天津的百年万国建筑群中，面积近1 000 m²的安里甘教堂实在神秘得不可思议。由于新中国成立后不久教堂就停止对外开放，其内部恢宏壮阔的建筑空间一直不为市民所知。

原安里甘教堂位于天津市和平区浙江路2号，在泰安道历史文化街区内。现存教堂重建于1936年6月，为天津市文物保护单位。建筑因长期缺乏管理，且受到自然、人为等综合因素的影响，导致内外檐皆受到一定程度的损害。

修缮改造后的安里甘艺术中心中，原教堂"主堂"将作为小型会演、艺术展览、展示等场地，原教堂"副堂"作为咖啡吧对外开放。总院文化遗产团队用打造样板、品牌的方式去推广文物建筑的保护及活化利用，让国人更多地了解有韵味的老建筑，在内部观演展示、办公会谈、庆典餐饮等活动中，将文物建筑融入市民的日常生活中，使其在烟火气息中升华，也让人们有条件沉浸在文物建筑的历史文化氛围中，逐渐提升人文素养及审美意识。

Among Tianjin's century-old building groups, the nearly 1 000 m² Anglican Church is incredibly mysterious. Since it was closed to the outside world shortly after the founding of the People's Republic of China, its magnificent interior architectural space has been hidden from the public.

The former Anglican Church is located at No.2, Zhejiang Road, Heping District, Tianjin, in the Tai'an Road historical and cultural block. The existing church was rebuilt in June 1936, which is a cultural relics protection unit in Tianjin. Due to the long-term lack of management and the influence of natural, man-made and other comprehensive factors, the interior and exterior eaves have been damaged to a certain extent.

In the renovated Anglican Art Center, the "main hall" of the original church will be used as a venue for mini-shows, art exhibitions and displays, while the "auxiliary hall" of the original church will be opened as a coffee bar. The cultural heritage team of the General Institute promotes the protection and activated utilization of cultural heritage buildings by creating models and brands, so that the Chinese people can know more about the old buildings full of charm. It integrates the cultural heritage buildings into the daily life of the citizens and sublimates them in exhibition, conference and celebratory catering. It also allows people to be immersed in the historical and cultural atmosphere of cultural relics buildings and gradually improve their humanistic quality and aesthetic consciousness.

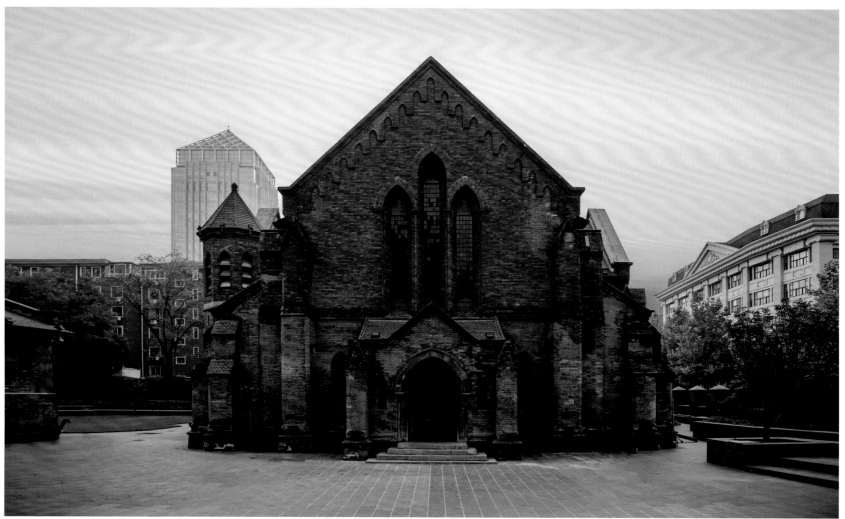

天津拓城建筑设计有限公司
Tianjin DC Design Co., Ltd.

拓城[DC DESIGN],意为"写意城市"。天津拓城建筑设计有限公司是在中国城市建设和开发领域从事综合性专业服务的工程实践咨询设计机构。业务涵盖广泛,涉及不同专业领域。承"艺术化解决问题"的设计法则,致力于提供创造性的专业设计方案。提供包括建筑设计、城市设计和城市更新等多领域的专业服务,涉及项目类型包括城市规划、居住类建筑、文旅类建筑、商业类建筑、办公类建筑、教育类建筑、文化类建筑、产业地产研发、城市更新及改造等,设计团队在中国市场前沿具有数十年的实践设计经验。

[DC DESIGN], meaning "freehand city". Adhering to the design principle of "artistic problem solving", we are committed to providing creative professional design solutions.Tianjin DC Design Co.,Ltd. is an engineering practice consulting and design institution engaged in comprehensive professional services in the field of urban construction and development in China. The business covers a wide range of industries and different professional fields. Providing professional services in many fields such as architectural design, urban design and urban renewal, involving project types including: urban planning, residential buildings, cultural tourism buildings, commercial buildings, office buildings, educational buildings, cultural buildings, industrial real estate research and development, urban renewal and transformation, etc., the design team has decades of practical design experience in the market frontier in China.

地址:天津市西青区海泰南道16号	Add: No.16, Haitai South Road, Xiqing District, Tianjin
邮编:300392	P.C: 300392
电话:15822556684	Tel: 15822556684
邮箱:dachingdesign@sina.com	Email: dachingdesign@sina.com

首创城商业综合体
Shouchuang City Commercial Complex

项目地点:天津　　Location: Tianjin
建筑面积:36 000 m²　　Building Area: 36 000 m²
用地面积:11 555 m²　　Site Area: 11 555 m²
设计时间:2020年　　Design Time: 2020

项目引入街区理念,在底层形成多变的商业街区空间。为了突出商业氛围,建筑体量采用"方盒子"作为设计语汇,整个建筑由大小不一的多个方盒子拼叠而成,形成戏剧性的商业建筑形象。

The project introduces the block concept, forming a changeable commercial block space on the ground floor. In order to highlight the commercial atmosphere, the building volume adopts "square box" as the design vocabulary, and the entire building is composed of multiple square boxes of different sizes, forming a dramatic commercial architectural image.

香江电商科创中心
E-commerce Technology Center

项目地点：山东 Location: Shandong
建筑面积：79 500 m² Building Area: 79 500 m²
用地面积：46 800 m² Site Area: 46 800 m²
设计时间：2017年 Design Time: 2017

香江电商科创中心集商业、办公、酒店等多重功能于一体。建筑设计突出东方文化，从传统建筑中汲取东方色彩，运用简洁大气的现代设计手法，展现东方语境下的商业氛围。

E-commerce Technology Center integrates multiple functions such as business, office and hotel; The architectural design highlights the oriental culture, draws oriental colors from traditional architecture, and uses simple and atmospheric modern design techniques to show the commercial atmosphere in the oriental context.

北师大附属学校
BNU Affiliated School

项目地点：天津	Location: Tianjin
建筑面积：118 200 m²	Building Area: 118 200 m²
用地面积：165 600 m²	Site Area: 165 600 m²
设计时间：2016年	Design Time: 2016

本项目以学生在校园内的活动为出发点，建筑设计中融入教学主街的现代教学理念，适当加入泛教育功能，营造灵活的素质教育公共空间。校园建筑尊重并延续传统校园的立面语言，同时适当进行创新，形成兼具传统与现代的校园建筑影像。

This project takes the students' activities on campus as the starting point, integrates the modern teaching concept of the main street of teaching into the architectural design, appropriately adds pan-educational functions, and creates a flexible public space for quality education. The campus building respects and continues the facade language of the traditional campus, while appropriately innovating to form a campus architectural image that combines tradition and modernity.

天津华厦建筑设计有限公司
Tianjin Huaxia Architecture Design Co., Ltd.

天津华厦建筑设计有限公司始创于1992年，总资产近3亿元，员工近500人，业务范围涉及建筑与规划、市政设计、施工图审查、工程监理、新型建材、房地产开发等。

拥有建筑行业建筑工程设计甲级、城乡规划编制乙级、市政行业乙级、房屋建筑工程监理甲级以及房屋建筑施工图审查(一类)等多项资质，通过GB/T19001—2008质量体系认证。

建筑与规划、市政设计业务拥有中高级专业技术职称的员工61人。其中一级注册建筑师8人、一级注册结构师4人、注册造价师3人、注册规划师5人、注册设备师4人。设计业务在写字楼、酒店、商厦、学校、医院、住宅等民用建筑领域和各类工业园区、大跨度钢结构等工业建筑领域以及规划设计领域业绩卓著，奖项云集，作品遍布华夏大地。

Tianjin Huaxia Architecture Design Co., Ltd. established in 1992, has total assets of nearly RMB 300 million and about 500 staff. Its business includes architectural design, city planning infrastructure design, construction drawing review, project supervision, new building materials and real estate development, etc.

We have qualifications of architectural design- grade A, urban and rural planning-grade B, infrastructure planning-grade B, construction project supervision-grade A and construction drawing review (type 1). We are certified by national qualification system GB/T19001-2008.

We have 61 staff with professional and technical titles at middle and senior levels in architecture, planning, and infrastructure design, including 8 certified class 1 architects, 4 certified class 1 structural engineers, 3 certified costevaluation engineers, 5 certified planners and 4 certified equipment engineers. We have made remarkable achievements in design and received numerous awards in the fields of civil architecture, including offices, hotels, commercial buildings, schools, hospitals and housing, as well as industrial parks, industrial buildings with large-span steel structures and city planning. Our projects are wide spread across China.

地址：天津市新技术产业园区(南开华苑)榕苑路16号鑫茂科技园中心楼三层/四层
电话：022-58693336-58598916
传真：022-58598916
邮箱：tjhxjzsjgs@126.com
网址：www.tj-huaxia.net

Add: 3rd / 4th floor, Central Building, Xinmao Science Park, 16 Rongyuan Road, Huayuan Industrial Park, Tianjin, China
Tel: 022-58693336　　58598916
Fax: 022-58598916
Email: tjhxjzsjgs@126.com
Web: www.tj-huaxia.net

东方市文化广场
Oriental Culture Square

项目地点：海南 东方
建筑面积：83 145 m²
用地面积：113 756.61 m²

Location: Dongfang, Hainan
Building Area: 83 145 m²
Site Area: 113 756.61 m²

本项目通过建筑整合的设计理念，为东方市营建出一个环境优美的室外文化广场，体现东方市的精神文明与物质文明风貌，体现文化与自然的融合性、建筑的时代性和前瞻性；通过统一的建筑基底，整合的建筑空间，形成一个建筑综合体。建筑形象大气磅礴、优雅庄重，具有强烈的视觉冲击力，为东方市的城市面貌增添色彩，成为东方市的新地标；通过空间梳理，一改传统博物馆等文化建筑的单一、单层、单调等印象，以综合多元化的设计理念，将各个文化场馆相互串联起来，保证文化场馆的长期运营及持久活力。

Through the design concept of architectural integration, this project builds an outdoor cultural square with beautiful environment for Dongfang City, which reflects the spiritual civilization and material civilization style of Dongfang City, reflects the integration of culture and nature, and reflects the contemporary and forward-looking architecture. Through the unified building base and integrated architectural space, an architectural complex is formed: the architectural image is magnificent, elegant and solemn, with strong visual impact, adding color to the urban appearance of Dongfang City, and becoming a new landmark of Dongfang City. Through spatial arrangement, the impression of traditional museums and other cultural buildings such as single, single-storey and monotonous is changed, and various cultural venues are connected with each other by comprehensive and diversified design concepts, so as to ensure the long-term operation and lasting vitality of cultural venues.

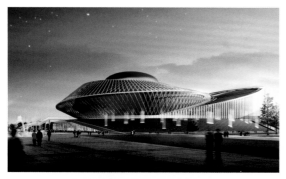

中新天津生态城中部片区32号地块小学、幼儿园项目
Primary School, Kindergarten Project at Plot No.32, Middle Area, Sino-Singapore Tianjin Eco-city

项目地点：天津
建筑面积：31 531.72 m²
用地面积：25 811.10 m²

Location: Tianjin
Building Area: 31 531.72 m²
Site Area: 25 811.10 m²

本项目位于天津市中新天津生态城中部片区。用地西至中天大道，南至华一路，东至规划用地边界，北至规划用地边界。规划用地总面积25 811.10 m²，总建筑面积31 531.72 m²，其中地上建筑面积18 060.00 m²，地下建筑面积13 471.72 m²，全部为中小学、幼儿园。用地现为荒地，地势平坦。周边规划道路交通便捷。用地西北部毗邻南开中学、东南侧为佳宅小区。总体规划充分体现生态环境的优美，与生态技术相结合，达到人、自然、生态的完美结合；建筑立面要求色彩质朴温暖，线条简洁明快又不失优美，取材朴实，用料简单；通过规划创新，解决接送孩子家长机动车停车造成道路拥堵的问题。参考生态城以往项目经验，在操场地下设置停车库，用于家长接送孩子临时停车。

This project is located in the central area of Sino-Singapore Tianjin Eco-City in Tianjin. Land to the west of Zhongtian Avenue, south to Hua Yilu, east to the boundary of the planned land, north to the boundary of the planned land. The total planning area is about 25 811.10 m², including a total construction area of 31 531.72 m², including an above-ground construction area of 18 060.00 m² and an underground construction area of 13 471.72 m². All for primary and secondary schools, kindergartens, land is now barren land, flat terrain. The surrounding planning road traffic is convenient; Adjacent to Nankai Middle School in the northwest of the site, the southeast side for the Jia Zhai community. The overall plan fully embodies the beauty of the ecological environment, and combines with ecological technology to achieve the perfect combination of human, nature and ecology; The building facade requires simple and warm colors, simple lines and elegant, simple materials, simple materials; Through planning innovation, to solve the problem of road wall caused by the parking of parents'vehicles to pick up their children. According to the previous project experience of Eco-City, a parking garage is set up under the playground for parents to pick up their children and park them temporarily.

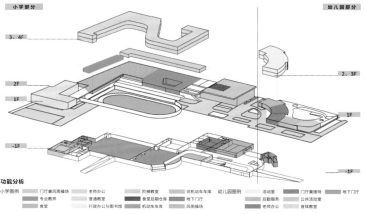

天津津兰国际黄金珠宝城
Tianjin Jinlan International Gold Jewelry City

项目地点：天津　　　　　　Location: Tianjin
建筑面积：353 861 m²　　　Building Area: 353 861 m²

本项目位于天津市西青区王兰庄，处于天津市南部，距离市中心约8 km。项目旨在打造一个金融—文化型街区。该项目为一个独立的商业区。其中包含中高层公寓。商业区是该项目中最具独特气息也最具吸引力的区域，1~4层的对外开放沿街店铺是当地人经营咖啡馆、酒吧、餐馆及店铺的绝佳选择。小商业区的外部是大型的商业区，在该区域内，店主和顾客可以畅游于销售民族或国际品牌商品的现代建筑中。景观设计以生态理念为核心，糅合文化内涵和现代功能，呈现一个光影交织、风月无边的诗意环境。设计结合场地特点进行适度堆坡，营造变化丰富的立体空间，以点、环、轴为元素，形成场景交错、收放自如的空间。贯穿始终的景观主轴如一条绿色丝带将几个建筑主体串联为一体，形成既相互独立又统一共生的绿色空间。

The project is located in Wanglanzhuang, Xiqing District, Tianjin. It is located in the south of Tianjin, about 8 km away from the city center. The project aims to turn it into a finance-cultural district. Divided into commercial and residential areas The project is a separate commercial area. It contains mid - to high-rise apartments. The business district is the most unique and attractive area in the project. The open shops along the street on the 1st to 4th floors are also the best choice for local individuals to run cafes, bars, restaurants and shops. Outside the small business district is a large commercial area where shopkeepers and customers can swim in modern buildings selling ethnic or international brands. The landscape design takes the ecological concept as the core, combines the cultural connotation and modern functions, presenting a poetic environment with light and shadow. According to the characteristics of the site, the design moderately piles the slope to create a three-dimensional space with rich changes. Points, rings and axes are used as elements to form a space with staggered scenes and free movement. The main axis of the landscape throughout, like a green ribbon, connects several building subjects as one, forming a green space that is both independent and unified.

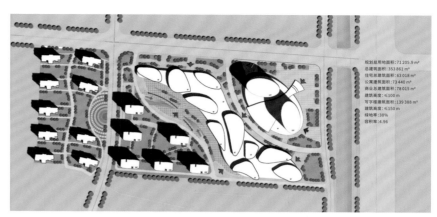

天津南开中学素质教育实验室
Quality Education Laboratory of Tianjin Nankai Middle School

项目地点：天津
Location: Tianjin

本项目是南开中学一系列素质教育实验室之一。南开中学设立素质教育实验室的目的，是为使学生在中学阶段利用选修课，较为全面地了解一些目前主流的领域。利用为南开中学建造素质教育实验室的契机，设计师大胆地将教育心理学、行为学研究成果与建筑设计相结合，根据青少年成长阶段的心理特点，以建筑手段营造三维空间，创造特殊的教育情境，以期达到改良教学方法、以更加直观的知觉体验推动学习动机的目的。它将成为向中学生介绍建筑学的窗口，以激发其对建筑的兴趣，让更多优秀的南开学子选择这个充满创意和挑战的职业，并将其作为终身事业为之奋斗，推动我国的建筑事业持续发展。

从建成后的使用效果来看，达到了预设的目的。根据南开中学授课教师的反馈，每一个首次进入房间的学生都在第一时间被整个空间氛围所吸引，学生会情不自禁地四处张望、四处游走、抚摸温暖的木材、欣赏精美的展板，全方位感受这件突破常规概念的作品。为了能使示范室更好地行使自己的职责，设计师还为南开中学提供了展板、电子课件、建筑模型等相关授课材料，使学生在被示范室本身吸引，想进一步了解建筑学的魅力时，可以得到更好的引导。

The architectural design demonstration studio of this project is one of a series of quality education laboratories in Nankai Middle School. The purpose of setting up quality education laboratories in Nankai Middle School is to enable students to make use of elective courses in middle school and have a more comprehensive understanding of some current mainstream fields. Taking Advantage of The Opportunity Of Building Quality Education Laboratory for NANkai Middle School, WE boldly combine the research results of educational psychology and behavior with architectural design, and create three-dimensional space and special educational situation by architectural means according to the psychological characteristics of teenagers' growth stage. In order to improve the teaching method, with more intuitive perceptual experience to promote the purpose of learning motivation. It will serve as a window to introduce architecture to middle school students, so as to stimulate their interest in architecture, and let more outstanding Nankai students choose this creative and challenging career, strive for it as a lifelong career, and promote the sustainable development of architecture in our country.

From the point of view of the use effect after completion, the preset purpose has been achieved. According to the feedback from the teachers of Nankai Middle School, every student who enters the room for the first time is immediately attracted by the atmosphere of the whole space. Students can't help looking around, walking around, touching the warm wood, admiring the exquisite display boards, and feeling this work that breaks through the conventional concept in all directions. In order to make the demonstration room better perform its duties, the designer also provided Nankai Middle School with exhibition boards, electronic courseware, architectural models and other relevant teaching materials, so that students can get better guidance when they are attracted by the demonstration room itself and want to further understand the charm of architecture.

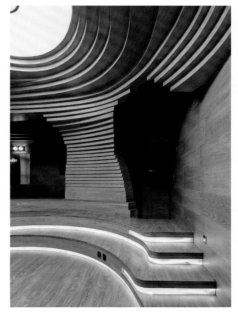

迁安青少年活动中心
Qianan Youth Activity Center

项目地点：河北 迁安	Location: Qianan, Hebei
建筑面积：24 818 m²	Building Area: 24 818 m²

本项目建筑体量具有地标性，同时体现了青少年建筑的活跃特征。项目基地紧邻滦河生态景观带，是生态理念向城市纵深发展的关键节点。建筑立面及造型元素着眼于自然，以"滦水之珠"的设计概念，充分呼应基地周边滦河水系和城市景观。

设计方法上，采取螺旋上升的方式，纯粹的几何形态突出了上部建筑体量，同时在入口处形成较好的半围合广场和开放平台。在形态上，基座逐渐抬升的方式，烘托了上部的建筑体量，使得其标志性得以增强，达到了"绿叶衬红花、相得益彰"的效果。同时，在空间使用上，亦可从小科学家广场进入一层的科技馆，形成良好的空间体验；立面流水状波纹和横向台阶相呼应，强化突出上部体量的漂浮感和地标性。

The project serves as a landmark, while reflecting the active character of the youth building. The project base is adjacent to Luanhe River Ecological Landscape Belt, which is a key node for the in-depth development of ecological concept to the city. The facade and modeling elements of the building focus on nature, and the design concept of "Luanhe River Pearl" fully echoes Luanhe River system and urban landscape around the base.

In terms of the design method, the pure geometric form in a spiral way highlights the upper volume of the building, while forming a good semi-enclosed square and open platform at the entrance. In terms of the form, the gradual lifting of the base accentuates the upper building volume and enhances its landmark feature, achieving the effect of "green leaves matching red flowers, complementing each other". At the same time, in terms of space use, you can also enter the Science and Technology Museum on the first floor from the Little Scientist Square, acquiring a good space experience. The flowing corrugation of the facade is echoed by the horizontal steps, which highlights the sense of floating and landmark feature of the upper volume.

新疆

新疆印象建设规划设计研究院（有限公司）
Xinjiang Impression Construction Planning & Design Institute (Co.,Ltd.)

新疆建设规划设计研究院，始创于2001年，2018年9月改制重组为新疆印象建设规划设计研究院（有限公司），至今已走过了21个春秋，现具有中国环境艺术设计甲级资质，建筑设计、城乡规划、风景园林、旅游规划等乙级综合设计资质。

2021年，在住房和城乡建设厅倡导下，为发挥创作优势，本院成立了"新疆印象建筑与文化研究中心"，已编印《界——印象在地作品集》上下册，收录作品遍及新疆14个地州，代表作有霍尔果斯中哈合作区国门、达坂城王洛宾音乐博物馆、石河子周恩来纪念馆新馆、昌吉回民小吃街、昌吉州博物馆、昌吉州市政务中心及科技馆、昌吉市规划展览馆、昌吉市滨湖河、吐鲁番老城印象、托克逊美食天下、阿克苏特色小镇美食街、焉耆西游古城、印象沙雅古城、柯坪古城博物馆及影剧院、伊犁那拉提草原山庄宾馆、石河子法院等。其中获得全国、自治区、乌鲁木齐市各类创作类设计奖40余项。单位和领军人物也获得了"中国特色文旅十大策划师""中国建筑设计领军人物""中国建筑优秀设计单位"等荣誉。

自2007年起，本院设计作品已入选《中国建筑设计年鉴》210幅，其中有10幅作品入选《2011年中国建筑优秀创意设计作品集》。2003年本院出版了新疆第一部规划专著《论城市的第一属性》，陆易农著。

本院一直主张"敬业、专业、乐业"的企业文化和"让特色从新疆印象开始"的企业愿景，努力打造一个集"跨文化策划、界印象设计、匠品质落地"三力合一的特色本土设计机构。

Xinjiang Architectural Planning and Design Institute, was founded in 2001 and restructured into Xinjiang Impression Construction Planning and Design Institute (Co., Ltd.) in September 2018. After 21 years of development, it now has the Grade A qualification of China environmental art design as well as Grade B comprehensive design qualification for architectural design, urban and rural planning, landscape architecture, tourism planning, etc.

In 2021, according to the proposal of the Ministry of Housing and Urban-Rural Development, in order to give full play to the creative advantages of the Institute, "Xinjiang Impression Architecture and Culture Research Center" was established. Two volumes of "Works of Impression in Xinjiang" have been compiled and printed, with the works covering 14 prefectures in Xinjiang. Representative projects include the entrance of Khorgos China-Kazakhstan Cooperation Zone, Wang Luobin Music Museum in Dabancheng, Zhou Enlai Memorial New Hall in Shihezi, the Hui People Snack Street in Changji, Changji Prefecture Museum, Changji Prefecture Municipal Service Center and Science and Technology Museum, Changji City Planning Exhibition Hall, Binhu River in Changji City, Turpan Old City Impression, Toksun Food World, Aksu Characteristic Town Food Street, Yanqi West Tour Ancient City, Impression Shaya Ancient City, Keping Ancient City Museum and Theater, Yili Narati Grassland Villa Hotel, Shihezi Court, etc. It has won more than 40 awards of various kinds of creative design in the country, autonomous region and Urumqi City. The unit and its leading figures have also won the honors of "Top Ten Planners of Cultural Tourism with Chinese Characteristics", "Leading Figures of Chinese Architectural Design", "Excellent Architectural Design Unit of China" and so on.

Since 2007, 210 design works of the Impression Institute have been selected into the "Yearbook of Chinese Architectural Design". Ten works have been selected into the "2011 Outstanding Creative Design Works of Chinese Architecture". In 2003, the first planning monograph of Xinjiang, On the First Attribute of Cities, by Lu Yinong, was published.

The design Institute has always advocated the corporate culture of "dedication, professionalism and contentment" and the corporate vision of "Let the characteristics start from Xinjiang Impression", striving to create a characteristic local design agency integrating "cross-cultural planning, community impression design and craftsman quality landing".

地址：乌鲁木齐市新市区苏州东街568号金邦大厦1513室
邮编：830011
电话：13325550916，18195884485
邮箱：1968675997@qq.com

Add: Room 1513, Jinbang building, 568 Suzhou East Street, Xincheng district, Urumqi
P.C: 830011
Tel: 13325550916, 18195884485
Email: 1968675997@qq.com

沙雅县·乌什喀特古城旅游景区规划设计
Planning and Design of Ushikat Ancient City Scenic Area in Shaya County

项目业主：沙雅县文化体育广播电视和旅游局
项目主持：王元新
主创人员：王元新、陈洋茹、张月磊、韦世武、许彦琴、袁泽
项目地点：新疆 阿克苏
用地面积：409 413.00 m²
设计时间：2021年10月—2022年10月

Project Owner: Shaya County Culture, Sports, Radio, Television and Tourism Bureau
Project Host: Yuanxin Wang
Chief Creators: Yuanxin Wang, Yangru Chen, Yuelei Zhang, Shiwu Wei, Yanqin Xu, Ze Yuan
Location: Aksu, Xinjiang
Site Area: 409 413.00 m²
Design Time: October 2021 to October 2022

沙雅县乌什喀特古城即"三重城"之意，是汉至晋代的古城，距今有2216年历史，被认为是汉代龟兹国的大城市之一。项目以乌什喀特古城（三重城）为原型，将消失的古城复建，复原古城建筑，重现历史，通过主题情境表现与故事策划演绎，植入千年古城主题文化体验产品，演绎一场丝绸之路上沙雅县乌什喀特古城的历史传奇。

在项目规划设计上，最大限度地使古城内城、宫城轮廓与乌什喀特古城遗址轮廓形态一致。内城平面呈不规则形，宫城平面略呈圆形。以城中的主轴线作为经纬线，城内所属的瓮城、宫殿、官署、街巷、佛寺等主要区域分列其间。古城中轴线上的广场分别以出土文物五铢、带扣及"汉归义羌长印"为设计元素。中轴线两侧的环路以"双龙戏珠"图案呈现，将各个功能区串联。

建筑设计上深入挖掘乌什喀特古城建筑特点，将沙雅地域文化元素融入其中，结合史料记载中古丝绸之路上驼铃声声、商旅不绝的美好画面，形成"大漠古驿"的整体印象。古城内建筑外立面采用仿生土、木材、石等原生建材，营造出具有千年古城风貌的集文化展示、参观、休闲、娱乐、旅游、影视等功能的开放式环境。

设计亮点：
——规划设计的交通流线为环形路网，把瓮城、内城、宫城、佛寺、外城有机地串联在一起，形成一个闭环。
——通过建筑外立面的再现，使古城看起来更有文化吸引力。

Ushikat Ancient City Scenic Area in Shaya County, meaning "Triple City", is an ancient city from Han Dynasty to Jin Dynasty, with a history of 2216 years. It is considered to be one of the major cities in Qiuci State of Han Dynasty. The project takes the ancient city of Ushkat (Triple City) as the prototype, and restores the ancient city to reproduce the history. Through the theme situation performance and story planning and interpretation, the project implants the cultural experience products with the theme of 1 000-year ancient city, and performs the historical legend of the ancient city of Ushkat in Shaya County on the Silk Road.

In the planning and design of the project, the outline of the inner city and the palace city should be consistent with the outline of the ruins of Ushkat ancient city to the greatest extent. The plane of the inner city is in irregular shape, while the plane of the palace is slightly circular. With the main axes of the city as longitude and latitude lines, Wengcheng, palace, government offices, streets, Buddhist temple and other main functional areas are arranged in the city. The square on the central axis of the ancient city is designed with the elements such as the excavated cultural relics Wuzhu coin, belt buckle and "Guiyi Qiangzhang Seal in Han Dynasty". The ring roads on both sides of the central axis is presented with a pattern of double dragon playing with bead, connecting each functional area in series.

In terms of architectural design, the architectural features of Ushkat ancient city are deeply explored, and the elements of Shaya regional culture are integrated into the design. Combined with the beautiful picture of camel bells and endless business trips on the ancient Silk Road recorded in historical materials, the overall impression of "ancient post in the desert" is formed. The external facade of the buildings in the ancient city uses bionic soil, wood, stone and other native building materials to create an open environment with the features of the ancient city for cultural display, visiting, leisure, entertainment, tourism, film and television.

Design highlights:
—In terms of traffic flow, ring road network is planned and designed to connect Wengcheng, inner city, palace, Buddhist temple and outer city together, forming a closed loop.
—The reappearance of the building facade makes the ancient city more culturally attractive.

柯坪古城博物馆影剧院
Kalpin Ancient City Museum Movie Theater

项目业主：柯坪县人民政府	Project Owner: Keping County people's Government
项目主持：孔伟雄	Project Host: Weixiong Kong
主创人员：王元新、张月磊、韦世武、黄俊华、赵国强	Chief Creator: Yuanxin Wang, Yuelei Zhang, Shiwu Wei, Junhua Huang, Guoqiang Zhao
项目地点：新疆 阿克苏	Location: Aksu, Xinjiang
总建筑面积：18 854 m²	Total Building Area: 18 854 m²
用地面积：15 000 m²	Site Area: 15 000 m²
设计时间：2022年7月	Design Time: July 2022

柯坪县历史悠久，境内拥有以古代齐兰古城为代表的70多处历史文化古迹。为了动态传承宝贵的历史文化资源，柯坪县拟建一座古城博物馆、影剧院两馆合一综合体，是个多功能的陈列展览、演艺会议、文化娱乐场所。

项目整体造型设计呈新疆大漠古城城郭形象，剧院外立面高仿柯坪齐兰古城旧址生土风貌，表现特有的在地建筑特色，其轮廓的残缺起伏，极具历史沧桑的厚重感，从正面入口较新古城博物馆的印象过度至背后老古城的形象，让时空在这里沉浸穿越了两千年。把建筑本身视为珍贵的艺术藏品得以收藏。

Kalpin County has a long history and has more than 70 historical and cultural sites represented by Qilan ancient city. In order to dynamically inherit valuable historical and cultural resources, Kalpin County plans to build a complex integrating the ancient city museum and the movie theater, which is a multifunctional place for exhibition, performance, conference and cultural entertainment.

The overall modeling design is the image of Xinjiang desert ancient city. The facade of the theater is vivid imitation of Qilan ancient city in Kalpin County, showing unique architectural characteristics. The incomplete and rolling outline has very heavy historical vicissitudes of life. From the front entrance of the new ancient city museum to the image of the old city behind, time and space here goes through two thousand years. The building itself is considered as a precious artwork to be collect.

丝路航田城市会客厅和玫瑰主题馆
Silk Road Aerospace City Reception Hall and Rose Main Pavilion

项目主持：王健宇、郝娜、张璟慧、王慧琴	Project Host: Jianyu Wang, Na Hao, Jinghui Zhang, Huiqin Wang
主创人员：王元新、张月磊、王策、石磊	Chief Creator: Yuanxin Wang, Yuelei Zhang, Ce Wang, Lei Shi
项目地点：新疆 伊犁	Location: Yili, Xinjiang
总建筑面积：17 616.79 m²	Total Building Area: 17 616.79 m²
用地面积：116 798.76 m²	Site Area: 116 798.76 m²
设计时间：2022年8月	Design Time: August 2022

项目主馆以航天科技为主题，采用了星系轨道的设计理念，以主馆为恒星、其他各功能空间为行星的空间关系为设计概念，将各个建筑物统一而协调地组合在一个整体里面。主入口结合北侧玫瑰种植基地设置一座玫瑰主题馆。建筑外观造型庄重简洁，其颜色白色为底，外形为一朵含苞待放的玫瑰花骨朵，侧面小窗造型呼应1号航天馆的科技性，加上褐色竖线条装饰，营造了鲜艳生动的田园氛围感，具有明确的标示性与可识别性，并与周边环境相协调。

The main pavilion of the project is themed on aerospace science and technology, so it adopts the design idea of galaxy orbit and the design concept of the main pavilion as the star and other functional spaces as the planet. All the buildings are integrated into a unified and coordinated whole. The main entrance to the building combines with the rose planting base on the north side to set up a rose theme pavilion. The modeling of the architectural appearance is solemn and simple, like a white budding rose flower. And the small window shape on the side echoes the sense of science and technology of Space Museum No.1 With the brown vertical line decoration, a vivid pastoral atmosphere is formed. It is clearly marked and recognizable, and harmonizes with the surrounding environment.

博乐城市会客厅
Bole City Reception Hall

项目业主：博乐中基地产开发有限公司
项目主持：陈捷、韩洪宇
主创人员：张月磊、王元新
项目地点：新疆 博乐
总建筑面积：23 498.5 m²
规划用地：22 351 m²
设计时间：2021年3月—2022年6月

Project Owner: Bole Zhongji Real Estate Development Co., Ltd.
Project Host: Jie Chen, Hongyu Han
Chief Creator: Yuelei Zhang, Yuanxin Wang
Location: Bole, Xinjiang
Total Building Area: 23 498.5 m²
Planned Area: 22 351 m²
Design Time: March 2021 to June 2022

  项目由画舫、戏舫、天宫合院、天宫酒店、天宫知夜等6大板块组成，集特色旅游、娱乐、餐饮美食、民宿酒店于一体，打造一处城市中心滨水活力区，为城市再添炫目风景线。为城市营造活力、健康、开放的核心主题，完善运动、休旅、会客、时尚消费等功能，满足百姓日益提升的多元、休闲、精神、文化的需求，全力提升滨水景观有机活力空间。

  The project is composed of six parts, namely, gaily-painted pleasure-boat, theatre boat, Tiangong Compound Courtyard, Tiangong Hotel, Tiangong Zhiye, etc., which integrates characteristic tourism, entertainment, food and beverage, and Minshuku hotel to create a dynamic waterfront area in the city center and add dazzling scenery. The project aims to create the core theme of vitality, health and openness for the city, improve sports, leisure, reception, fashion consumption and other functions, meet the people's increasing demand for diversity, leisure, spirit and culture, and make every effort to improve the waterfront landscape organic vitality space.

达坂城（大果院子）疆风民宿设计
Dabancheng (Daguo Courtyard) Xinjiang Style Minshuku Design

项目业主：乌鲁木齐达坂城旅游开发公司	Project Owner: Urumqi Dabancheng Tourism Development Company
项目主持：刘生清	Project Host: Shengqing Liu
主创人员：王元新、张月磊、王策、袁泽	Chief Creator: Yuanxin Wang, Yuelei Zhang, Ce Wang, Zhe Yuan
项目地点：新疆 乌鲁木齐	Location: Urumqi, Xinjiang
总建筑面积：7 550 m²	Total Building Area: 7 550 m²
用地面积：42 443 m²	Site Area: 42 443 m²
设计时间：2022年9月	Design Time: September 2022

大果院子是新疆达坂城火车站站前合院建筑，能让人体验不一样的疆风民宿，体验不一样的花海水街，满足人们追求"一生有一个院子"的心愿。项目主题定位为生态文化旅游商业综合体。这里是一个多民族建筑博物馆聚落，一个多民族民宿聚落，一个新疆人的院子，一个开在建筑中的花园，一个城市建筑师的"心院"，一个会"变脸"的文化艺术长廊，一个新疆特产线上线下的交易平台，一个展示达坂城洛宾文化的窗口，一个多民族婚纱摄影的网红打卡地，一个体验新疆风情的窗口，一个讲好新疆故事的基地。

Daguo Courtyard is the courtyard building in front of Dabancheng Railway Station in Xinjiang, which can let people experience different Xinjiang style Minshuku and different flower sea along water street. It is the greatest wish to pursue a yard in life. The theme positioning is ecological culture tourism commercial complex. It aims to create a multi-ethnic architecture museum settlement, a multi-ethnic homestay yard settlement of Xinjiang people, a garden opened in the building, a waiting place in the wetland garden, a city director architect's courtyard, a changing culture and art corridor, an online and offline trading platform for Xinjiang specialty, a window for Dabancheng Luobin culture, a internet celebrity place for multi-ethnic wedding photography, a window to experience Xinjiang flavor, a cultural base for telling a good story of Xinjiang.

兵团103团一连乡村振兴风貌提升
The Rural Revitalization of Company 1, Regiment 103

项目业主：兵团第六师103团	Project Owner: 103rd Regiment, 6th Division, Bingtuan
项目主持：阿余拉哲、魏久华、解文超	Project Host: Ayula Zhe, Jiuhua Wei, Wenchao Xie
主创人员：王策、张月磊	Chief Creator: Ce Wang, Yuelei Zhang
项目地点：新疆 五家渠	Location: Wujiaqu, Xinjiang
用地面积：6 000 m²	Site Area: 6 000 m²
设计时间：2022年9月	Design Time: September 2022

兵团第六师103团一连乡村振兴示范项目重点围绕广场和民宿的打造。设计师利用原有农作物晒场空地，打造一处集会展、演绎、百姓纳凉、夜晚休闲于一体多功能"亮剑"休闲广场。广场两侧是虚实透景休闲长廊，是军垦文化元素和乡村本地材料的有机结合。广场中央是一个巨大中国结灯笼造型的地景，寓意军民团结一家亲。民宿设计集农家乐、旅游住宿、文化沉浸式体验等功能进行重点打造，对民居住房外立面改造提升，立面造型植入了军垦怀旧记忆的红砖元素符号。

以职工自筹资金和团场适当鼓励补贴的方式，示范性地打造农文旅民宿。民宿建筑外观采用朴素的具有军垦怀旧记忆的红砖建筑风格，力求集农家乐、旅游民宿、文化沉浸等功能于一体。

通过一系列的乡村振兴文化挖掘和屯垦风貌打造，改变和提高农场职工的生产生活环境。

The demonstration project of rural revitalization of Regiment 103 of Division 6 focuses on the construction of squares and Minshuku. By using the original open space of crop sunning ground, the designers create a multi-functional recreational square where people can exhibit, perform, enjoy the cool, and have leisure activities at night. On both sides of the square is the leisure corridor, which is an organic combination of army reclamation cultural elements and rural local materials. In the middle of the square is the landscape equipped with a huge Chinese knot in the shape of a lantern, which symbolizes the unity of the army and the people. The Minshuku design focuses on integrating agritainment, tourism Minshuku with cultural immersion. The facade of residential houses is transformed and upgraded, adopting the red brick element symbol with nostalgic memory of army reclamation.

With funds raised by the staff themselves and appropriate encouragement subsidies from the regiment, it aims to build rural tourism Minshuku. The architectural appearance adopts the simple red brick architectural style with nostalgic memory of army reclamation, striving to integrate agritainment, tourism Minshuku with cultural immersion.

It uses a series of rural revitalization culture mining and cultivation to change and improve the production and living environment for farm workers.

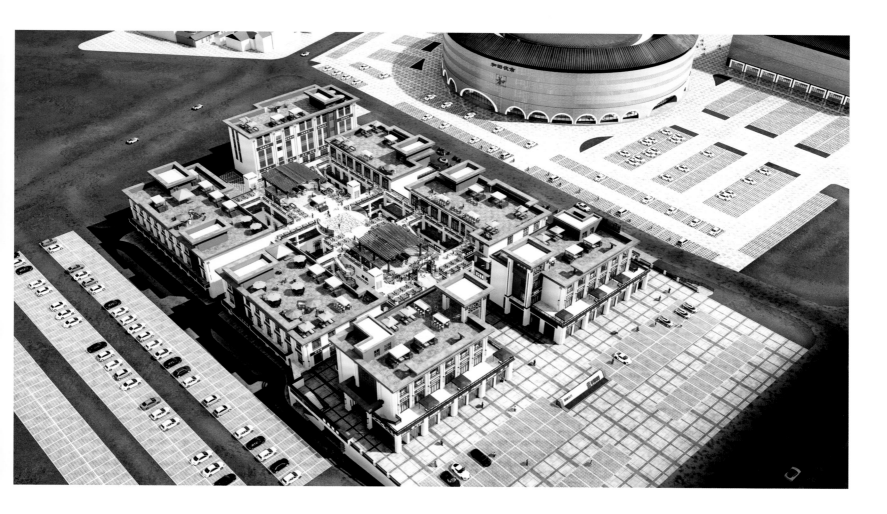

和田团城·谜街
Hetian Tuancheng · Maze Street

项目业主：和田中基地产开发有限公司	Project Owner: Hotan Zhongji property Development Co., Ltd.
项目主持：陈捷、韩洪宇	Project Host: Jie Chen, Hongyu Han
主创人员：张月磊	Chief Creator: Yuelei Zhang
项目地点：新疆 和田	Location: Hetian, Xinjiang
总建筑面积：22 000 m²	Total Building Area: 22 000 m²
用地面积：14 000 m²	Site Area: 14 000 m²
设计时间：2020年1月—2022年2月	Design Time: January 2020 to February 2022

团城·谜街项目位于和田团城景区东南侧，在北京东路的核心地带，是和田传统的流金淌银之地。项目占地14 000 m²，总建筑面积为22 000 m²，设计定位为：团城·谜街——新疆特种旅游集散中心。这里是西域神秘之旅的序幕，"尼雅之谜、公主堡传奇、西市笙歌"……这里是正在发生的故事，这里是偶遇后的再出发，这里是精神之旅，是观光、度假旅游的再升华。团城·谜街依托和田的历史文化，进行深度发掘，将22 000 m²的文旅街区整体打造为和田地区名优博览、自助旅游集散分中心、丝路驿站文化夜市；团城·谜街是集文化体验、文化创意、旅游观光、休闲度假为一体，服务与设施一流、参与性和体验性极高的综合性和田旅游目的地，包含60余家独立餐厅、7家文化主题客栈、100余家文博名优店、10多个文化展示体验广场及完善的配套服务设施。

团城·谜街采取国际上最为流行、最为成功的街铺混业格局，双首层建筑设计，公主堡立体的广场动线，为人们提供了良好旅游购物动线，通过1部扶梯、3部垂直电梯、8部人行步梯，连廊、广场，打通一、二层的商业脉络，所有商业层面的无缝连接，八面贯通，构建便捷畅达的多维立体商业空间。

The project of Tuancheng · Maze Street is located in the southeast of Tuancheng Scenic Spot in Hetian, at the core area of Beijing East Road, the traditional luxury land in Hetian. The project covers an area of 14 000 m², with a total construction area of 22 000 m². The design is positioned as: Tuancheng · Maze Street - the special tourism distribution center of Xinjiang. Here is the prelude to the mystery tour of the Western Regions - "the Mystery of Nyaya, the Legend of Princess Castle, and the Western City Playing and Singing"...... Here is the story of what is happening, the departure after a chance encounter, the spiritual journey, the sublimation of sightseeing, vacation tourism. Relying on the historical and cultural heritage left by Hotan, Tuancheng · Maze Street is deeply explored, and the cultural and tourism block of about 22 000 m² is built into the cultural and tourism exhibition center, self-service tourism distribution center and Silk Road post cultural night market in Hotan. Tuancheng · Maze Street is a comprehensive Hetian culture tourist destination with first-class services and facilities, high participation and experience which integrates cultural experience, cultural creativity, tour and leisure, including more than 60 independent restaurants, 7 culture-themed inns, more than 100 famous and excellent shops, more than 10 cultural exhibition experience squares and complete supporting service facilities.

Tuancheng · Maze Street adopts the most popular and successful mixed-industry layout in the world, with the double first-floor building design and the three-dimensional square line of Princess Castle, providing people with a good travel and shopping line. Through 1 escalator, 3 elevators, 8 pedestrian stairs, connecting corridors and squares, all commercial levels and commercial veins on the first and second floor are seamlessly connected, creating a convenient and accessible multi-dimensional commercial space.

沙雁洲景区旅游基础设施建设项目——胡杨学院
Tourism Infrastructure Construction Project OF Yanzhou Scenic Area — Huyang College

项目地点：新疆 阿克苏　　Location: Aksu, Xinjiang
主创人员：王洋　　Chief Creator: Yang Wang
总建筑面积：3 917 m²　　Total Building Area: 3 917 m²
设计时间：2022年5月（建设中）　　Design Time: May 2022 (Under Construction)

建筑功能涵盖游客中心、餐厅、旅游接待、商品售卖、办公及员工宿舍。建筑造型大气又不失变化，屋顶造型层次丰富，建筑颜色与周围环境融合，同时也与周边现状建筑风格统一。

　　The architectural functions include tourist center, restaurant, tourist reception, merchandise sales, office and staff dormitory. The architectural style is magnificent with changes. The roof shape is rich in layers and the architectural color is integrated with the surrounding environment, but also with the current surrounding architectural style.

沙雅县沙雁洲景区游客服务中心
Shayanzhou Tourist Service Center in Shaya County

项目地点：新疆 阿克苏　　Location: Aksu, Xinjiang
主创人员：王洋　　Chief Creator: Yang Wang
总建筑面积：950 m²　　Total Building Area: 950 m²
设计时间：2019年3月　　Design Time: March 2019
竣工时间：2019年11月　　Completion Time: November 2019

景区出入口布置在建筑中间，两侧设置功能用房，同时将公共区域与办公区域有机地分隔开。建筑屋顶造型形似一只展翅高飞的大雁，与景区特点相呼应，整体造型简约大气，同时也与周围环境融为一体。

　　The entrance and exit of the scenic area are arranged in the middle of the building, and functional rooms are set on both sides, while the public area and office area are separated organically. The roof of the building is shaped like a flying goose, echoing the characteristics of the scenic area. The overall shape is simple and magnificent, but also can integrate with the surrounding environment.

泰和大酒店及民宿综合开发利用建设项目
Taihe Hotel and Minshuku Comprehensive Development and Utilization Construction Project

项目地点：新疆 伊犁　　Location: Yili, Xinjiang
主创人员：王洋、罗莎　　Chief Creator: Yang Wang, Sha Luo
设计时间：2021年6月（建设中）　　Design Time: June 2021(Under Construction)

项目主要分为3个区域——酒店区、商业区、民宿区，整体风格采用新中式风格，同时在建筑外立面融入蒙古族风格元素。规划把酒店布置在交通位置较好的交叉路口处，并沿道路布置商业建筑，满足本地块商业需求的同时也可为周边居民提供便利。民宿建筑位于地块内侧，保持安静的同时可以遥望北侧雪山。

The project is mainly divided into three areas - hotel area, commercial area and Minshuku area. The overall style of the project adopts new Chinese style, while local Mongolian style elements are integrated into the facade design of the building. According to the plan, the hotel is arranged at the intersection with good traffic location, and commercial buildings are arranged along the road to meet the commercial needs of the local block and provide convenience for the surrounding residents. The Minshuku building is located in the inner part of the plot, which can enable people to keep quiet and look at the snow mountain on the north side.

拉萨市设计集团有限公司
Lhasa Municipality Design Group Co., Ltd.

　　拉萨市设计集团有限公司成立于1978年，现隶属于拉萨市城市建设投资经营有限公司。集团是西藏自治区2019年第一批国家级高新技术企业，现有工程设计建筑行业（建筑工程）甲级、风景园林工程专项乙级、市政行业乙级、城乡规划编制乙级、工程勘察岩土工程（勘察、设计）、工程测量乙级资质，具备完整的ISO9001标准质量管理体系。目前，设计集团员工共约160人。经过30余年的资源积淀与整合，设计集团在大型城市综合体、教育、医疗、办公、酒店、商业、旅游、司法、公安、产业园区、大型住宅小区、景观园林、城市改造更新、市政及道桥建设等领域建成了一大批有社会影响力的项目，并深耕传统建筑文化保护与研究，塑造了优质的品牌，享有良好的社会信誉，在业内拥有众多战略合作伙伴，在西藏自治区建筑设计单位中综合实力位居前列。拉萨市设计集团在驻藏兴城这条路上始终贯彻以人为本的理念，秉承"追求卓越设计、打造精品工程、竭诚服务客户、共创美好未来"的宗旨，扎根雪域高原，以建设美丽家园为己任，同时为了更好地参与城市建设，设计集团以研究为推力，设立了"青藏极地建筑文化研究中心"，集团始终坚持传承和发扬传统文化，努力将拉萨市设计集团打造成为文化和品质并重的行业领航者。

　　Lhasa Municipality Design Group Co., Ltd. was founded in 1978, is a branch under the Lhasa Urban Construction Investment Management Co., Ltd. The Group is the first batch of state-level high-tech enterprises in Tibet Autonomous Region. Currently, the Group's qualifications in the fields are including Level A in Construction Engineering, Special Grade Level B in Landscape Engineering, Grade B in City Design, Grade B in Urban and Rural Planning, Level B in Geotechnical Engineering Survey and Design & Engineering Survey with a complete ISO9001 standard quality management system. At present, our Group has about 160 employees. After more than 30 years of accumulating and integrating of resources, the design group has built a large number of projects with visible social impact in such as large urban complexes, education, medical care, office space, hotel, commerce, tourism, justice and public security, industrial parks, large residential quarters, landscape gardens, urban transformation and renewal, city and road-bridge construction. These projects are deeply cultivated the protection of traditional architectural culture of the region and combined with extensive research, created high-quality brands and enjoy positive social reputation. Alone the years, we have established many strategic partners in the industry and ranks in the forefront of architectural design units in the Tibet Autonomous Region. In its mission in "stationing in Tibet and bringing prosper to the city", Lhasa Municipality Design Group Co., Ltd has always implemented the people-oriented concept, adhering to the purpose of "pursuing excellence in designing, creating high-quality projects, whole-heartedly serving customers and creating a better future", we root ourselves on the Plateau and takes building a beautiful home as our main responsibility. In order to better participate in urban construction, our Group has set up the Qinghai-Tibetan Architectural Research Center with research as the thrust, and always adhere to inheriting and carrying forward traditional culture and strive to build Lhasa Municipality Design Group Co., Ltd. an industrial leader with equal emphasis on culture and quality.

地址：拉萨市城关区纳金路74号　　　Add: No.74 Najin Rd, Chengguan District, Lhasa
电话/传真：0891-6231076　　　　　Tel(Fax): 0891-6231076
网址：www.xzlssjy.com　　　　　　　Web: www.xzlssjy.com

西藏非遗千工坊设计项目
Design Project for Tibet Intangible Cultural Heritage Qiangong Workshop

项目地点：西藏 拉萨	Location: Lhasa, Tibet
建筑面积：5 163.67 m²	Building Area: 5 163.67 m²
用地面积：4 000.13 m²	Site Area: 4 000.13 m²
建筑功能：文化建筑	Architectural Function: Cultural Architecture
项目状态：竣工	Project Status: Completed
设计时间：2019 年	Design Time: 2019
设计团队：王慧林、周立军、陈宝、陈伟、林辉、洛桑达瓦、多庆巴珠等	Design Team: Huilin Wang, Lijun Zhou, Bao Chen, Wei Chen, Hui Lin, Luosangdawa, Duoqingbazhu, etc.

　　博物馆不仅以展品展示非遗文化，还应有非遗文化体验场所，而新建的千工坊正是非遗传承人工作及游客体验的地方，由19栋单体、8个组团构成，每个单体、组团错落有致，互相间有灵活的步行路径进行串联，流线曲折贯通，一步一景。整体形态运用藏式建筑中与环境融为一体的山地建筑布局手法，体现地域民族建筑的文脉延续。建筑采用当地石材，整体色调及门窗样式均采用藏式传统风格，具有浓郁的藏式风味，符合片区的整体风貌。千工坊具有最佳的视线通廊，不仅能与神圣的布达拉宫遥相呼应，还能欣赏整个拉萨市区自然而独特的风情，并且在晚上还能一睹拉萨市独特的夜景之美。

　　The museum not only displays the intangible cultural heritage with exhibits, but also should be a place to experience the intangible cultural heritage. The new Qiangong Workshop is a place where the non-inheritors work and tourists experience. It is composed of 19 single buildings and 8 group buildings, each of which is well-arranged and connected with each other by flexible walking paths. The overall form uses the mountain building layout technique integrated with the environment in Tibetan architecture, reflecting the continuation of the cultural context of regional ethnic architecture. The building is made of local stone. The overall tone and the style of doors and windows are all in traditional Tibetan style, which is with strong Tibetan flavor and in line with the overall style of the area. The Qiangong Workshop has the best view corridors, which not only echo the sacred Potala Palace, but also enable people to enjoy the natural and unique scenery of the whole Lhasa city and the unique beauty of Lhasa city at night.

达孜叶巴康养小镇城市设计
Urban Design for the Health Care Town in Yeba Village, Dagze County

项目地点：西藏 拉萨
规划总建筑面积：615 200 m²
规划用地面积：313.29 hm²
项目状态：规划中
设计时间：2020年
项目主创：马扎·索南周扎 多庆巴珠
设计团队：李昆、旦增索朗、杨大华、普布顿珠、索朗旺久、旦增曲珠、边巴次仁、普布伦珠等

Location: Lhasa, Tibet
Total Planned Building Area: 615 200 m²
Planned Site Area: 313.29 hm²
Project Status: Planning
Design Time: 2020
Project Chief Creator: Mazha · Sonanzhouzha, Duoqing Bazhu
Design Team: Kun Li, Danzengsuolang, Dahua Yang, Pubudunzhu, Suolangwangjiu, Danzengquzhu, Bianbaciren, Pubulunzhu, etc.

以藏医文化为核心,打造最具西藏生活美学的宜居样本。设计突出雪域高原疗养、藏医养生疗养、康养社区、特色公园等核心特色活动功能,打造雪域高原度假旅游康养活力印象地。期许为居民提供一个设施完备、环境优美的健康活力智慧社区,为游客提供一个智慧化、现代化、特色化的集疗养、旅居、购物、游玩等功能于一体的藏医文化体验目的地。中心建筑方案特色:场地位于小镇核心区域,建筑整体造型取自藏式庄园,反映出藏式建筑的悠久历史,宗堡般坚固挺拔层层退台的建筑造型,营造出整个藏医养生文化展示中心的雄伟气势。设计高低不同的两进院落。退台与院落相叠,营造丰富的充满地域文化的院落空间。整个建筑大实大虚,底部沉稳厚重,顶部构造精细、结构轻盈,且顶部视野开阔,独具西藏传统建筑文化特色,也是整个康养小镇的标志性建筑之一。

With Tibetan medicine culture as the core, the project aims to create a livable sample full of Tibetan living aesthetics. The design highlights the core features of the snow plateau rehabilitation, Tibetan medicine health rehabilitation, health care community and characteristic park, trying to create a tourism health care place with impressive vitality in the snow plateau. It is expected to provide residents with a healthy, vibrant and intelligent community with complete facilities and beautiful environment, and provide tourists with an intelligent, modern and distinctive Tibetan medicine cultural experience destination integrating rehabilitation, residence, shopping and amusement. The features of the central architecture are as follows: the site is located in the core area of the town, and the overall shape of the building is taken from the Tibetan manor, reflecting the long history of Tibetan architecture. The architectural shape of standing tall and straight like a fort and retreating layer by layer creates the majestic atmosphere of the whole Tibetan medical health culture exhibition center. Two courtyards with different levels are designed. The terrace and courtyard overlap to create a rich courtyard space full of regional culture. The bottom of the whole building is solid and heavy while the top is fine and light in structure. There is a broad vision at the upper part. It has unique characteristics of traditional Tibetan architecture culture, and is also one of the landmark buildings of the whole health care town.

西藏文化艺术创作园区
Tibetan Culture and Art Creation Park

项目地点：西藏 拉萨
建筑面积：26 020 m²
用地面积：43 929 m²
建筑功能：文化建筑
项目状态：竣工
设计时间：2018年
设计团队：多庆巴珠、洛桑达瓦、普布顿珠、李昆、于梦璇等

Location: Lhasa, Tibet
Building Area: 26 020 m²
Site Area: 43 929 m²
Architectural Function: Cultural Architecture
Project Status: Completed
Design Time: 2018
Design Team: Duoqingbazhu, Luosangdawa, Pubudunzhu, Kun Li, Mengxuan Yu, etc.

西藏文化艺术创作园区位于慈觉林西山脚下,距离拉萨市中心约3.6 km。文化艺术创作园主要由唐卡博物馆、艺术家创作室、学校及附属功能的公共空间、服务设施组成。园区内建筑立面设计充分考虑拉萨传统建筑的立面颜色构成,采用藏式传统建筑颜色,从藏式服饰上采集颜色与构图形式,形成丰富、变化的立面效果。

项目靠近山体,像典型西藏建筑一样强调与山和天空的对话。形体方面,借鉴布达拉宫的形象,与山石呼应,建筑不是单一的体块,而是由多个斜墙体块组成的有机的聚合体,即是模拟天然山体的形态,也像山间村落的集合,除了营造契合唐卡艺术收藏和展示的空间,本项目也着重打造精神体验空间,将游览过程与唐卡文化和西藏文化的精神追求相结合,通过内部观展路径与外部拓展路径的组合,启发游客感受和领悟其文化特色。另外,建筑细部尊重当地风格的传承,用现代建筑语汇重现传统艺术特色,例如主入口门头、立面窗花、中庭扶手栏杆等,借鉴拉萨传统装饰样式,将其样式简化,并用现代材料和工艺重新表达,使本项目成为独具地域特色与时代特征的标志性建筑。

The Tibetan Culture and Art Creation Park is located at the foot of the West Mountain of Cijuelin Village, about 3.6 km from downtown Lhasa. The Tibetan Culture and Art Creation Park consists of public space and service facilities of Thang-ga Museum, artists' studio, school and their ancillary functions. The facade design of buildings in the park fully considers the facade color composition of traditional Lhasa buildings. Traditional Tibetan architectural colors are adopted while colors and composition forms are collected from Tibetan clothing to form a rich and changing facade effect.

The project is close to the mountain and emphasizes the dialogue with the mountain and sky like typical Tibetan architecture. In terms of image, the project draws on the topological image of the Potala Palace, which echoes the rocks. It is not a single volume, but an organic aggregate composed of several slanted wall blocks, which simulates the form of a natural mountain and also resembles a collection of mountain villages. In addition to creating a space suitable for the collection and display of Thang-ga art, this project also focuses on creating a spiritual experience space. The tour process is combined with the spiritual pursuit of Thang-ga culture and Tibetan culture, and through the combination of internal viewing path and external expansion path, tourists are inspired to feel and understand its cultural characteristics. Moreover, the architectural details respect the inheritance of local style, using modern architectural vocabulary to reproduce traditional artistic features. Traditional and simplified Lhasa decoration style is adopted in the design of the door head of the main entrance, facade window sash, and atrium handrail, etc. with modern materials and techniques, so that the project becomes a symbol of unique regional characteristics and characteristics of the times.

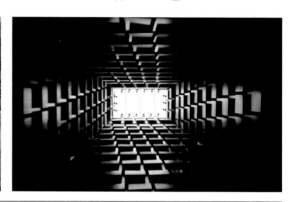

布达拉宫广场装饰柱提升改造项目
Potala Palace Square Decoration Column Upgrading Project

项目地点：西藏 拉萨
项目状态：竣工
设计时间：2021 年
项目主创：多庆巴珠
设计团队：李昆、旦增索朗、边巴次仁、杨大华、普布伦珠等

Location: Lhasa, Tibet
Project Status: Complete
Design Time: 2021
Project Founder: Duoqing Bazhu
Design Team: Kun Li, Danzengsuolang, Bianbaciren, Dahua Yang, Pubulunzhu, etc.

此次设计将自然元素与建筑形体相融汇,将民族文化与建筑手法相结合;使装饰柱完美地融入周边环境,成为美观的景观构筑物;与布达拉宫的建筑特点相协调,成为衬托布达拉宫的辅助构筑物,既能蕴含地域建筑文化精粹又能表现多民族文化融合的内涵。

The design integrates natural elements with architectural form while combining national culture with architectural techniques. The decorative column is perfectly integrated into the surrounding environment, becoming a beautiful landscape structure. In harmony with the architectural features of the Potala Palace, it becomes the auxiliary structure that sets off the Potala Palace. It can not only contain the essence of regional architectural culture but also express the connotation of multi-ethnic cultural integration.

设计理念:"四方五和"。
1. 取土为基,砌石为墙,与布达拉宫呼应,环境相宜之和;
2. 取自然五色,结建筑之美,自然敬畏之和;
3. 攒尖汉阙为顶,敦厚藏式为本,汉藏交融之和;
4. 56根椽子木,搭接成拱,民族团结之和;
5. 四方稳定之形,成四面八方,国泰民安之和。

Design concept:
1. Take earth as the base, masonry as the wall, echo with the Potala Palace, in harmony with the environment;
2. Take the five colors of nature, combine the beauty of architecture and the awe of nature;
3. Combine the watchtower in the Han Dynasty with architecture in Tibetan style;
4. Fifty-six rafter wood in the form of the arch means national unity;
5. Boxy buildings garrison in all directions means peace and prosperity.

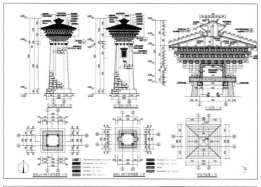
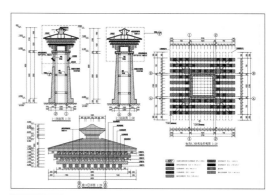
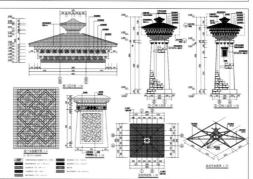
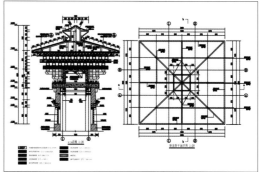
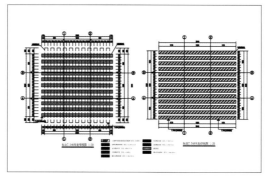
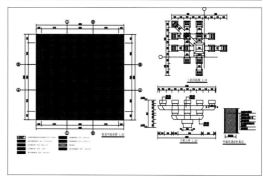

西藏技师学院
Tibet Institute of Technology

项目业主：西藏自治区人力资源和社会保障厅	Project Owner: Department of Human Resources and Social Security of Tibet Autonomous Region
项目地点：西藏 拉萨	Location: Lhasa, Tibet
建筑功能：教育建筑	Building Function: Educational Building
建筑面积：128 100 m²	Site Area: 249 084 m²
用地面积：249 084 m²	Building Area: 128 100 m²
设计时间：2018年	Design Time: 2018
项目状态：竣工	Project Status: Completed
项目负责人：王慧琳	Project Leader: Huilin Wang
设计团队：多庆巴珠、鲁海、旦增索朗、普布顿珠、郭霄雯	Design Team: Duoqing Bazhu, Hai Lu, Danzeng Sorang, Pubu Dunzhu, Xiaowen Guo

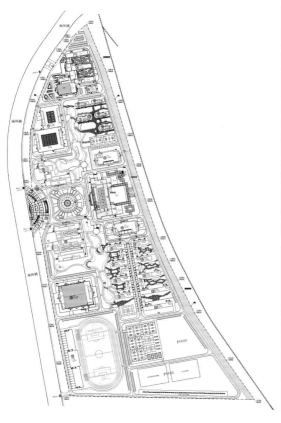

项目位于拉萨市柳梧新区（柳东大桥以北、拉日铁路以西、南环快速通道以东）南环路与拉日铁路交叉口南侧，占地面积为249 084 m²。本设计方案有别于一般大专院校的简单地将生活、教学、办公截然分区的设置，而是更多考虑技工技师教学学生年轻、实训时间长等教学特点，按照产业（即第二产业专业、第三产业专业）分区设置，将学生的日常生活、学习有效地设置在可高效管理的范围内；在产业分区的基础上，再进行教学、实训、生活等功能的细分设置。如此，既动静分离、减小相互干扰，又方便高效管理，以提升教学和实训的效率。

The project is located on the south side of the intersection of the South Ring Road and Lasa-Rikaze Railway in Liuwu New District of Lhasa City(north of Liudong Bridge, west of Lasa-Rikaze Railway, east of the South Ring Road Express Road) covering an area of about 380 mu, about 250 000 m². Different from the simple partition of life, teaching and office in general colleges and universities, the design pays more attention to the teaching characteristics of young students and long training time. The zones of students' daily life and learning are effectively set within the scope of efficient management according to the industry (namely the secondary industry, tertiary industry) partition setting; On the basis of industrial zoning, the subdivision of teaching, practical training, life and other functions is set up. In this way, it not only achieves the dynamic-static separation to reduce mutual interference, but also realizes convenient and efficient management, so as to improve the efficiency of teaching and training.

昆明新正东阳建筑工程设计有限公司
Kunming Xinzheng Dongyang Construction Engineering Design Co., Ltd.

昆明新正东阳建筑工程设计有限公司(以下简称"正东设计")成立于2007年1月,具有住房与城乡建设部正式颁发的建筑工程甲级资质证书。多年的耕耘和付出,正东设计造就了一支240人的具有较高管理水平和技术水平的专业化团队,至今已承接了400余项工程设计任务。公司设计作品曾多次获得国家级、部级、省级优秀设计奖。

公司注重创新,有一支30多的人专业建筑设计、规划设计和室内设计方案创作团队,并成立了BIM研发组等新技术项目团队。

在工程设计技术和质量上,正东设计一贯坚持精益求精,建设了完善的技术质量控制体系,并在数百项工程中贯彻落实。

近10年,基于房地产设计市场,公司由生产导向型,逐步转型为客户导向型设计公司,为各类大型地产开发企业提供全面的技术咨询服务以及综合解决方案。

正东设计秉持"专业创造价值"的核心理念,坚持技术专业、管理专业、态度专业,通过专业的服务为顾客创造价值。

Kunming Xinzheng Dongyang Construction Engineering Design Co., Ltd. (briefly referred to as "Zhengdong Design") was established in January 2007, it has A-level Design Qualification Certificate of Constructional Engineering officially issued by the construction department of the country. With many years' cultivation and endeavors, Zhengdong Design has built up a professionalized team with approximately 240 persons that have high management and technical level. Up to now, the company has undertaken design tasks of more than four hundred projects. The design works of the company have gained national level and ministerial level excellent design awards for many times. The company focuses on innovation and has a professional scheme creation team with 30 persons on building, planning and interior design, and established technical project teams like BIM R&D group etc. Zhengdong Design consistently persists keeping improvement on engineering design technology and quality, established complete technical quality control system and implemented on hundreds of projects. In recent 10 years, based on real estate design market, the company gradually transited from production oriented type to customer oriented type design company, and provides overall technical consulting service and comprehensive solution for various large real estate development enterprises. Zhengdong Design adheres to the core conception of "specialty creates value", persists technical professionalism, management professionalism and attitude professionalism, creates value for customers through professional services.

地址:云南省昆明市盘龙区北京路俊发中心10楼
电话:0871-65713781
邮箱:kmzd111@hotmail.com
网址:www.yndesign.net

Add: 10F, Junfa Center, Beijing Road, Panlong District, Kunming, Yunnan
Tel: 0871-65713781
Email: kmzd111@hotmail.com
Web: www.yndesign.net

御府中央小区(A2-1/3地块)
Yufu Central Community (Plot A2-1/3)

项目地点:云南 昆明
建筑面积:73 141.79 m²
用地面积:8 834.12 m²
容积率:1.80
设计时间:2015年4月
竣工时间:2017年9月
主持建筑师:苏晓战

Location: Kunming, Yunnan
Building Area: 73 141.79 m²
Site Area: 8 834.12 m²
Plot Ratio: 1.80
Design Time: April 2015
Completion Time: December 2019
Architect: Xiaozhan Su

御府中央小区(A2-1/3地块)位于昆明滇池度假区滇池路。项目为多层公共建筑,由昆明新正东阳建筑工程设计有限公司与赛艾建筑设计咨询(上海)有限公司联合设计。

本项目充分利用基地自然线型的形状,打造一个独一无二的城市地标。建筑总体分为4个塔楼,不同的独立塔楼相互连接,形成整个地块的一个连续"脊椎",在这条脊椎中折叠出4个独立的空间。设计运用4个元素来形成这些空间的装饰特征,脊椎沿着建筑的长度方向向前起伏延伸。脊椎部分采用铝板和玻璃为设计元素,通过玻璃和铝板的交叉设计,形成体量的虚实对比。采用浅色玻璃、七巧板、玻璃幕墙为设计元素,以体现办公建筑庄重、现代的建筑特质。项目采用BIM技术进行建模分析,实现建筑的集成化信息处理方式,高效、便捷地进行建筑设计、设备管理、成本管控。本项目获云南省2021年度优秀工程设计(建筑设计)三等奖。

Yufu Central Community (plot A2-1/3) is located in Dianchi Road, Dianchi Resort, Kunming. The project is a multi-storey public building jointly designed by Kunming Xinzheng Dongyang Architectural Engineering Design Co., Ltd. and SAA Architectural Design Consulting (Shanghai) Co., Ltd.

The project makes full use of the shape of the base natural linear, creating a unique city landmarks, construction in general can be divided into four towers, the towers of different independent are interconnected to form the whole plot of a continuous spine, fold in the spinal cord out of the four independent space, design the adornment of the four elements are used to form the space characteristics, The spine rises and falls along the length of the building. Aluminum plate and glass are used as the design elements for the spine part. Through the cross design of glass and aluminum plate, the virtual and real contrast of the volume is formed. The remaining four gems adopt light-colored glass, tangram and glass curtain wall as design elements to reflect the solemn and modern architectural characteristics of the office building. At the same time, the project adopts BIM technology for modeling and analysis. To realize the integrated information processing mode of architecture. Efficient and convenient construction design, equipment management, cost control.

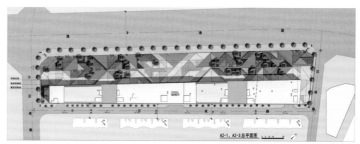

磨黑法庭及配套项目
Mohei Court and Ancillary Projects

项目地点：云南 普洱
建筑面积：775.38 m²
用地面积：911.78 m²
容积率：0.85
设计时间：2014年9月－2016年11月
竣工时间：2018年10月
主持建筑师：饶红

Location: Pu'er, Yunnan County, Pu'er, Yunnan
Building Area: 775.38 m²
Site Area: 911.78 m²
Plot Ratio: 0.85
Design Time: September 2014 to November 2016
Completion Time: October 2018
Principal Architcet: Hong Rao

本项目是一个小型乡镇法庭及配套项目。乡镇法庭是我国民主与法制建设的重要场所，量大面广，但因乡镇法庭地处偏僻，项目规模很小，所以一直未得到应有的重视。本项目的综合效益及意义在于：虽然项目小，地处偏僻，但公司并没有等闲视之，而是通过分析项目所处的具体环境、用地条件及周围建筑的语境，深入了解乡镇法庭的各种具体需求和工作方式，既摒弃了常见的、因循守旧的"衙门式"法庭建筑形式，也不拘泥于传统建筑形式，而是在逻辑分析的基础上，采用"因地制宜、化整为零、和而不同、守正出新"的设计理念和创作手法。经过精心推敲，反复比较，积极思考，运用新思维、新理念、新方法，造就了一个功能合理、使用方便、环境优美、造型独特、极富特色的"小而美"的乡镇法庭。建成后该项目成为当地标志性的景点，得到了业主及社会各界的广泛赞誉，用业主的话来说可谓"好评如潮"。项目发表于《云南建筑》《建筑技艺》杂志。本项目的意义还在于为量大面广的乡镇小型建筑设计提供了更多的可能性。本项目获云南省2021年度优秀工程设计（建筑设计）二等奖。

Mohei Court and ancillary projects is a small township court and supporting project. The township court is an important place for the building of democracy and the rule of law in our country, with a large volume of people. However, because of the remote location of the township court and the small scale of the project, it has not received the attention it deserves. The overall benefit and significance of this project lies in the fact that, despite its small size and remote location, we did not wait to analyse the specific environment in which the project is located, the conditions of the site and the context of the surrounding buildings, and to gain an in-depth understanding of the various specific needs and ways of working of the township courts. We have abandoned the common, old-fashioned "courthouse" form of courtroom architecture, and have not adhered to the "traditional architectural" form. Instead, based on a logical analysis, the design concept and creative approach is "to adapt to local conditions, to transform the whole into a piece, to harmonise but differ, and to keep the righteousness new". After careful deliberation, repeated comparisons and active thinking, new thinking, new ideas and new methods were applied. The result is a township court that is functional, easy to use, beautiful and unique in shape. After completion, it became a local landmark attraction and was widely praised by the owner and the community, and in the words of the owner, was "well received" and published in Yunnan Architecture and Architecture Technique magazines. The significance of this project also lies in the fact that it offers more possibilities for small-scale architectural design in a large number of towns and villages.

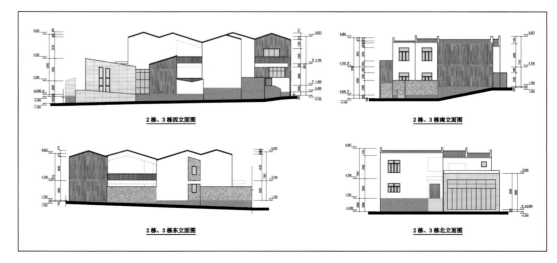

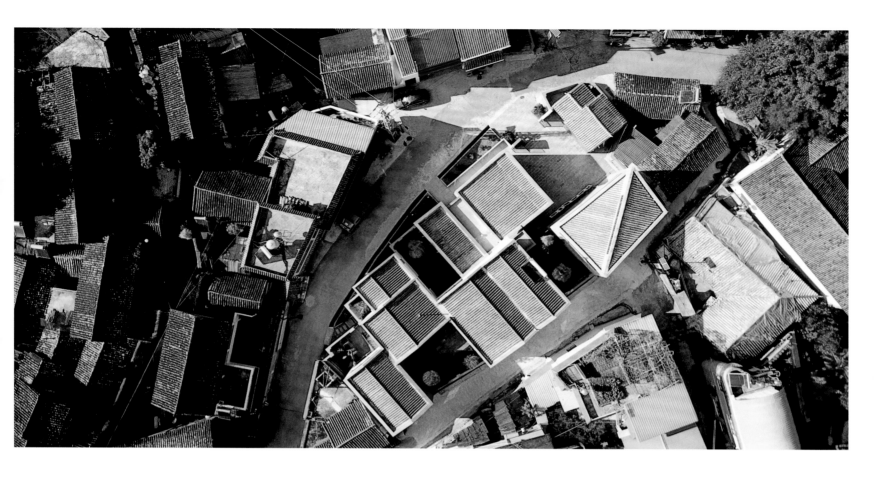

俊宏誉园
Junhong Yu Yuan

项目地点：云南 普洱	Location: Pu'er, Yunnan
建筑面积：113 947.64 m²	Building Area: 113 947.64 m²
用地面积：43 441 m²	Site Area: 43 441 m²
容积率：1.93	Plot Ratio: 1.93
设计时间：2012年11月	Design Time: November 2012
竣工时间：2017年12月	Completion Time: December 2017
主持建筑师：饶红	Principal Architect: Hong Rao

俊宏誉园位于普洱市茶城大道，区位条件好，交通方便，但地形复杂，高差大，有限高要求。

本项目是一个集住宅、公寓、办公及社区配套于一体的住区项目。因地制宜化解高差、恰到好处利用地形、力求得体体现风貌是本项目的设计特色。

本项目规划设计方案科学合理，逻辑清晰。设计师通过对现场地形地貌的反复勘查，与甲方的充分交流，对项目的特点及地形地貌进行了逻辑分析；精心设计，反复推敲，进行多方案比较；因地制宜，做到了充分合理地利用地形高差。建筑设计布局合理，风格独特，造型得体，清新雅致，得到了规划部门及与会专家的高度评价。施工图设计各专业紧密配合，力求实现经济节能，设计大样细节表达清晰。施工过程中设计师全程跟踪，经常走场解决问题，使项目具有较高的完成度。项目建成后，不仅得到了项目参与各方的一致认可，而且得到了规划部门、各验收部门、业主及社会大众的普遍赞誉，已经成为当地的标杆项目和体现"地方风貌"的范例并发表于《云南建筑》。本项目获云南省2021年度优秀工程设计（住宅与住宅小区设计）二等奖。

Junhong Yuyuan is located on Cha Cheng Avenue in Pu'er City, with good location and convenient transportation, but complex topography, large height difference and limited height requirements.

The project is a residential project that integrates houses, flats, offices and community facilities. The design of this project is characterised by the appropriate use of the terrain and the appropriate use of height differences.

The planning and design of this project is scientific and logical. Through repeated surveys of the topography of the site and full communication with the A-party, the characteristics and topography of the project were logically analysed. Careful design, repeated deliberation and multi-proposal comparison were carried out. The project has made full and reasonable use of the terrain height difference according to the local conditions. The building design is reasonable in layout, unique in style, decent in shape and fresh in elegance, and has been highly evaluated by the planning department and the attending The construction drawings were designed in close collaboration with all disciplines, striving for economy and energy conservation, and the design details were clearly expressed. During the construction process, the whole process is tracked, and the problems are often solved on site, so that the project has a high degree of completion. After the completion of the project, it has not only received the unanimous approval of all parties involved in the project, but also the general praise of the planning department, the acceptance department, the owner and the public. It has become a local benchmark project and an example of 'local style' and is listed in Yunnan Architecture.

朗韵花园（A2地块）
Langyun Garden (Plot A2)

项目地点：云南 昆明	Location: Kunming, Yunnan
建筑面积：90 835.55 m²	Building Area: 90 835.55 m²
用地面积：15 375.41 m²	Site Area: 15 375.41 m²
容积率：1.20	Plot Ratio: 1.20
设计时间：2018年3月	Design Time: March 2018
竣工时间：2019年12月	Completion Time: December 2019
主持建筑师：吕郅、郭瑞	Principal Architect: Zhi Lv, Rui Guo

　　朗韵花园（A2地块）位于云南省昆明市东白沙河片区。本项目为住宅项目，场地高差较大，自北向南50 m高差，为保证住户视野的通透性，采用错台方式处理场地高差，形成疏密有致、高低错落的建筑形态。

　　本项目为昆明新正东阳建筑工程设计有限公司与汇张思建筑设计事务所（上海）有限公司联合设计，建筑外立面风格采用中式风格，采用云南一颗印的手法修饰立面，将项目平面与立面完美融合，并采用BIM技术进行建模分析，实现建筑的集成化信息处理方式，高效、便捷地进行建筑设计、设备管理、成本管控。本项目获云南省2021年度优秀工程设计（住宅与住宅小区设计）三等奖。

　　Langyun Garden (Plot A2) is located in the East Baishahe area of Kunming, Yunnan Province. The project is a residential project. The elevation difference of the general layout site is large, with an elevation difference of 50 m from north to south. In order to ensure the permeability of the residents' field of vision, the staggered platform is used to deal with the elevation difference of the site. The shape of the general layout is dense and scattered.

　　The project is jointly designed by Kunming Xinzheng Dongyang Architectural Engineering Design Co., Ltd. and huizhangsi architectural design firm (Shanghai) Co., Ltd. the building facade style adopts Chinese style, and the facade is decorated with the method of Yunnan one seal. The project plane and facade are perfectly integrated, and BIM Technology is used for modeling and analysis. Realize the integrated information processing mode of architecture. Efficient and convenient architectural design, equipment management and cost control.

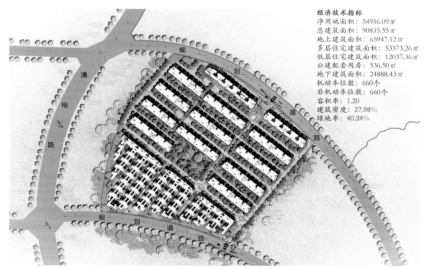

北京中鸿建筑工程设计有限公司昆明分公司
Kunming Filiale of Zhong Hong Holding Group

北京中鸿建筑工程设计有限公司昆明分公司（简称"中鸿昆明"）成立于2010年底，现配置有规划、场地、建筑、结构、设备、技术管理等专业，团队架构完善。中鸿昆明团队致力于建筑全周期设计管控，擅长山地、公共、住宅建筑及特色产业园区、综合文旅项目等设计，自成立以来累计完成的设计项目建筑面积约9 000 000 m²，其中紫云青鸟文化创意博览园、中国（云南）自由贸易区昆明片区指挥部等众多已建成项目获得业主及合作伙伴一致好评。团队通过不同项目的经历与沉淀，探索出了独有的项目全过程设计、管控心得。对于项目最终呈现的期待以及团队对自身的要求将始终推动中鸿昆明秉持初心，努力创造更多务本、尽心的设计作品。

Kunming Filiale of Zhong Hong Holding Group is established in 2010. The professional team of this filiale has a complete professional service from planning, architecture, structure, facility, to technical management especially excels in the project of mountainous building structure, public facility, industrial parks and comprehensive entertainment. The accumulated built-up area since the establishment is about 9 000 000 m², These works are received positive feedbacks. The whole company which now has a unique design language persists in the concept of prudent attitude and ardently affection towards this industry for the aim of presenting an anticipated perfection on every project.

地址：云南省昆明市彩云北路3567号大都二期五号楼1001-1005号
邮编：650217
电话：13678725633（童） 0871-63816228-800
传真：0871-63816228-801
邮箱：862177342@qq.com

Add: No.1001-1005, Building 5, Dadu Phase II, No.3567 Caiyun North Road, Kunming, Yunnan
P.C: 650217
Tel: 13678725633 (Tong) 0871-63816228-800
Fax: 0871-63816228-801
Email: 862177342@qq.com

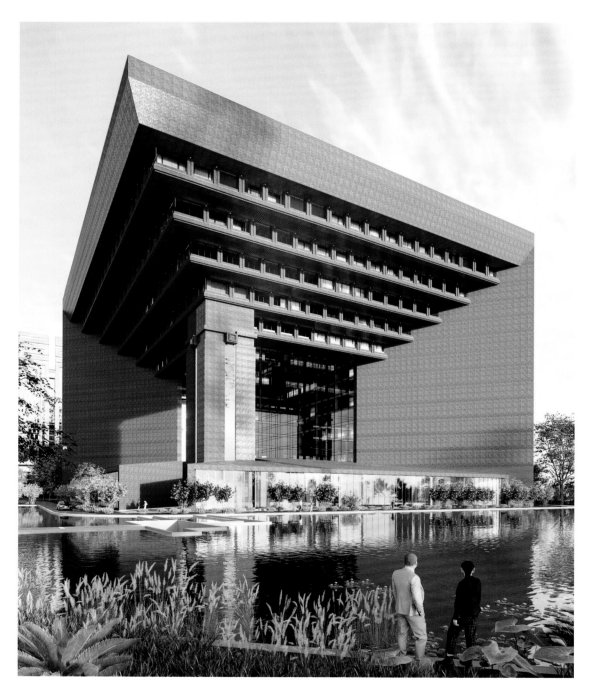

北辰兰苑
Beichen Lanyuan

设计师：张羽、张力、顾伟
项目地点：云南 保山
建筑面积：339 430 m²
用地面积：12.8 hm²
容积率：1.61

Designers: Yu Zhang, Li Zhang, Wei Gu
Location: BaoShan, YunNan
Building Area: 339 430 m²
Site Area: 12.8 hm²
Plot Ratio: 1.61

项目位于云南省保山市，主要功能为企业总部办公、酒店、部分住宅，总部办公、酒店功能区与居住功能区互相独立。建成后公共区域还将承担一部分城市功能，满足企业的经营与发展需要。项目设计结合企业"城市建设者"的独特身份及保山市特有的悠久汉文化历史，以古建筑中"斗拱"构件为原型，通过现代建筑语汇的提炼与抽象，展现中国传统建筑文化的特征。

The project is located in Baoshan City, Yunnan Province. The main functions are the headquarters office, hotel and some residences. The headquarters office, hotel function area is independent from the residential function area. After its completion, the public area will also take on part of the urban functions to meet the needs of enterprise management and development. The design of the project combines the unique identity of the enterprise as an "urban builder" and the unique long history of Han culture in Baoshan City. Taking the "bracket" component in ancient architecture as the prototype, the design displays the characteristics of traditional Chinese architectural culture through the extraction and abstraction of modern architectural vocabulary.

汉营走马古镇
Hanying Zouma Ancient Town

设计师：吕建芸、顾伟、赵杏花	Designers: Jianyun Lv, Wei Gu, Xinghua Zhao
项目地点：云南 保山	Location: BaoShan, YunNan
建筑面积：215 885 m²	Building Area: 215 885 m²
用地面积：33.7 hm²	Site Area: 33.7 hm²
容积率：0.58	Plot Ratio: 0.58

汉营走马古镇项目以"蜀汉"文化为主线，用蜀汉文化诠释建筑景观节点，用走马场文化构筑项目独特建筑符号，以"休闲度假、文化演艺、古镇体验"为主题，实现古镇观光、文化体验、休闲度假及生态人居等功能，构建一个宜居、宜旅、宜商、宜游、体现多元文化包容性的文化型精品旅游小镇。

The project takes "Shu Han" culture as the main line, interprets architectural landscape nodes with Shu Han culture, constructs unique architectural symbols with stable culture, and takes "leisure vacation, cultural performance, ancient town experience" as the theme to realize functions of ancient town sightseeing, cultural experience, leisure vacation and ecological habitat. It aims to build a cultural boutique tourism town that is suitable for living, traveling, business and tourism and reflects the inclusiveness of multiculturalism.

曼德勒·缪达经济贸易合作区
Mandalay· Myuda Economic and Trade Cooperation Zone

设计师：张羽、张力、顾伟	Designers: Yu Zhang, Li Zhang, Wei Gu
项目地点：缅甸 曼德勒	Location: Mandalay, Myanmar
建筑面积：141 000 m²	Building Area: 141 000 m²
用地面积：36.79 hm²	Site Area: 36.79 hm²
容积率：0.38	Plot Ratio: 0.38

项目位于缅甸曼德勒，是"一带一路"项目建设的一部分，承载着制造、加工、物流、金融服务等重要功能。作为"万塔之邦"的曼德勒传承着深厚的佛教文化，项目将佛塔的剪影抽象成核心区建筑的外立面元素，作为建筑的外遮阳系统，用一种低成本的绿色方式缓解当地强烈日照带来的炎热问题，同时也与佛教文化相得益彰。

Located in Mandalay, Myanmar, the project is a part of the Belt and Road Initiative, carrying important functions of Myanmar's manufacturing, processing, logistics, and financial services. Mandalay, a city with tens of thousands of pagodas, inherits profound Buddhist culture. The project abstracts the silhouette of pagodas into the facade elements of the buildings in the core area as the exterior shading system of the buildings, which alleviates the hot problem caused by the intense sunshine in a low-cost and green way, and at the same time, it is in harmony with the Buddhist culture.

近水书院
Jinshui Academy

设计师：张羽、顾伟、赵杏花
项目地点：云南 文山
建筑面积：2 240 m²
用地面积：0.49 hm²
容积率：0.45

Designers: Yu Zhang, Wei Gu, Xinghua Zhao
Location: WenShan, YunNan
Building Area: 2 240 m²
Site Area: 0.49 hm²
Plot Ration: 0.45

项目位于云南省文山州普者黑景区旁。近水书院是云上水乡·近水镇整体用地北部一块四面环水的孤岛，独特的地理位置赋予了书院独特的周边环境。业主是酷爱书法的一位雅士，希望借此书院修性养心，交友论道。设计借"书中藏万宝"的理念，打造外表质朴而内在丰富的内向型建筑空间，借"众妙之门"的寓意，将建筑形式固化为入口空间的石墙与飞虹。

The project is located near Puzhehei Scenic Spot in Wenshan Prefecture, Yunnan Province. Jinshui Academy is an isolated island surrounded by water in the north of Yunshang Water Town · Jinshui Town. Its unique geographical location gives the academy a unique surrounding environment. The owner is fond of calligraphy, hoping to cultivate his original nature and make friends. With the concept of "ten thousand treasures hidden in books", the design creates an introverted architectural space that is simple in appearance but rich in interior. With the meaning of "the door of many wonderful things", the building adopts the form of the stone wall and bridge in the entrance space.

浙江

浙江大学建筑设计研究院有限公司
The Architectural Design & Research Institute Of Zhejiang University Co., Ltd.

浙江大学建筑设计研究院有限公司始建于1953年，前身为浙江大学建筑设计室，国家重点高校中最早成立的甲级设计研究院之一。

以"营造和谐、放眼国际、产学研创、高精专强"为办院方针，以"高品位的文化、高宽远的视野、高效能的管理、高素质的人才、高精专的技术、高质量的作品"为发展战略，以"平衡建筑"学术理论指导设计实践，多年来始终坚持走创作路线和精品路线，在各个领域均有大量的优秀作品问世。历年来共获得近1 500项国家、部、省级优秀设计奖、国际设计奖、优质工程奖及科技成果奖等。

始终秉承"求是、创新"的企业精神和"精益求精、崇尚完美"的质量方针，竭诚为社会各界提供优质的设计服务，努力实现"设计创造共同价值"的核心价值观。

Founded in 1953, The Architectural Design & Research Institute of Zhejiang University Co., Ltd. grew out of Architectural Design Studio of Zhejiang University. It is one of the earliest Class-A design institutes established among national key universities.

Guided by the principle of "Harmonious Environment, Global Vision, Design integrating Teaching & Research & Creation, Aiming High & Professional & Powerful", implementing the development strategy of "High-grade Culture, Long & Wide Vision, High-efficiency Management, High-caliber Personnel, Highly professional Technique and High-quality Works", and applying the academic theory of "Balance Architecture" as design guidelines, UAD has been persevering in the creation of innovative, high-end projects. A large number of outstanding projects in various disciplines have been realized. UAD has been credited by over 1 500 awards of various types over the years, including excellent design awards at national, ministerial and provincial level, apart from international design awards, Premium-quality project awards, Science & technology achievement awards and so on.

UAD has always been adhering to the enterprise spirit of "Truth seeking & Innovation" and following "Striving for Excellence & Advocating Perfection" as quality guideline. UAD will continue to provide premium design and services for all sectors of the community, and stick to the idea of "Sharing Value through Design".

地址：中国浙江杭州天目山路148号浙江大学西溪校区东一楼	Add: East 1st Floor, Xixi Campus, Zhejiang University, No.148, Tianmushan Road, Hangzhou, Zhejiang, China
邮编：310028	P.C: 310028
电话：0571-85891018	Tel: 0571-85891018
传真：0571-85891080	Fax: 0571-85891080
邮箱：uad1953@zuadr.com	Email: uad1953@zuadr.com
网址：www.uad.com.cn	Web: www.uad.com.cn

大禹纪念馆
Da Yu Memorial Hall

设计师：董丹申	Designer: Danshen Dong
项目地点：浙江 绍兴	Location: Shaoxing, Zhejiang
建筑面积：27 913 m²	Building Area: 27 913 m²
用地面积：14.2 hm²	Site Area: 14.2 hm²
容积率：0.242	Plot Ratio: 0.242

大禹纪念馆试图从更加广阔的场所中，通过外物的参照寻求建筑内在的线索，轴线的意义，既串联起了历史与文脉，也限定了建筑的朝向与布局。设计在场地原有的基础上劈山理水，并通过地景的设计，重塑建筑与自然的关系，寻求建筑体量与自然风貌的和谐平衡。中心冥想厅保留缅怀和共鸣的感受；共生院落场所体现王权和民权的融合。

The Da Yu Memorial Hall reorganize the mountain and water layout on the original site and reinvent the relationship between architecture and nature with proper landscape design so as to hit a harmonious balance of architectural volume and natural features.The Central Meditation Hall retains the remembering and resonating sensation; The Coexisting Courtyard integrates the royal power and civil rights.

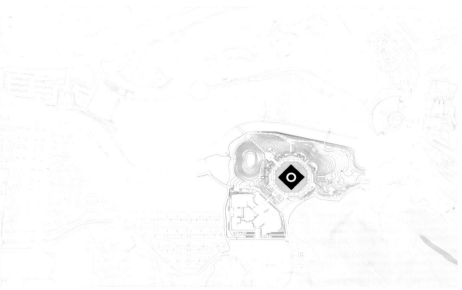

桑洲清溪文史馆
Sangzhou Qingxi Culture & History Museum

设计师：吴震陵	Designer: Zhenling Wu
项目地点：浙江 宁海	Location: Ninghai, Zhejiang
建筑面积：1 691 m²	Building Area: 1 691 m²
用地面积：0.29 hm²	Site Area: 0.29 hm²
容积率：0.54	Plot Ratio: 0.54

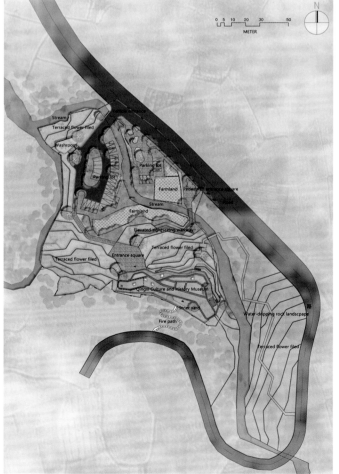

浙江大学紫金港校区文科类组团一（人文社科大楼）
Humanities & Social Sciences Building Cluster 1 in Zijingang Campus of Zhejiang University

设计师：陆激	Designer: Ji Lu
项目地点：浙江 杭州	Location: Hangzhou, Zhejiang
建筑面积：114 694 m²	Building Area: 114 694 m²
用地面积：4.76 hm²	Site Area: 4.76 hm²
容积率：2.0	Plot Ratio: 2.0

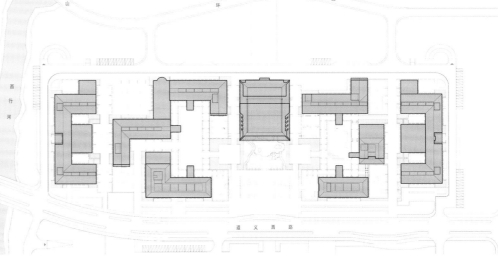

普陀山正门客运中心
Mount Putuo Main Gate Passenger Transport Center

设计师：沈晓鸣	Designer: Xiaoming Shen
项目地点：浙江 舟山	Location: Zhoushan, Zhejiang
建筑面积：11 778 m²	Building Area: 11 778 m²
用地面积：4.8344 hm²	Site Area: 4.8344 hm²
容积率：0.24	Plot Ratio: 0.24

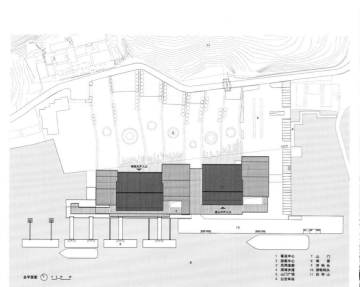

杭州职业技术学院学生食堂
Canteen of Hangzhou Vocational & Technical College

设计师：钱锡栋
项目地点：浙江 杭州
建筑面积：86 966 m²
用地面积：6.01 hm²
容积率：0.75

Designer: Xidong Qian
Location: Hangzhou, Zhejiang
Building Area: 86 966 m²
Site Area: 6.01 hm²
Plot Ratio: 0.75

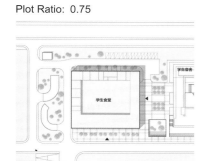

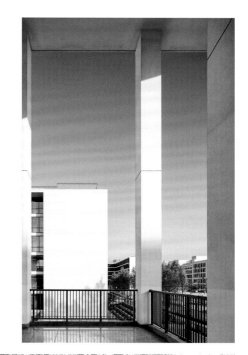

杭州英蓝中心
Hangzhou Yinglan Center

设计师：王昕洁
项目地点：浙江 杭州
建筑面积：224 199 m²
用地面积：4.381 hm²
容积率：2.7

Designer: Xinjie Wang
Location: Hangzhou, Zhejiang
Building Area: 224 199 m²
Site Area: 4.831 hm²
Plot Ratio: 2.7

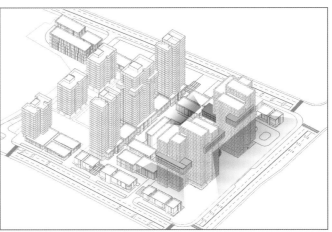

钱投邻居中心（望江店）
Qiantou Neighborhood Centre of Wangjiang

设计师：孙啸野
项目地点：浙江 杭州
建筑面积：17 690 m²
用地面积：0.55 hm²
容积率：1.73

Designer: Xiaoye Sun
Location: Hangzhou, Zhejiang
Building Area: 17 690 m²
Site Area: 0.55 hm²
Plot Ratio: 1.73

浙江大学紫金港校区管理学院大楼
School of Management Building, Zijingang Campus, Zhejiang University

设计师：邱文晓	Designer: Wenxiao Qiu
项目地点：浙江 杭州	Location: Hangzhou, Zhejiang
建筑面积：56 596 m²	Building Area: 56 596 m²
用地面积：2.32 hm²	Site Area: 2.32 hm²
容积率：2.27	Plot Ratio: 2.27

杭州电子科技大学第一实验楼
The Primary Laboratory Building of Hangzhou Dianzi University

设计师：吴雅萍	Designer: Yaping Wu
项目地点：浙江 杭州	Location: Hangzhou, Zhejiang
建筑面积：34 291 m²	Building Area: 34 291 m²
用地面积：1.57 hm²	Site Area: 1.57 hm²
容积率：1.8	Plot Ratio: 1.8

嘉兴学院梁林校区扩建工程二期
Jiaxing University Lianglin Campus Expansion Project Phase II

设计师：滕美芳	Designer: Meifang Teng
项目地点：浙江 嘉兴	Location: Jiaxing, Zhejiang
建筑面积：230 836 m²	Building Area: 230 836 m²
用地面积：39.78 hm²	Site Area: 39.78 hm²
容积率：0.56	Plot Ratio: 0.56

超重力离心模拟与实验装置
国家重大科技基础设施
Centrifugal Hypergravity and Interdisciplinary Experiment Facility (CHIEF) Base

项目地点：浙江 杭州	Location: Hangzhou, Zhejiang
建筑面积：34 560 m²	Building Area: 34 560 m²
用地面积：5.9 hm²	Site Area: 5.9 hm²
容积率：0.39	Plot Ratio: 0.39

中国电子科技集团科技园谷雨村
China Electronics Technology Group Science and Technology Park Guyu Village

项目地点：河北 涞水	Location: Laishui, Hebei
建筑面积：139 265 m²	Building Area: 139 265 m²
用地面积：14.6 hm²	Site Area: 14.6 hm²
容积率：0.85	Plot Ratio: 0.85

义乌全球数字自贸中心
Yiwu Global Digital Free Trade Center

项目地点：浙江 义乌	Location: Yiwu, Zhejiang
建筑面积：1 300 000 m²	Building Area: 1 300 000 m²
用地面积：39.13 hm²	Site Area: 39.13 hm²
容积率：2.5	Plot Ratio: 2.5

浙医二院柯桥未来医学中心
Keqiao Future Medical Center, the Second Affiliated Hospital Zhejiang University School of Medicine

项目地点：浙江 绍兴	Location: Shaoxing, Zhejiang
建筑面积：500 000 m²	Building Area: 500 000 m²
用地面积：23.87 hm²	Site Area: 23.87 hm²
容积率：1.6	Plot Ratio: 1.6

杭州市城建设计研究院有限公司
Hangzhou Architectural & Civil Engineering Design and Research Institute Co., Ltd.

杭州市城建设计研究院有限公司建制于1952年，2003年改制组建成为有限责任公司，是一家具有70年光辉历史的勘察设计甲级设计院。拥有建筑行业（建筑工程）甲级、市政行业甲级、风景园林工程设计专项甲级、工程咨询资信甲级、公路行业（公路）专业乙级、城乡规划乙级、工程勘察专业类（岩土工程）设计乙级及施工图审查、建筑节能评估、消防验收评估等资质。

公司配套完善，技术力量雄厚，现有员工近700人，其中教授级高级工程师30余人，各类注册师160余人；荣获詹天佑奖、新中国成立60周年百项经典暨精品工程奖、国家勘察设计行业优秀设计奖、市级科技进步奖等多项奖项，以及省、市级优秀勘察设计奖300多项。

公司系浙江省勘察设计行业协会副会长兼民营设计企业分会会长单位、中国勘察设计协会民营企业分会副会长单位，公司连年获得国家高新技术企业、全国勘察设计行业创优型企业、全国勘察设计行业优秀民营设计企业、首批全国建筑设计行业诚信单位等多项荣誉称号。

Hangzhou Architectural and Civil Engineering Design and Research Institute Co., Ltd. was established in 1952 and restructured in 2003 to become a limited liability company. It is a Class-A design institute with a glorious history of 70 years. The qualifications we owned are: Class-A in construction industry (construction engineering), Class-A in municipal industry, Class-A in landscape engineering design, Class-A in engineering consulting credit, Class-B in highway industry (highway), Class-B in urban and rural planning, Class-B in engineering investigation professional design (geotechnical engineering), and including construction drawing review, building energy conservation assessment, fire acceptance assessment and other qualifications.

The company has perfect supporting and strong technical force. Nowadays, our company has nearly 700 employees, more than half of them have the technical title of engineer or above, including over 30 professor-level senior engineers and 160 registered professionals. We has won many awards, such as Zhan Tianyou Award, one hundred Classic and Boutique Engineering Works Award for the 60th anniversary of the founding of New China, National Excellent Design Award for Exploration and Design industry, municipal Science and Technology Progress Award, and nearly 300 provincial and municipal excellent exploration and design awards.

The company is the vice president unit of Zhejiang Province Exploration and Design Industry Association and president unit of private enterprise branch of design, vice president unit of China Exploration and Design Association Private Enterprise Branch. The company has been awarded many honorary titles such as National high-tech enterprise, National Exploration and Design Industry Excellent Enterprise and Excellent Private Design Enterprise, first batch of National architectural design industry integrity unit and so on.

电话：0571-87924760
传真：0571-87917516
邮箱：hcj1952.com
网址：www.hcj1952.com

Tel: 0571-87924760
Fax: 0571-87917516
Email: hcj1952.com
Web: www.hcj1952.com

余杭区文化艺术中心一期工程
Phase I Project of Yuhang District Culture and Art Center

合作单位：HENNING LARSEN ARCHTECTS A/S香港分公司
项目地点：浙江 杭州
建筑面积：82 547 m²
用地面积：52 954 m²
容积率：0.592
设计时间：2014年1月—2016年3月
竣工时间：2019年5月
项目功能：集剧院、展览等功能为一体的复合型建筑
主持建筑师：杨书林
设计团队：王小红、姚帅、克劳德·戈德弗罗伊、金天德、卢江、姜锡敏、周建、金樟其

Cooperation Unit: HENNING LARSEN ARCHTECTS A/S Hong Kong Branch
Location: Hangzhou, Zhejiang
Building Area: 82 547 m²
Site Area: 52 954 m²
Plot Ratio: 0.592
Design Time: January 2014 to March 2016
Completion Time: May 2019
Function: Multi-functional building (Theatre, Exhibition) or (Theatre, Exhibition)
Principal Architect: Shulin Yang
Design Team: Xiaohong Wang, Shuai Yao, Claude Godefroy, Tiande Jin, Jiang Lu, Ximin Jiang, Jian Zhou, Zhangqi Jin

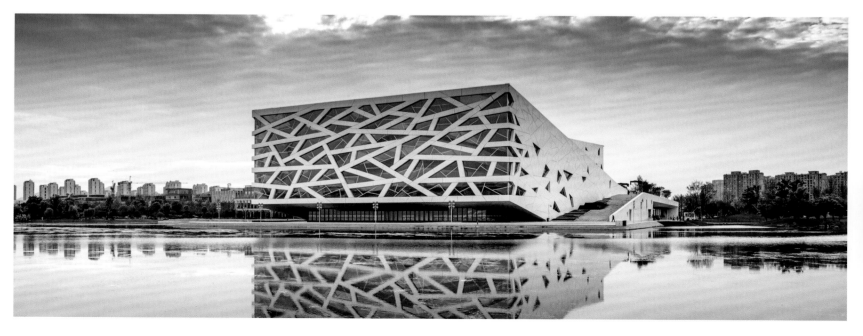

总平面图

  总平面图布局兼顾各方视觉效果,从北向南依次抬升布置展览馆、架空城市广场、小剧院及大剧院,并使建筑伸展至水面,各功能和谐统一。架空城市广场连接剧院和展览馆,成为市民登高望远的好场所。

  造型设计灵感源自冰裂纹图案,提取自剪纸和余杭滚灯元素,不规则外墙采用白色UHPC板材幕墙系统,小剧院采用可开启镜框式舞台设计,堤岸高筑的造型是对良渚文明的致敬、传承和发展。

  The layout of the general plan takes into account the visual effects of all parties. From north to south, the exhibition hall, the overhead city square, the small theater and the grand theater are raised and arranged in sequence, and the buildings are stretched to the water surface, and all functions are harmonious and unified.The overhead city plaza connects the theater and the exhibition hall, making it a good place for citizens to climb up and look into the distance.

  The shape design is inspired by the ice crack pattern, which is extracted from paper-cut and Yuhang rolling lanterns. The irregular exterior wall adopts a white UHPC sheet curtain wall system. The small theater adopts an openable mirror frame stage design. The shape of the high embankment is a tribute to the inheritance and development of Liangzhu civilization.

浙江和泽医药研发生产项目
Heze Pharmaceutical R & D and Production Project, Zhejiang

设计团队：王海平、边琦、张杭、李旻聪、徐若晨、黄杰、罗玄、周永忠、周敏峰、
　　　　　徐晓雪、张益华、杨燕
项目地点：浙江 杭州
建筑面积：100 150 m²
用地面积：25 710 m²
容积率：3.0

Design Team: Haiping Wang, Qi Bian, Hang Zhang, Mincong Li, Ruochen Xu, Jie Huang, Xuan Luo, Yongzhong Zhou, Minfeng Zhou, Xiaoxue Xu, Yihua Zhang, Yan Yang
Location: Hangzhou, Zhejiang
Building Area: 100 150 m²
Site Area: 25 710 m²
Plot Ratio: 3.0

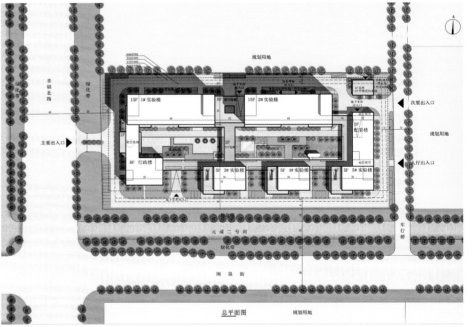

设计在展现自己独特性的同时与周边环境紧密结合，从城市设计的角度出发，无论是建筑的布局、设计手法还是建筑立面都能与周边的项目相辅相成，共同将医药港小镇打造为一个有机的"创新服务综合体"。

空间结构规划上打破传统单一的厂房功能布局模式，遵循项目地域特征和企业独特性的原则，打造一个现代新型的原创医药生产示范基地，推进中国医药产品研发创新体系的建设。

While showing its uniqueness, the design is closely combined with the surrounding environment. From the perspective of urban design, both the architectural layout, design techniques and architectural facade can complement the surrounding projects, and jointly build the medical port town into an organic "innovative service complex".

In terms of spatial structure planning, we should break the traditional single plant function layout mode, follow the principles of regional characteristics of the project and enterprise uniqueness, create a modern new original pharmaceutical production demonstration base, and promote the construction of China's pharmaceutical product R & D innovation system.

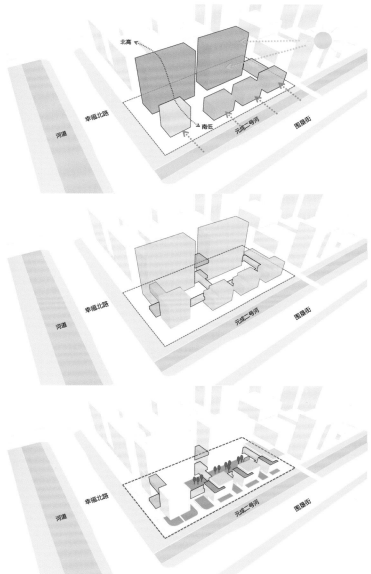

499

杭政工出【2018】17号地块创新型产业用房（LOFT49创意城市先行区）
Innovative Industrial Housing in No. 17 Plot [2018] of Hang Zheng Gong Chu (LOFT49 Creative City Pilot Zone)

项目地点：浙江 杭州
建筑面积：62 247.50 m²
用地面积：15 925.00 m²
容积率：2.3
绿地率：20%
设计团队：任丹东、金天德、彭展华、李万荣、张志强、叶再利、张益华、徐晓雪、聂剑飞

Location: Hangzhou, Zhejiang
Building Area: 62 247.50 m²
Site Area: 15 925.00 m²
Plot Ratio: 2.3
Greening Rate: 20%
Design Team: Dandong Ren, Tiande Jin, Pang Bob, Wanrong Li, Zhiqiang Zhang, Zaili Ye, Yihua Zhang, Xiaoxue Xu, Jianfei Nie

"LOFT 49创意城市先行区"位于京杭大运河旁,是杭州城市更新总体规划下的新型创意办公建筑。

设计优先探索开放空间的设计,功能空间次之。6 000 m²的"三角中央公园"位于毗邻住宅区的北侧边缘,为休闲花园、活动空间和聚会场所。由此产生的剩余场地将建筑塑造成一个"L-Block"。空中花园、退让露台、观景平台和公共走廊等负空间通过体量减法截断延长的建筑体量,沿着200 m朝南的步行街界面上形成标志性的城市门窗,并与邻近的工业遗产产生互动连接。

造型上紧贴运河工业设计文化,简朴明快,色彩庄重,重视质量和功能。建筑下部以竖向玻璃幕墙为主,上部以陶板结合深色系统玻璃窗形成统一有序的立面机理。横与竖、虚与实的对比穿插中营造活泼、谐和气氛,造型简朴统一中体现细腻的变化,展现工业设计力量感的同时融入创新型产业对美学的精致诉求。

"Loft 49 creative city pilot area" is located next to the Beijing Hangzhou Grand Canal. It is a new creative office building under the master plan of Hangzhou urban renewal.

Design gives priority to exploring the design of open space, followed by functional space. The 6 000 m² "triangle Central Park" is located on the north edge of the adjacent residential area, which is a leisure garden, activity space and gathering place. The resulting remaining site shapes the building into an "L-Block". Negative spaces such as sky gardens, retreating terraces, viewing platforms and public corridors cut off the extended building volume through volume subtraction, forming iconic urban doors and windows along the 200 m south facing pedestrian street interface, and interacting with adjacent industrial heritage.

The shape is close to the industrial design culture of the canal, simple and lively, solemn in color, and attaches importance to quality and function. The lower part of the building is mainly vertical glass curtain wall, and the upper part is ceramic plate combined with dark system glass window to form a unified and orderly facade mechanism. Horizontal and vertical, virtual and real contrast interspersed to create a lively and harmonious atmosphere, simple and unified modeling reflects delicate changes, shows the power of industrial design, and integrates the delicate aesthetic appeal of innovative industries.

郑州海洋生物博物馆
Zhengzhou Museum of Marine Life

设计师：杨书林、陈岳峰、袁子翀、杨燕、姜锡敏、金樟其
项目地点：河南 郑州
建筑面积：39 989.98 m²
建筑高度：39.95 m
容积率：1.999

Designers: Shulin Yang, Yuefeng Chen, Zichong Yuan, Yan Yang, Ximin Jiang, Zhangqi Jin
Location: Zhengzhou, Henan
Building Area: 39 989.98 m²
Building Height: 39.95 m
Plot Ratio: 1.999

本工程为改扩建项目，原部分水族馆拆除，在基地东侧新建极地馆，西侧新建小丑鱼世界和鲸豚表演馆。项目总用地面积为16 349.40 m²，总建筑面积为39 989.98 m²（含原水族馆保留部分7 092.65 m²），是全亚洲竖向垂直高度最高的单体海洋馆。设计提取原有建筑的弧形母题，平立面以折线或弧线为延续，一则为建筑立意，隐喻建筑仿如风帆；二则柔化建筑边缘，减少建筑对边界的压迫感。充分利用格式塔心理学及视觉美学的基本原理，将建筑形体进行分离，新老参观流线有机交融，达到整个项目新老建筑的协调融合。

This project is a reconstruction and expansion project. The original part of the aquarium will be demolished, and a new polar pavilion will be built on the east side of the base, and the west side of the newly built is the clownfish world and whale and dolphin performance pavilion. The total land area of the project is 16 349.40 m², and the total construction area is 39 989.98 m² (including 7 092.65 m² of the original aquarium), which is the single aquarium with the highest vertical height in Asia. The design extracts the "arc" shape motif of the original building, and the plane and facade are continued by polylines or arcs. First is the intention of architecture, which metaphors that architecture is like sails. The second is to soften the edge of the building and reduce the pressure of the building on the boundary. Make full use of basic principles of Gestalt psychology and visual aesthetics to separate the building form and organically blend the new and old visiting streamline, so as to achieve the coordination and integration of the new and old buildings of the whole project.

瓶窑鼎胜轻智造产业园
Pingyao Dingsheng Light Intelligent Manufacturing Industrial Park

项目地点：浙江 杭州
建筑面积：1 270 000 m²
用地面积：255 809 m²
容积率：3.5
绿化率：20%
主创建筑师：吴正群、李成、金利益
设计团队：周永忠、张鑫、王坚烽、蒋美丽、陈道南

Location: Hangzhou, Zhejiang
Building Area: 1 270 000 m²
Site Area: 255 809 m²
Plot Ratio: 3.5
Greening Rate: 20%
Chief Architects: Zhengqun Wu, Cheng Li, Liyi Jin
Design Team: Yongzhong Zhou, Xin Zhang, Jianfeng Wang, Meili Jiang, Daonan Chen

项目地块位于杭州市余杭区瓶窑新城核心位置，总用地面积为255 809 m²（约383.7亩）。周边建设有中法航空大学、盒马鲜生杭州区域供应链运营中心，北区预计建设地铁12号线站点，可预期未来产业氛围浓厚。

项目定位：打造为集工业生产、货运仓储、工业办公、生活配套等功能于一体的标志性高档现代综合工业园区，建立轻智造物联网及航空服务产业集聚地。

The project site is located in the core of Pingyao new town, Yuhang District, Hangzhou, with a total land area of 255 809 m² (about 383.7 mu). The project is surrounded by the China France Aviation University and Hangzhou regional supply chain operation center of freshhema. The North District is expected to build a subway line 12 station, which can be expected to have a strong industrial atmosphere in the future.

Project positioning: to build a landmark high-end modern comprehensive industrial park integrating industrial production, freight storage, industrial office, living facilities and other functions, and to establish a light intelligent Internet of things and aviation service industry cluster.

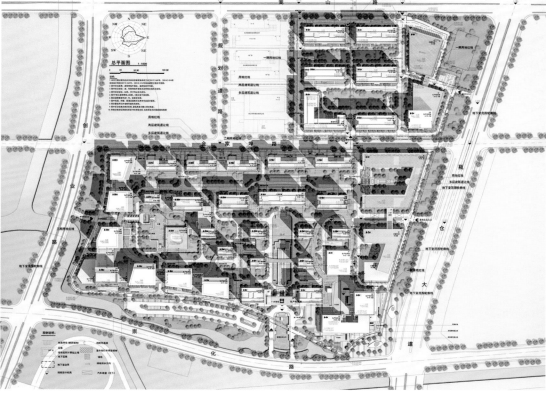

设计理念:不仅提供一个产业的容器,而是融合多种功能于一体,塑造一个综合的城市社区。功能的丰富、尺度的合宜、密度的适中是着眼点。整片255 809 m²的土地分3期启动建设,具有极为丰富的功能业态。本次介绍为1、2期,在生产车间、员工宿舍的基础上,重新规划了总体布局,并在东南转角瓶窑入口位置设计多个中心轴线,打造了标志建筑——综合中心大楼。

大楼主体功能及建筑定位:对外主要有企业文化展示、来宾接待等功能,各个楼层拥有不同的展示空间,大楼前的大型广场也为城市提供休闲娱乐和活动交流的场所,以此树立高档鲜明积极的园区形象;对内主要为员工提供会议中心及生活交流平台,以此来打造融多功能于一体的标志性高档现代综合工业园区。

Design concept: it is not limited to providing an industrial container, but integrating multiple functions to create a comprehensive urban community. Rich functions, appropriate scales and moderate density are the focus. The whole 255 809 m² of land will be constructed in three phases, with extremely rich functional formats. This introduction is for phase I and phase II. On the basis of the existing production workshops and staff dormitories, the overall layout has been replanned, and several central axes have been designed at the southeast corner and the entrance of the bottle kiln, creating a landmark building - the comprehensive center building.

The main functions and building positioning of the building: for external personnel, there are mainly corporate culture display, guest reception and other functions. Each floor has different display spaces. The large square in front of the building also brings places for leisure, entertainment and activity exchanges to the city, so as to establish a high-end bright and positive image of the park; Internally, it mainly provides conference center and life exchange platform for employees, so as to create a landmark high-end modern comprehensive industrial park with multiple functions.

本案通过内退的手法设计了大尺度的城市广场,并通过大跨度架空空间的设立,延伸广场至内部建筑空间,围绕建筑设计多个下沉广场,通过点、面、线相结合的设计手法,打破了大空间的空旷感和视觉的单一感,运用垂直高差的手法分隔空间,以取得空间和视觉效果的变化。采用下沉式广场贯穿整个中心大楼,一直延伸到地块中部,演变成中庭,同时也为中心区域增加立体景观,并获得空间视觉盛宴。

In this case, a large-scale city square is designed by retreating, and the square is extended to the internal architectural space through the establishment of a large-span overhead space. Multiple sunken squares are designed around the architecture. Through the design techniques combining points, surfaces and lines, the sense of emptiness and visual oneness of the huge space are broken, and the vertical height difference is used to separate the space, In order to achieve the change of space and visual effect, the sunken square is used to run through the whole central building, extending to the middle of the plot and evolving into a atrium. At the same time, it also increases the three-dimensional view for the central area and obtains the visual feast of space.

浙江

城 市 建 设 专 家

华汇集团是以城市建设事业为发展领域，以工程设计咨询为核心，从事工程建设全过程服务和投资的平台企业。现为中国勘察设计协会副理事长单位、民营企业分会会长单位，国家高新技术企业，蝉联"中国工程设计企业60强"、"中国十大民营工程设计企业"。

华汇集团通过数字化赋能和平台生态圈打造，成功服务于全国31个省区上百个城市，历年来各类作品获国际、部、省、市级奖项500多项，拥有授权专利超百项，在建筑、市政道桥、绿色节能技术、数字化技术、全过程咨询、工程总承包、城市更新等领域具有行业专业领先地位。

华汇集团以"共同的企业 共同的事业"为核心价值观，率先出台《共同富裕行动纲领（2021—2025）》；华汇积极承担企业公民责任，依托"浙江华汇建设美好生活基金会"和"夏禹智库"，打造公益乡村振兴技术专家支持平台，助力绍兴打造"新时代共同富裕地"。

Huahui Group is a platform enterprise with urban construction as the development field and engineering design consulting as the core, is a platform enterprise engaged in the whole process of engineering construction services and investment. As the vice chairman of China Engineering& Consulting Association, the President Unit of private enterprises branch, Huahui is a national high-tech enterprise. Huahui has been successively ranked as "Top 60 Engineering Design Enterprises in China" and "Top Ten Private Engineering Design Enterprises in China".

Through the creation of digital empowerment and platform ecosystem, Huahui has successfully served hundreds of cities in 31 provinces and regions across the country. Over the years, various works have won more than 500 international, ministerial, provincial and municipal awards and more than 100 authorized patents. In the fields of Planning, Architecture, Municipal Road& bridge, Intelligent transportation, Ecological environment, green energy-saving technology, Digital technology, Whole Process Engineering Consulting, Project General Contracting, Urban operation, Urban renewal, and other fields, Huahui has a leading position.

With "common enterprise and common cause" as its core values, Huahui took the lead in issuing the Action Program for Common Prosperity (2021-2025). Huahui actively assumes the responsibility of corporate citizenship. Relying on "Zhejiang Huahui Building a Better Life Foundation" and "Xia Yu Think Tank", Huahui builds a technical expert support platform for public welfare rural revitalization and helps Shaoxing build a "common prosperity in the new era".

地址：浙江省绍兴市解放大道177号　　Add: No.177 Jiefang Road, Shaoxing, Zhejiang, China
邮编：312000　　　　　　　　　　　　P.C: 312000
电话：0575-88208005　　　　　　　　Tel: 0575-88208005
传真：0575-88208005　　　　　　　　Fax: 0575-88208005
邮箱：mkt@cnhh.com　　　　　　　　Email: mkt@cnhh.com
网址：www.cnhh.com　　　　　　　　Web: www.cnhh.com

绍兴羊山攀岩中心
Yangshan Rock Climbing Center

设计师：黄会明、王清、金烽　　Designers: Huiming Huang, Qing Wang, Feng Jin
项目地点：浙江 绍兴　　　　　　Location: Shaoxing, Zhejiang
建筑面积：9 000 m²　　　　　　　Building Area: 9 000 m²
用地面积：2 hm²　　　　　　　　Site Area: 2 hm²
容积率：0.45　　　　　　　　　　Plot Ratio: 0.45

绍兴柯桥羊山攀岩中心获2022年杭州亚运会组委会"建筑特色奖",晶莹剔透的"蚕茧"造型诠释了亚运地域文化和攀岩运动特点。攀岩中心与羊山公园和谐共生,升旗位置与公园主峰巧妙融合,"天人合一"的无界场馆使赛后利用有无限可能。建筑总体呈现了融入环境的自然之美、刚柔并济的运动之美、包容开放的人文之美和低碳智慧的绿色之美。

Yangshan Rock Climbing Center has won the prize of Architectural Excellence Award granted by Organization Committee of Hangzhou Asian Games 2022, the glittering cocoon-like shape exhibits both specific culture of the Asian Games and the characteristics of rock climbing sport. The climbing center and Yangshan Park harmoniously coexist while the flag-raising position is vividly integrated with the main peak of the Park, which makes the post-race utilization possible. The center generally presents the beauty which is immersed into the environment naturally, the beauty of the sports with strength and tenderness, and the beauty of diversified humanity as well as the beauty of low-carbon lifestyle.

"鲁镇·社戏"演艺中心
Luzhen Village Opera Performance Center

设计师：黄会明、王清、金涛
项目地点：浙江 绍兴
建筑面积：9 000 m²
用地面积：2 hm²
容积率：0.45

Designers: Huiming Huang, Qing Wang, Tao Jin
Location: Shaoxing, Zhejiang
Building Area: 9 000 m²
Site Area: 2 hm²
Plot Ratio: 0.45

"鲁镇·社戏"演艺中心是大型影画剧定制剧场，位于国家AAAA级旅游景区绍兴柯岩风景区。通过视觉延展、形体差异、交错转折、高差台地、化整为零处理，项目设计大大削弱了剧场建筑的体量感，从而缓和了建筑与鲁镇原有小尺度环境的冲突，并营造和呈现了"现代山水意向"，使建筑嵌入鉴湖，融于古镇。主入口传统戏台是建筑与剧情矛盾冲突的营造。

Luzhen Village Opera Performance Center is a giant film and photographic customized theater, the project locates in National AAAA Scenic Spot Shaoxing Keyan Scenic Spot. Through visual expansion, figure difference, exchange and turning, tableland with height difference and rounded to zero treatment, the volume sense of the theater building has been reduced, which has eased the conflict between the giant building and the small size of Luzhen and created and presented the "Modern Mountain and Water Image" as well as insert the building into the scenic view of Jian Lake.

南溪书舍
Nanxi Book Collection

设计师：黄会明、王清、梁伟
项目地点：浙江 绍兴
建筑面积：2 000 m²
用地面积：0.5 hm²
容积率：0.39

Designers: Huiming Huang, Qing Wang, Wei Liang
Location: Shaoxing, Zhejiang
Building Area: 2 000 m²
Site Area: 0.5 hm²
Plot Ratio: 0.39

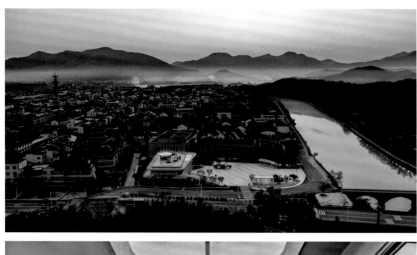

南溪书舍试图从无序、繁杂的广阔场所中,通过建筑、广场、流线的精心组织,重塑山水、民居、场馆、道路的秩序,从而形成一个以小型公共文化综合体为中心的村镇空间,大大增强了村民的凝聚力。立体多维的灰空间与交通空间交错重叠,建筑内外界限模糊,建筑空间多元,活动多样。该项目是乡村文化生活的建筑探索实践。

Nanxi Book Collectio tries to recreate the orders for the mountain, the waters, the civil residence, the venues and the streets based on the disordered, complicated large spaces, through architectures, squares and the delicate organization of streamline to form a village space with the center of small size public cultural comprehensive integration, which greatly increase the integrative force of the villagers. Cubic and multiple dimensional grey space staggers and overlaps with the traffic space, the building boundary inside and outside the building is vague, the architectural space are multiple-elements and activities are various. This project is a trial experiment for the building exploration of village culture life.

设计作品索引
Index of Design Work

办公、商务、金融建筑

银科金融中心	2
临港新片区创新魔坊	6
星扬西岸中心	10
广州三七互娱全球总部大厦	23
大参林医药集团荔湾运营中心	24
亚洲金融大厦暨亚洲基础设施投资银行总部	69
永久办公场所	
中国美术出版大厦（人民美术出版社新址）	73
嘉铭东枫产业园	78
中坤广场改造项目	83
长阳航天城科技园项目	87
重庆环球金融中心	122
珠江城	154
中葡商贸中心	163
广州荔胜广场	172
合肥天玥广场	174
合肥政务新区天珑广场	174
合肥复星文化金融创新城	175
上海虹桥万创中心	196
悍高星际总部	214
金颖科技（惠州）磁性元器件制造项目	215
云米互联科技园	216
中南高科·仲恺高端电子信息产业园	217
五象总部大厦	226
开阳县冯三镇乡村振兴新华生态产业示范园	239
合肥金融港中心	252
黄石市科技创新中心	254
中电光谷智造中心·武汉阳逻	256
哈尔滨工业大学建筑设计研究院科研楼	268
寒地建筑科学研究中心	274
苏州城发建筑设计院设计研发中心	314
中建东北院总部大厦	323
润友科技长三角（临港）总部项目	327
烟台天马中心	338
潍坊滨海经济开发区综合商务中心	339
新源智慧建设运行总部A座	347
中美清洁能源研究中心	349
成都时代天境	416
青岛创智产业园一期工程	437
香江电商科创中心	441
天津津兰国际黄金珠宝城	447
磨黑法庭及配套项目	472
超重力离心模拟与实验装置国家重大科技基础设施	493
瓶窑鼎胜轻智造产业园	504

产业园区规划设计

成都芯谷IC研发及产业基地项目二期	258
天津欧微优科创园（中电科创园）	260
长沙中电软件园二期扩大项目	262
枫桥工业园	300
大兆瓦风机新园区项目	307
黄桥总部经济园	320
潇河新城项目	342
山西科技创新城科技创新综合服务平台（一期）项目	352
曼德勒·缪达经济贸易合作区	479
中国电子科技集团科技园谷雨村	493

城市更新改造

920街坊项目	114
长嘉汇弹子石老街	118
解放碑至朝天门步行大道品质提升	120
金山国际商务中心一期改造升级项目（城市更新）	127
科苑小学改扩建工程	177
南宁市"三街两巷"项目金狮巷银狮巷保护整治改造（二期）工程设计	222
南宁书画院装修改造工程	224
潘祖荫故居修缮整治工程	309
包钢宾馆	330
太原武宿国际机场三期改扩建工程航站区工程	346
太原工人文化宫大修改造工程 B段	350
钟楼街项目	354
第一实验室（太原第一机床厂改造项目）	356
解放路沿线城市风貌更新	358
上海五角场万达改造	392
中国川西林盘非遗（马椅子）工坊更新	410
中国两弹城	412
烟台市所城里、朝阳街（一期）历史文化街区保护性改造	438
兵团103团一连乡村振兴风貌提升	456
布达拉宫广场装饰柱提升改造项目	466

城市规划、景观规划

嘉定新城远香湖地区城市设计	28
连云港新港城几何中心总体空间规划及核心区城市设计	37
东台高新区邻里中心	39
江苏淮海科技城创智产业园	42
上海金山冰雪海养生公园	48
绍兴上虞·东关古运河滨水地区城市设计研究	56
雄安新区容东片区C组团	66
太原万达沙河澜山销售物业大区景观设计	110
成都杜甫草堂片区规划	135
白云山柯子岭门岗及周边景观整治工程	144
贵阳永乐湖国家水利风景旅游城市综合体	244
常州市三江口公园	311
长株潭绿心中央公园总体城市设计方案	428
杨柳青大运河国家文化公园（元宝岛）	432
沙雅县·乌什喀特古城旅游景区规划设计	450
达孜叶巴康养小镇城市设计	462

城市综合体

印尼星迈黎亚综合体	17
越南胡志明市莲花东	18
广州丰盛101超高层综合体	22
中海国际中心与环宇荟办公商业综合体	34
中华药港核心区建筑设计	38
绛县涑水河新区（一期）	108
建设项目总承包（EPC）项目	
成都文殊坊二期	134
重庆江北嘴紫金大厦	136
成都金牛区商业综合体	137
石阡县楼上村特色田园乡村·乡村振兴集成示范试点	242
伸马商业综合体	278
首创城商业综合体	440
博乐城市会客厅	454
北辰兰苑	476

工业建筑

上海微小卫星工程中心卫星研制项目	86
星空标准厂房项目	90
智能制造基础件产业集群建设项目（一期）	91
中国航发沈阳发动机研究所JG12项目	92
迈百瑞生物医药（苏州）有限责任公司生物医药创新中心及运营总部	100
北京大兴国际机场临空经济区（廊坊）物流港	101
海口综合保税区高端食（药）材加工标准厂房项目	101
贵州义龙新区海庄、花月光伏电站EPC项目（新能源）	132
陕西正通煤业矿井水净化处理工程	133
广州110KV猎桥变电站	140
翁源县气象防灾减灾业务技术用房	161
茅台集团贵定昌明玻璃瓶厂建设项目	232
贵州茅台酒股份有限公司"十四五"酱香酒习水同民坝一期建设项目	234
黄山小罐茶超级工厂	302
骊住建材（苏州）有限公司新工厂	308
中安创芯半导体及电子设备制造项目	353
浙江和泽医药研发生产项目	498

公共建筑

天津生态城图书档案馆	76
南京江北新区市民中心	178
横琴汽车营地（露营乐园）项目二期勘察设计施工总承包	200

绥化市图书馆	285	龙华区实验学校小学部设计采购施工总承包（EPC）	203	温州云顶草上世界度假酒店	55
潇河国际会议中心项目	345			资溪·冠合开元芳草地乡村度假酒店	55
西安丝路国际会议中心	361	高圳车革命传统教育基地建筑设计	206	禅泉酒店	150
深圳龙岗万达广场	386	逢简小学牌楼修复加固项目	211	保利汕尾金町湾A006地块项目（希尔顿逸林酒店）	152
成都金融城双子塔	398	顺德实验中学提质扩容设计	213	黄山雨润涵月楼酒店	369
南沙青少年宫	401	广西大学大学生创新实验中心大楼	220	九华山涵月楼度假酒店	370
中建科技成都绿色建筑产业园研发中心	405	茅台学院图书馆	228	安吉柏翠姚良度假酒店	371
皓月·江屿城	408	中共贵州省委党校（贵州行政学院）	236	崇明长兴岛开心农场酒店	372
天津市梅江地块医疗和公共卫生服务组团新建工程	433	中国环境管理干部学院新校区	264	希尔顿格芮酒店二期	373
		佳木斯大学实训楼AB栋	282	西安云山湖酒店	374
黄骅市群众艺术公园和"四馆一中心"项目	434	江湾学校	283	泰和大酒店及民宿综合开发利用建设项目	459
"兴文石海"世界级旅游目的地游客中心	436	绥化市廉政教育中心	284		
杭州英蓝中心	488	哈尔滨麦硕国际教育社区	287	**居住建筑**	
郑州海洋生物博物馆	502	软件谷学校	296	沈阳万科·东第	13
		中国科学技术大学苏州研究院仁爱路校区（166、188地块）	312	武汉华润半岛九里	14
纪念馆、古建筑保护				长春嘉惠九里	15
金陵大报恩寺遗址博物馆	292	中国中医科学院大学	312	郑州康桥东麓园	15
状元楼	332	昆山杜克大学二期	313	深圳勤诚达·云邸	16
山西省省情（方志）馆	351	中国常熟世联书院	313	沈阳万科新都心	16
红楼梦酒坊遗址馆	407	玉成实验小学	319	沈阳汇置·峯	18
大禹纪念馆	483	淄博市环理工大学创业创新带	380	天津金隅·云筑	19
		淄博美达菲国际学校项目	381	济南市中万科城	20
交通建筑		成都市蜀都小学	414	北京万科观承别墅	21
东莞CBD示范性停车楼	25	河北工程大学新校区规划设计	429	建控·江山赋	40
成都博览城北站交通枢纽	44	河北工程大学新校区图书馆	429	华润清河橡树湾（海淀区清河镇住宅及配套工程）	80
上海金桥车辆段上盖物业综合开发	46	雄安新区启动区高教园区图书馆	430	海德园	88
北京新机场东航机务维修区项目	85	邢台新能源职业技术学院设计	431	绛县涑水河新区（一期）建设项目总承包（EPC）项目	108
南粤古驿道梅岭驿站	142	河北工业大学北辰校区化工、海洋学院教学实验楼	433		
粤澳新通道（青茂口岸）	162	北师大附属学校	442	唐山·水山樾城一期、二期	111
日照市奎山综合客运站及配套工程设计项目	334	中新天津生态城中部片区32号地块小学、幼儿园项目	446	沈阳远洋上河风景项目	113
淄博火车站北城市设计项目	382			重庆荣昌棠悦府	138
青岛胶东国际机场	403	天津南开中学素质教育实验室	448	西永翰粼天辰	138
贵安站	418	迁安青少年活动中心	449	铜梁原乡溪岸	139
川藏铁路拉萨至林芝铁路站房设计	420	西藏技师学院	468	重庆博翠宸章	139
玉磨线站房	422	浙江大学紫金港校区文科类组团一（人文社科大楼）	485	贵阳中天未来方舟	175
重庆西站	424	杭州职业技术学院学生食堂	487	截流河（沙井段）、蚝乡路等项目拆迁安置房建设工程可行性研究和全过程设计	205
普陀山正门客运中心	486	浙江大学紫金港校区管理学院大楼	490		
		杭州电子科技大学第一实验楼	491	高地天域	246
教育建筑		嘉兴学院梁林校区扩建工程二期	492	安顺经开区领秀山水	248
中国人民大学附属中学丰台学校	82			阳光城·未来悦（一期）A组团	249
礼贤小学	89	**酒店建筑**		黎平天玺湾	250
北京工商大学良乡校区二期	96	东方美谷JW万豪酒店	4	顺海绿洲一期建设项目	251
东营市河口第一中学	102	日本北海道温泉度假酒店	26	华润·置地公馆	281
广州黄埔区委党校	156	青岛海天中心	31	旭辉铂悦犀湖	317
澳门大学珠海新校区	160	天目湖 WEI 精品酒店	35	垂直森林－山水林居	328
广东外语外贸大学北校区校门建筑和门前广场项目	165	安吉吾想园酒店	50	宜兴凤凰山康养谷	375
		高二高姆山度假酒店	51	御府中央小区（A2-1/3地块）	470
		杭州婚庆文化基地	52	俊宏誉园	474
		磐安县乌石村乡村度假酒店	53	朗韵花园（A2地块）	475
		磐安樱花谷野生酒店	54		

旅游、休闲建筑

柬埔寨西哈努克市海龙湾度假中心	41
九寨沟景区沟口立体式游客服务中心	64
博湖博物馆及游客中心项目	106
丹寨旅游小镇	124
曼飞龙国际养生度假区	169
湛江玥珑湖养生体验中心	173
柬埔寨七星海红树林度假区（星月海岸）一期	202
佛山市顺德北滘碧江美食聚集区	212
雷山县大塘农业观光园	240
西昌邛泸景区游客中心	402
沙雁洲景区旅游基础设施建设项目——胡杨学院	458
沙雅县沙雁洲景区游客服务中心	458
汉营走马古镇	478

商业建筑

前海世茂中心	8
杭州苕溪公园商业街	27
郑州格拉姆国际中心	36
宝应新城吾悦广场	94
中关村科技园区丰台园产业基地东区三期	99
广东塑料交易所仓储中心/沟通之弧	168
苏州丁家坞精品酒店	303
中银大厦	305
星湖大厦	306
仁恒仓街商业广场	316
大连恒隆广场	325
大连中心·裕景	326
潇河新城酒店项目	343
合肥银泰中心二期项目	390
万萃城二期(1)M1组团项目	394
绿地济南CBD项目	396
达坂城(大果院子)疆风民宿设计	455
和田团城·谜街	457
钱投邻居中心(望江店)	489
义乌全球数字自贸中心	493

体育建筑

国家跳台滑雪中心	62
冬奥会环境建设项目（二期）	112
重庆市青少年活动中心	128
大连体育中心体育场、体育馆	270
第十三届全国冬季运动会冰上运动中心	272
黑河学院综合体育馆	280
青岛市民健身中心	290
苏州北部文体中心	304
北山四季越野滑雪场项目	324
武夷新区体育中心	427
绍兴羊山攀岩中心	506

文化艺术建筑

中国第一历史档案馆	67
清华大学图书馆	68
中国工艺美术馆、中国非物质文化遗产馆	70
南海艺术中心	148
雁塔文化艺术中心概念规划方案	166
江西省文化中心	171
西藏非物质文化遗产博物馆	184
碧江牌坊	210
深圳清真寺	294
南昌汉代海昏侯国遗址博物馆	298
永州市一宫两馆四中心	340
浦东美术馆	362
宁波院士中心	364
淄博市城市馆、美术馆建设项目	376
淄博市"一馆两中心"项目	378
成都露天音乐公园	404
天津第二工人文化宫公园改造提升（二宫）	432
宝鸡大剧院	435
南宫文化中心·尚小云大剧院	436
天津市安里甘艺术中心	439
东方市文化广场	445
丝路航田城市会客厅和玫瑰主题馆	453
西藏非遗千工坊设计项目	461
西藏文化艺术创作园区	464
近水书院	480
桑洲清溪文史馆	484
余杭区文化艺术中心一期工程	494
南溪书舍	510

乡村振兴规划与设计

南江华润希望乡村规划及建筑设计项目	116
官田兵工小镇	266
树山村改造提升工程	310

医院、医疗建筑

北京美中宜和妇儿医院	97
北京市顺义区域医疗中心	98
重庆市人民医院	129
国家呼吸医学中心一期工程	159
宁波市杭州湾医院	180
上饶市立医院三江总院建设工程设计	204
贵州医科大学附属医院第三住院综合楼	230
哈尔滨市儿童医院松北院区、市妇幼保健院松北院区、市红十字中心医院松北院区	277
哈尔滨市血液中心	286
深圳大学学府（深圳大学总医院）	335
东营市人民医院急诊急救中心暨内科病房综合楼	336
山东省立医院儿科综合楼及辅助保障设施建设项目	337
浙医二院柯桥未来医学中心	493

影院、剧院建筑

澄海音乐厅	172
新疆大剧院	182
中国扬州运河大剧院	366
成都城市音乐厅	400
柯坪古城博物馆影剧院	452
"鲁镇·社戏"演艺中心	508

展览、展示建筑

南京国际展览中心	43
国家会展中心（天津）	72
北京亦创国际会展中心（世界机器人大会永久会址）	74
中国·红岛国际会议展览中心	75
2019北京世界园艺博览会—央视动画馆设计	104
重庆市规划展览馆迁建EPC项目	130
广州空港中央商务区项目会展中心	146
深圳改革开放主题公园项目	188
莲花山公园展示中心及公共卫生间	
智谷AI科技中心	192
顺德勒流龙眼历史人物纪念馆	209
防城港市堤路园（园博园）项目（园博园主展馆）	218
第十届江苏省园艺博览会园博览园主展馆	288
太仓天镜湖展示中心	318
超级碗	331
一号高炉展厅设计	333
潇河国际会展中心项目	344
太忻一体化展示馆	348
西安丝路国际会展中心	360
西安丝路国际会展中心（一期）	361
鑫创科技展示中心	384
武汉绿地天河国际会展城	388